Satellite Newsgathering

Jonathan Higgins

Focal Press
OXFORD AUCKLAND BOSTON JOHANNESBURG MELBOURNE NEW DELHI

Focal Press
An imprint of Butterworth-Heinemann
Linacre House, Jordan Hill, Oxford OX2 8DP
225 Wildwood Avenue, Woburn, MA 01801–2041
A division of Reed Educational and Professional Publishing Ltd

A member of the Reed Elsevier plc group

First published 2000
Reprinted 2002

British Library Cataloguing in Publication Data
Higgins, Jonathan
Satellite newsgathering
1. Artificial satellites in telecommunication
I. Title
621.3'825

Library of Congress Cataloguing in Publication Data
A catalogue record for this book is available from the Library of Congress

ISBN 0 240 51551 X

For more information on all Butterworth-Heinemann publications please visit our website at www.bh.com

Composition by Genesis Typesetting, Rochester, Kent
Printed and bound by Antony Rowe Ltd, Eastbourne

Satellite Newsgathering

Contents

Preface

Satellite newsgathering (SNG) is essentially the process that delivers 'live' and 'breaking' news to the viewer and listener as it happens, providing a window through which we can all learn about news events as they happen. It is a very powerful tool of modern newsgathering and, when used effectively, it can be staggering in terms of the way a situation can be so vividly conveyed to an audience.

The process of newsgathering is a not a subject that most people consider as they take in a television newscast or a radio news bulletin. Yet a higher proportion of the world population now obtain their 'fix' of daily news from radio or television than from newspapers, and consequently they rely on the processes of electronic newsgathering to deliver information from the scene. Satellite newsgathering is just one of a number of tools used, and it involves the use of very advanced technology in sometimes the most primitive of conditions.

A war virtually anywhere in the world can be brought 'live' instantaneously onto the television screen in the living room – yet just twenty years ago this was impossible. Film cameras were still then largely the key tools of television newsgathering, as electronic newsgathering (ENG) cameras were still not commonplace, except in certain markets such as the US. Traditional electronic outside broadcast cameras were too large and heavy for newsgathering operations. There were reports from the scene of a story on occasions but, because they were on film, there was a delay because of the transportation, processing and editing involved.

Arthur C. Clarke first proposed the concept of artificial communications satellites in 1945. Through the 1950s and early 1960s, a race ensued between the US and Russia to develop rocket technology to offer improved military capability, and in addition to attempt to place the first man on the Moon – though this second objective was more political than military. The development of communications satellites occurred in parallel and as part of the overall scheme to exploit space for the 'betterment of mankind'. (Incidentally, the word 'satellite' comes from the Latin *satelles*, meaning 'an attendant'.)

The use of satellites alone to deliver news material back to the major news networks really began with the Vietnam War in the late 1960s – Vietnam has been dubbed the first 'TV war'. The US networks needed a way to obtain video from overseas that was faster than by air transportation, and the relatively new communications satellites provided the means for doing so – though the stories were still at least a day old before they were reported on television with pictures. The first significant use of satellite technology for news occurred in 1968 during the Tet offensive in the Vietnam War. But the passage of time does not always improve matters, as the Falkland Islands conflict between Britain and Argentina of 1982 illustrates: it was not well reported in television news terms. There were no satellite links available and, due to other hindrances imposed by the British government, many of the stories were not well covered on television with visual background until film pictures arrived, typically up to 48 hours or more later. The Persian Gulf conflict in 1990, and the ensuing Persian Gulf War in 1991, changed that forever. SNG was 'made' by the situation in the Gulf, and no major international conflict has since gone by without the use of SNG.

The change has crept up on the public almost unnoticed. For instance, today's television news bulletins are a complex weave of studio material, sophisticated electronic graphics, video (from tape or computer 'server'), and live reports delivered by terrestrial microwave links and satellite uplinks from innumerable scenes of news stories. Radio bulletins are similarly a complex mix of different sources, put together to recreate the situation in the listener's mind and convey the key information of the story.

But the focus of this book is the use of satellite uplinks for newsgathering and, although that typically means television news, we will not be restricting our discussion to television only, but also considering the part that satellite delivery systems from the field play in radio news reporting. Radio has had the advantage of the availability of the telephone since the advent of reports from the scene; though in the early days of radio news, it was common to merely read news items from newspapers or news agency wire services. Even now, many radio reports from difficult locations are delivered via the telephone – often a satellite telephone.

The process of satellite transmission is essentially that of a radio transmitter transmitting a focused beam to an artificial satellite orbiting the Earth, which in turn transmits another focused beam back to a radio receiver at a different point on the Earth's surface. However, the process of SNG goes beyond solely the equipment used, to how communication satellites are operated in orbit, regulatory affairs, transportation, safety and logistics. Not least, it is about working in a wide range of conditions, from city streets to battlefields, in all climates and where time is of the essence.

For some time now I've felt that there was a need for a book that can be used as a primer for anyone wanting to know more about satellite newsgathering. As an intrinsic part of most news bulletins, whether on television or radio, the use of SNG is now taken for granted. Yet even for those involved in the broadcast news business, it can be very difficult to find out about many of the aspects of the use of SNG equipment and the challenges posed for those who are responsible, either directly or indirectly, for its deployment or operation in the field.

In this book, those interested in this subject should gain an understanding of the technical and practical considerations in specifying and operating these systems around the world, and an insight into both the satisfaction gained in doing so successfully and the challenges occasionally faced. It is intended to give a practical treatment of this specific application of satellite communications engineering, and the use of mathematics has been virtually restricted to the appendices for those who are interested.

A note of caution – the term 'SNG' is often used erroneously in a generic sense to describe any mobile satellite uplink that may be used for coverage of a sports fixture, a local or national event, or an international event such as the Olympic Games. This book is specifically about using mobile satellite uplinks for the gathering of news, and nothing else. Throughout, the focus remains on news-gathering and the use of satellite technology in the field to serve this purpose.

In this book, I have aimed to treat the whole subject of SNG in a 'holistic' fashion. There is a flow and intertwining of the different themes in the following chapters that puts this subject beyond a strictly technical treatise on satellite communications engineering (of which there are numerous in print). This book covers a range of issues, and hopefully can act as a handbook for both professionals working in the field as well as those with an incidental interest. The most expert people in the use of SNG have strong editorial, technical and logistical skills – be they an uplink technician, producer, or reporter – and so this book covers the subject in a similarly multi-disciplined way.

Finally, I want to thank the many colleagues and friends across the industry who have freely given their time to look at drafts of chapters, made suggestions, supplied information, provided photographs and illustrations, and generally encouraged me in this endeavour. I also want to thank Margaret Riley and Jenny Welham at Focal Press, who patiently waited for the manuscript while several deadlines passed, and guided me in the writing of this, my first book.

Not least, I thank my wife Annie, who has constantly supported and encouraged me from the very beginning when I first decided to write this book, and has put up with countless lost evenings, weekends (and holidays!) throughout.

Jonathan Higgins
January 2000

Acknowledgements

My thanks to all my colleagues who supplied information, debated issues and generally aided me in the writing of this book. In particular I would like to acknowledge the observations of the following on specific parts of the text: my colleagues Richard Penman, Andy Woodhouse and Tim Barrow; Derek Tam and Marco Franken (INMARSAT); Peter V.F. Beardow (7E Communications); Steve McGuiness (Advent Communications); Dick Tauber (CNN); Professor Steven Livingston (The George Washington University); and the staff at INTELSAT.

Thanks are also due to Bill Andrews (MSAS Global Logistics), Richard Wolf and Wes Gordon (Wolf Coach), and the staff at Thomson Broadcast Systems, who made available information vital to the accuracy of the text.

I would like to express my appreciation to the following colleagues for their excellent photographs taken on location: Paul Szeless (particularly the front cover), Simon Atkinson, Martin Cheshire and Harvey Watson.

Thanks also to A.R. Lewis and R. Penman for their help with Appendix B.

The author thanks INMARSAT for allowing him to quote extensively from *Inmarsat-B High Speed Data Reference Manual* in Chapter 5.

I also wish to record my thanks to the following companies and organizations who kindly supplied illustrations and information: Advent Communications, BT Broadcast Services, Continental Microwave, CPI Satcom Division, EDAK, Marconi Applied Technologies (formerly EEV), ESA-CNES-ARIANESPACE, EUTELSAT, FCC, Frontline Communications, Glensound Electronics, Hughes Space & Communications, INMARSAT, INTELSAT, Livewire, NDS, Nera, PanAmSat, Telecast Fiber Systems, Thomson Broadcast Systems, Toko America, Vertex Communications, Vocality International, Wahlberg & Selin and Wolf Coach.

1

From the beat of the distant drum. . .

In the hotel conference room in Dharhan, Saudi Arabia, the US general turned away from the overhead projection screen, completing the briefing to the assembled international media and inviting questions. Across the world, millions of people watching and listening via satellite had been drawn into the story told by the general's words and the pictures projected onto the screen from the on-board 'smart bomb' cameras. The world had watched graphic pictures illustrating the latest round of sorties of Operation Desert Storm on Iraq, which had taken

Figure 1.1 'Satellite dish farm' on a crazy golf course at a hotel in Dharhan, Saudi Arabia, 1991

place only hours before. It was January 1991, and this briefing in the Persian Gulf conflict was yet another example of satellite newsgathering giving audiences across the world the sense of 'being there' – and this power had also become a political force.

This type of scene occurred on numerous occasions in early 1991, with 'live' coverage from Riyadh and other Iraqi targets in the Middle East of attacking Scud missiles. The use of satellite newsgathering is now so powerful that no-one can be in any doubt that war is fought as much on television as on the battlefield – the political impact of which has been dubbed the 'CNN effect'[1,2]. It was for this reason that the Iraqi government wanted the Western media present in Baghdad during this period to bring to the world live coverage of the night-time Allied attacks. As a tool of the media, satellite newsgathering (SNG) is of such significance and immediacy that no major international conflict can ever now be reported without its use. While SNG brings everything from reports on the pile-up on the local freeway, to live 'vox-pops' on the street, to national political events, it is in the arena of major international news events that its remarkable power to evoke the sense to the audience of 'being there' is felt the most. It brings these news events into the home and the workplace, delivering the impression to audiences that they are viewing through a 'window on the world'. The development and use of SNG is the culmination of a long history of gathering and distributing news and, to put SNG into context, we are going to take a journey through the history of telecommunications, from the early methods of communicating news to today's sophisticated electronic newsgathering process.

1.1 News by drum, horn, shout, fire and smoke

The early communication of news is naturally closely linked with the development of spoken language over 100 000 years ago. Spoken news sent by messenger was essential for the survival of early peoples – by warning of attack or flood, informing of the whereabouts of sources of food, or the proclamation of the birth and death of leaders.

The earliest technologies used to send news over distances (without sending a human messenger) revolved around the use of drums, horns, birds, shouting, beacon fires, mirrors and smoke. In the sixth century BC the Persians could send news from the capital of the Persian Empire to the provinces by means of a line of men shouting one to another, positioned on hills. This kind of transmission was up to thirty times faster than using messengers. In the first century BC, the Gauls could call all their warriors to war in just three days using only the human voice in a similar manner.

However, in many societies the drum and the horn were the quickest and easiest way of relaying information by sound waves, by using a string of local 'repeaters' to cover large distances. (It is interesting to consider the parallel with modern telecommunications, which still relies on 'repeaters' and 'waves', although at a much higher frequency than sound.)

The recording and sending of complex messages is tied to the development of writing. Telegraphy (from the Greek 'writing in the distance') describes a communication system able to convey signals that represent coded letters, numbers and signs of writing, and has very ancient origins. The use of hieroglyphics and symbols on clay tablets brought the ability to send more complex information by messenger, and the first alphabet appeared around 1500 BC. In 105 AD T'sai Lun (China) first developed paper, and the use of paper spread across Asia to Egypt in the third and fourth centuries AD.

The beginning of the development of a technological telegraph system came in the late eighteenth century. In 1791, Claude Chappe (France) invented the optical telegraph, which was a mechanical semaphore system based on towers. In 1792, he showed his invention to the French Legislative Assembly, which adopted it officially, and in 1794 the first message sent via this semaphore system brought news of French victory over the Austrians. Napoleon was reported to have been impressed by this, and soon semaphore systems were established between most of the cities of France, followed by prominent cities in Italy, Germany, Russia and the US, and many remained in use until the 1860s, when they were superseded by the electric telegraph.

1.2 Dots and dashes

The true revolution in telegraphy came in the eighteenth and early nineteenth centuries with the discovery of electricity. The work on electricity at various times over this period of Stephen Gray (UK), Pieter van Musschenbroek (Holland), Ewald Georg von Kleist (Germany), Luigi Galvani (Italy), Alessandro Volta (Italy), Andre Ampere (France), Christian Oersted (Denmark) and Michael Faraday (UK) all contributed to the development of the electric telegraph.

The invention of the electric telegraph was the first significant technological leap in the conveyance of news. As with so many inventions, a number of people were working on the same idea at the same time (as we shall also see later with both the telephone and television), and came up with various types of equipment. In 1830, Joseph Henry (US) transmitted the first practical electrical signal, and in 1832 Pawel Schilling (Russia) had constructed the needle telegraph in St Petersburg, in which electricity flowed through a magnetic spool and moved a magnetic needle, thus demonstrating the first working electromagnetic telegraph.

In 1835 William Cooke (UK) and Charles Wheatstone (UK) demonstrated an electric telegraph system, based on Cooke's observation of a demonstration of Schilling's system which he had seen a few years earlier. Their system used electric currents to deflect needles that pointed to letters of the alphabet laid out on a grid. They developed their system and built a telegraph line from London to Slough in 1843. In 1845, their telegraph system made history when it aided the capture of a murder suspect, who was seen boarding the train for London. Police used the telegraph to send details to Paddington railway station in London, and the police captured the suspect as he got off the train.

However, the first practical electric telegraph that was widely adopted was based on a relatively simple design, developed by Samuel Morse (US) in 1837. Morse had seen Cooke and Wheatstone's telegraph demonstrated in London. His system used a key (a switch) to make or break the electrical circuit, a pair of wires joining one telegraph station to another distant station, and an electromagnetic receiver or sounder that, upon being turned on and off, produced a buzzing noise. These electrical impulses were received as a series of dots and dashes that could be translated into letters and hence words – thus came the invention of the Morse code. Morse patented his telegraph system in 1840.

In 1844 Morse demonstrated the use of the telegraph to send a news story over his first long distance telegraph circuit under construction alongside the railway line from Baltimore in Maryland to Washington DC, a distance of about 60 km (37 miles). A telegraph operator at Annapolis Junction, Maryland, heard news of the announcement of the presidential nominee for the Whig party at a convention in Baltimore from a messenger on a Washington-bound train passing through Annapolis Junction. The telegraph line did not yet extend the full distance from Washington to Baltimore, but the telegraph operator was able to signal the news to Morse in Washington – the message via the train arrived in Washington an hour later. A few weeks later, on completion of the telegraph line, Morse sent the famous words from the Bible, 'What hath God wrought!' on his telegraph from the Supreme Court chamber in the US Capitol Building in Washington to the Mount Clare railroad depot in Baltimore.

This was completely revolutionary, and newspapers in the US and Europe quickly seized upon the idea of using the telegraph as a means of gathering news from distant locations. Telegraph lines sprung up alongside railway tracks – for these provided a convenient path across the country – so that by 1852 there were 37 000 km (23 000 miles) of lines across the USA. By 1867 the Western Union Telegraph (which had an almost monopolistic position in the US) had over 74 000 km (46 000 miles) of lines, and by 1895 over 300 000 km (189 000 miles.)

In 1865, twenty participating countries signed the first International Telegraph Convention, and thus the International Telegraph Union (ITU) was established. (The birth and significance of the ITU is examined in Chapter 6.) However, up until around 1880, the land-line telegraph was mostly used for short-distance metropolitan communication.

1.2.1 Wiring the world

To really take advantage of its potential, the telegraph needed to reach around the world. In 1866, after eleven years of attempts to lay a cable across the Atlantic, the first submarine cable between Newfoundland and Ireland came into service. By 1879, there was a combination of routes, both overland and under the sea, linking London with India, China, Java and Australia. By 1890, Brazil was linked by submarine cable with the US and Europe. In 1880 a cable linking the east coast of Africa was completed, linking the Yemen in the Arabian Gulf via Mozambique to South Africa, and by the late 1880s Europe was linked to the west coast of Africa via Portugal. By 1892, there were ten transoceanic telegraph links. In 1902, the first transpacific telegraph cable began operating, stretching from Vancouver, British Columbia, through the Fiji Islands and Norfolk Islands to Australia, before landing in New Zealand.

1.2.2 The news and the telegraph

The use of the telegraph for newsgathering grew, as the following few examples illustrate. The telegraph played an important role in disseminating news from the battlefronts during the American Civil War, feeding detailed reports to the newspapers. The telegraph was instrumental in disseminating news of Abraham Lincoln's assassination, for within twelve hours of his shooting in 1865, newspapers in every major American city had printed the story. In an article entitled 'The Intellectual Effects of Electricity' in *The Spectator*, London, 9 November 1889, the following observation was made:

> 'With the recording of every event, and especially every crime, everywhere without perceptible interval of time the world is for purposes of intelligence reduced to a village. . .All men are compelled to think of all things, at the same time, on imperfect information, and with too little interval for reflection. . .The constant diffusion of statements in snippets, the constant excitements of feeling unjustified by fact, the constant formation of hasty or erroneous opinions, must, in the end, one would think, deteriorate the intelligence of all to whom the telegraph appeals.'

Some areas of the press were obviously not as impressed by this new tool for newsgathering as others. Even after the invention of the telephone, the telegraph remained the predominant method of long-distance communication for some considerable time, because of the extensive network that had become established and the relatively low cost of sending messages. The use of Morse code, of course, was essential to the early development of radio.

1.3 Bell ringing

The development of the telephone began soon after the invention of the electric telegraph, and although attributed to Alexander Graham Bell (US), others also had a hand in its development. In 1854, Charles Bourseul (France) wrote about transmitting speech electrically and, in 1861, Philip Reis (Germany) devised a telephone system, which unfortunately, although tantalizingly close to reproducing speech, only poorly conveyed certain sounds.

However, it was Bell who filed a patent for a telephone on 14 February 1876, only hours before another inventor, Elisha Gray (US), also filed his patent for a telephone system. Even though neither man had actually built a working telephone, Bell made his telephone operate three weeks later using ideas outlined in Gray's patent and not described in his own patent – an issue of some dispute at the time. Bell was heavily dependent on the electrical knowledge of his assistant, Thomas Watson (US). On 10 March 1876, the first complete sentence was heard on the telephone when Watson was waiting in another room to receive the call. Bell was setting up his equipment when he accidentally spilled battery acid on his trousers and called to his assistant, 'Mr Watson, come here. I need you.' These were the first words transmitted by telephone.

Bell's invention used a continuously-varying (analogue) electrical signal to transmit a person's voice over a line, unlike the telegraph's 'dots' and 'dashes'. As a transmitter, the telephone converts sound waves into analogue signals by reacting to variations in air density caused by the sound waves. These variations in air density are converted into an electrical current of varying force and, at the receiver, the sequence is reversed whereby the analogue signals are converted back into sound waves.

Telephones were developed as personal communication devices, as usually only one person transmitted and received messages with one other person at a time; so from the point of view of newsgathering, the telephone did not offer a significant advantage over the telegraph. Although the messages could be instantly understood, without the need for decoding Morse, the use of the telephone for newsgathering lagged behind the telegraph until well into the first part of the twentieth century. However, for messages to be telegraphed or telephoned in the late nineteenth century, a physical link had to be established and maintained between the points of communication. Remove the need for the physical link, and the potential for communication expanded greatly.

1.4 Wireless

In 1894, Guglielmo Marconi (Italy) began to investigate the laws of electricity and magnetism determined by previous experimenters. He built upon the work of Heinrich Hertz, who in 1887 produced radio waves, and the other work on radio waves by Oliver Lodge (UK) and Edouard Branly (France).

Specifically, Marconi believed that there was a way to use electromagnetic waves for the purpose of communication, and he began to conduct his own experiments to test these ideas. These experiments led to the first elementary radio set.

In 1894, Marconi's first radio transmissions travelled only a few hundred metres. Marconi successfully sent the Morse code letter 'S' to a farmer who owned the adjacent property through the use of a basic transmitter and receiver. He noted that ground obstacles such as vegetation and hills weakened the signal slightly, but otherwise the experiment was a huge success. For the following eighteen months, Marconi continued to experiment in seeking to improve upon his previous successes, during which time he successfully demonstrated the sending of a telegram by radio wave over 20 km (13 miles). Marconi obtained his first patent for his radio device in 1896 and, from then on, worked to improve and to commercialize his design. The Marconi Wireless Telegraph Company was formed in 1898, backed by British investors, to develop Marconi's system for maritime applications and to continue experimentation, and in 1899, Marconi wireless systems were fitted to three British battleships and two US battleships.

Marconi's wireless telegraph helped make the headlines in 1899 when he placed transmitters on two steamships in New York harbour and telegraph operators aboard reported the progress of an international yacht race. The ship-to-shore signals were received at stations in New York and then relayed across the US and the Atlantic by land-line telegraph. This was the first use of wireless telegraphy (radiotelegraphy) to cover 'breaking' news.

On 12 December 1901, Marconi achieved his furthest radio transmission to date – across the Atlantic Ocean, from St Johns in Newfoundland to Poldhu in Cornwall. This advancement was of great impact to worldwide communications, as the potential for transmitting information and news across the world instantaneously was realized.

Over the following twenty years, the Marconi Wireless Telegraph Company established such a dominance in marine communications, and conducted its business in such a manner, that considerable problems were caused which began to encroach on what a number of countries regarded as issues of sovereignty. In both the US and the UK, in the years up to and during the First World War, wireless became the preserve of the respective navy departments of government. After 1912 (directly as a result of the sinking of the ocean liner *Titanic*), a number of countries, including the UK and the US, made it a condition that all vessels had to be fitted with wireless telegraphy equipment. The particular details of this aspect of wireless are not relevant to our exploration of the development of broadcast news, save to say that in the US it eventually led to the establishment of a company who became dominant in US broadcasting – RCA.

American Telephone & Telegraph (AT&T), Westinghouse Electric, Western Electric Company, Marconi Wireless of America and General Electric were all US companies who were crucial to the war effort in their manufacture of wireless telegraphy equipment for the Allies. However, because of US government concern over the dominant position of the Marconi Company (due to its foreign roots), there was a determination to end the grip of the Marconi Company on US radiotelegraphy. This led to encouragement by the US government to form a consortium of US companies in order to challenge the dominance of the Marconi Company in the US.

In this aspect of the development of wireless in the US, one name is prominent – David Sarnoff (US). Sarnoff was not an inventor, but he was a visionary and an entrepreneur. He started working for Marconi Wireless of America in 1906, and there is an apocryphal story that in 1912, while on duty working as a wireless telegraphy operator, he picked up the signals from the ships near to the doomed *Titanic* and relayed news of survivors to waiting relatives for the following 72 hours. Another story is that he wrote a memo to his boss, E.J. Nally, in 1916: 'I have in mind a plan of development that would make radio a "household utility" in the same sense as the piano or phonograph. The idea is to bring music into the house by wireless.' Nally is said to have rejected Sarnoff's idea.

In 1919, General Electric bought out Marconi Wireless of America to form the Radio Corporation of America (RCA), with the aim of developing the US market for radio equipment. McNally, the President of Marconi America, and Sarnoff, by then the General Manager, transferred to the same positions in RCA. Within two years, RCA had taken significant holdings in AT&T and Westinghouse, to become the dominant commercial force in the US broadcasting industry. Sarnoff, the ultimate broadcasting marketeer, later became President of RCA in 1930.

The development of wireless was the subject of innumerable patent wrangles (and similar patent-ownership struggles continued on in the development in television up until 1939), but with the formation of RCA, combining a 'patent-pool' of all the important technologies, the issues were largely resolved.

1.5 Radio is born

Marconi's wireless relied on the use of Morse code, but voice transmission over the airwaves had been a dream of many, including Reginald Fessenden (Canada). Marconi did not believe there was much future in voice wireless, preferring to concentrate on improving radiotelegraphy. (The last maritime transmission of Morse code was on 31 January 1997.)

Fessenden believed that electromagnetic waves could be altered with the voice superimposed, using the same principles that made Bell's telephone work, and in 1900 he first proved that it was possible. Fessenden succeeded in being the first to broadcast a voice message, on Christmas Eve 1906, from Massachusetts, US, when radio operators on ships in the Atlantic were startled to hear a human voice emitting from their Morse telegraph equipment. Many operators heard Fessenden make a short speech, play a record and give a rendition of 'O Holy Night' on his violin. Radio broadcasting was born.

The technique of voice transmission that Fessenden had developed lacked power and amplification, and in 1906 Lee De Forest (US) patented the 'Audion' amplifier (the triode electron tube) which was an improvement on the thermionic valve (vacuum tube) invented by John Ambrose Fleming (UK) in 1904. De Forest, like many other pioneers of radio, was a controversial figure, and in 1910 he attempted (unsuccessfully) the first live broadcast of an opera, sung by Caruso, from the Metropolitan Opera House in New York. On the night of the US presidential election in 1916, De Forest also transmitted the returns of the elections. (Unfortunately, he announced the wrong candidate as the winner, as he closed down at 11 p.m. before the final results were in.) This transmission arguably made him the first broadcast journalist.

Although Fessenden's work and De Forest's invention made voice radio possible and practicable, it was not until after the First World War that public and commercial broadcasting began. Throughout this period, radio was still seen primarily as point-to-point communication – a 'wireless' telephone. The notion of 'broadcasting' to a large audience of listeners was not seen as practical or of much interest. Radio at that time was used mostly for commercial shipping purposes, but land-based amateur operators began to appear as electronic technology developed, 'broadcasting' to other enthusiasts listening on crystal radio sets. Most of these operations were very tentative affairs – just single operators transmitting recordings and talking – but in 1920 in the US alone it is estimated that radio experimenters spent over US$2 million on radio parts. In many other countries, groups of amateur enthusiasts were communicating with each other, and laying the foundations for the establishment of radio broadcasting in their respective countries.

The beginning of true public broadcasting is generally attributed to station KDKA in Pittsburgh, Pennsylvania, US. It is still a matter of historical debate as to whether it was KDKA or one of a number of other stations in the US which was first in broadcasting, and much turns on the definition of 'broadcasting'. An amateur enthusiast, Frank Conrad (US), who worked for Westinghouse Electric, built a transmitting station in his garage at his home in East Pittsburgh, Pennsylvania. In his experimental 'broadcasts' in 1919, he read items from the newspapers but discovered that when he substituted a phonograph for his microphone, a large number of listeners (who had built their own crystal radio sets) wrote or telephoned requests for more music. When Conrad became swamped with these requests, he decided to broadcast a regular programme to satisfy his listeners. Having exhausted his own collection of records, he borrowed records from a local music store in exchange for mentioning the store on the air – the first radio advertisement. All of these concepts of broadcasting – the station, the audience, the programmes and a means to pay for the programmes – came about through Conrad's experiments. A local department store promoted Conrad's broadcasts in one of their newspaper advertisements to sell radios, and promptly sold out of them. As a direct result, Westinghouse took notice and, seeing the potential of broadcasting for selling radios they could manufacture, decided to set up its own station and use its defunct war radio manufacturing facilities to produce simple receivers.

On 2 November 1920, station KDKA went on air for the first time with a 4-hour news programme – bringing the returns of the US elections relayed over the telephone from the wire service of the local newspaper, culminating in the announcement that Harding had been elected as US President. This is generally considered as marking the birth of the broadcasting industry. (Shortly after this, Westinghouse merged its radio interests with RCA.)

During the latter part of the 1920s, networks of affiliated radio stations were formed in the US, as part of the natural growth of a new industry. In 1926 at RCA, Sarnoff established the National Broadcasting Company (NBC), and in 1928, William Paley formed the Columbia Broadcasting System (CBS) from the struggling United Independent Broadcasters company. In 1943, the independent Blue Network Company was formed from the Blue Network of NBC (one of its two networks) as a result of a federal anti-trust ruling against the dominance of NBC. The Blue Network Company changed its name in 1945 to the American Broadcasting Company (ABC). Other networks, such as the Mutual Broadcasting System (1934), the DuMont Network (1946) and the Liberty Broadcasting System (1949), were all formed but were eventually overwhelmed by the dominance of the 'Big Three' networks and their well-established network of affiliate stations.

A large number of stations around the world began broadcasting through the 1920s, many using wireless equipment developed during the First World War. In 1920, in Chelmsford in the UK, the Marconi Company's station 2MT famously transmitted an experimental broadcast of a concert by the opera singer Nellie Melba. In 1922, 2MT went on air with scheduled broadcasts, followed by the British Broadcasting Company's station, 2LO, in London later that year. The Marconi station became integrated into the BBC, who bought Marconi transmitters. In 1927, the British Broadcasting Company became the British Broadcasting Corporation, and by 1930 covered most of the populated areas of the UK with their broadcasts.

As a vast continent as well as a nation, Australia was a natural market for radio. The first 'broadcast' in Australia in 1919 was by AWA (Amalgamated Wireless Australasia, an amalgamation of Marconi and Telefunken in 1913), when the Australian National Anthem was broadcast. Radio station 2SB in Sydney was the first to go on air with scheduled programmes in 1923. A string of other stations went on air through the 1920s, and in 1929 the Australian Broadcasting Company (ABC) was formed. By 1932, there were fifty-five stations across the country.

1.5.1 Early news bulletins

Through the 1920s in the UK, the BBC developed its programming of entertainment and music, but news was only broadcast after 7 p.m. to avoid upsetting the newspapers. Radio news bulletins consisted of announcers (in full evening dress) reading brief items of news, usually from the newspapers or prepared from news agency wire services. There was little or no 'actuality' (news from the scene itself) and the amount of news conveyed was roughly equivalent to brief headlines. Nevertheless, the newspapers were increasingly hostile to radio news, and the BBC held off objections from the press by beginning its evening news bulletins with the statement[3]: 'Here is the last news bulletin for today. General situation section. The following items of news indicate the general situation in the country. Copyright in these items is reserved outside the British Isles.'

This awkward relationship with the press was repeated in other countries as radio quickly established itself as the predominant medium, with the newspapers continually fighting rearguard actions as they jealously fought to hold on to their monopoly of news before it was heard on the radio. The newspapers considered that they had the primary right to disseminate news, and many newspaper groups sought to control the situation by buying up radio stations.

In the US, radio stations, as in other places, rarely had their own systems of newsgathering, but produced news bulletins from news agency wire services or newspapers. Disputes raged when radio stations not owned by local newspapers used material from newspapers. This developed into the 'press-radio wars' from around 1931 to 1935, when the news agency wire services were forced to stop providing news material to radio stations that threatened sales of newspapers. Obviously, radio could broadcast prime news agency wire service stories before the newspapers could print them.

CBS is of particular interest, as from the beginning it established a stronger interest in news than NBC. It later developed the 'eyewitness news' concept, with the emphasis on a report from the scene, either from a reporter in the studio who had just returned from the scene or, where possible, a live report by whatever means (telephone or radio).

In response to the embargo of the news agencies, the networks developed their own newsgathering systems, bypassing the news agencies, even covering major stories overseas as the following example illustrates. In 1936, the Spanish Civil War was raging, and an NBC news correspondent was caught in the middle of a battle. He and an engineer established a telephone line at an abandoned farmhouse between the Loyalist and Rebel armies, and reported while sounds of artillery shells exploding could be heard during the live broadcasts, which were relayed to the US by short-wave radio.

Occasionally the 'live' broadcasting of news events in the US occurred as well. The first famous instance was in 1925, when a Chicago radio station transmitted live coverage of the Scopes 'monkey' trial from Dayton, Tennessee, where a teacher named Scopes had been charged with breaking a Tennessee state law prohibiting the teaching of the Darwinian Theory of Evolution. Dayton was 800 km (500 miles) from Chicago, and the broadcasts were carried using 'open' telephone lines. In 1930, a Colombus, Ohio, station transmitted very dramatic live reports of a fire at the Ohio State Penitentiary, with the sounds of the fire and the screams of three hundred inmates dying in the background. But such reporting of major news stories was unusual. Newsgathering was not the first priority of the mainly entertainment-oriented radio networks, though they did provide coverage of important speeches, political conventions and election results (as we have seen).

The battle for disseminating news between the newspaper publishers on one side and the broadcasters on the other, continued in several countries until the Second World War. Meanwhile, in the UK, the BBC generally avoided the same type of confrontation with the newspapers. Live reporting from the scene of a story happened only very occasionally for special events, such as at the funeral of King George V (1936), and the Coronation of King Edward VIII later that year (which, incidentally, was the first 'live' television transmission on the BBC's experimental TV service). The abdication speech of Edward VIII on 11 December 1936 was also carried 'live' by radio from Windsor Castle.

One of the more hilarious incidents of early BBC 'live' reporting of a news event was the Review of the Fleet, one evening in 1937. The commentator had plainly enjoyed the spirit of the occasion too much before he went on air – the broadcast was a shambles as the commentator incoherently stumbled on, while no-one in London seemed to have the courage to pull the plug!

1.5.2 Getting the story back

From the 1920s radio stations relied on telephone lines, mobile transmitters and, later on, short-wave radio transmissions from abroad, to receive reports from remote locations. In the US, the networks, with more money to invest in new technology, started using hand-held portable high-frequency (microwave) transmitters for such events as party conventions. This gave the reporters increased mobility to roam the convention floors, and their comments and interviews, picked up by temporary control rooms set up at the convention centres, were then relayed by telephone lines to the network. For many years, neither NBC nor CBS allowed recorded material in their news bulletins or entertainment programmes, as they took pride in their live presentations, believing audiences would otherwise object. In any case, recording breaking news was impossible with the technology available in the 1920s.

Recording in the field became possible when the technology had developed further. The arrival of a recorder using steel tape, the 'Blattnerphone', in 1931 meant that material could now be recorded – but the equipment was large, heavy and impractical for newsgathering in the field. Editing metal tape required more the skills of a welder than a radio technician. From 1933 onwards, both the BBC in the UK and the US networks pursued the use of acetate disc recorders, but the machines were still not very practical for use in the field. Meanwhile, the Germans were developing the use of cellulose acetate tape impregnated with magnetic particles with the 'Magnetophon' system. In 1937, a disc recording made by a reporter sent to cover the arrival of the German airship the *Hindenburg* in New Jersey, and its destruction in the subsequent terrible explosion, was both the first time a recording had been broadcast on a radio network and also the first transatlantic broadcast from the US to Europe.

An example of how the BBC responded to a major UK news story in 1937 was the appalling East Anglian floods in which many died, where a BBC correspondent was sent to report on the disaster. His reports were not to be transmitted live, but instead a mobile recording team in a truck was also sent, who recorded the correspondent's words and the location sounds onto acetate discs, and then sent the disc recordings back via the guard on the London train.

1.5.3 The Second World War

During the Second World War, radio was the focus as a disseminator of news on both sides. Leaders on all sides used radio to make speeches to rally morale, and radio was considered vital by all sides as a way of ensuring that the war effort was sustained. The use of radio in such a political manner, of course, was fitting for the time. Listeners in the US and the UK kept abreast of the war listening to live-on-the-scene reports from Europe, North Africa and the Far East. The BBC developed the use of portable 'midget' disc recorders, weighing under 15 kg (30 lb), for recording reports from the battlefield, which were then flown back to London for transmission to give an element of 'actuality'. The Americans used wire recorders, using steel wire with a magnetic coating, which although more portable than the traditional disc recorders, still weighed over 22 kg (50 lb). They were still too large to be moved easily from one location to another, and some American correspondents jealously eyed the midget disc recorders used by the BBC correspondents. However, whatever the type of recorder used, without them many reports and sounds of the war would not have been heard by radio listeners, because military censors (and practical considerations) prohibited live broadcasts from the battlefield. After the war, continuing development work by the Americans using the acetate tape system pioneered by the Germans, led to the development of 6 mm (¼ in) 'reel-to-reel' tape recorders in the late 1940s. This recording format remained in common use until the development of the professional 'compact cassette' units in the 1990s.

1.5.4 The threat of television

From 1920 until the late 1940s, radio broadcasting enjoyed an explosive growth. But by the 1950s, with the development of television, radio stations found themselves fighting a similar battle for survival as the newspapers had decades earlier. Unfortunately for radio in the US, television was fast becoming the dominant medium as advertisers and audiences were drawn away from radio. The calibre of programming on radio had reached a peak, but many foresaw the demise of radio in the face of the 'trivialities' of television.

1.6 Television

Television (from the Greek 'seeing at distance') was a concept that, as with the history of radio, had seriously been discussed since the middle of the nineteenth century, and had some overlap with the development of wireless. The roots of television can be traced back to 1817, with the discovery of the light-sensitive element selenium by Jons Berzelius (Sweden). The first development was in 1839, when Edmond Becquerel (France) discovered a relationship between the voltage of a metal-acid battery and exposure to light. In 1878 Willoughby Smith (UK) showed that current flowing through selenium crystals changed with fluctuations in light falling on the crystals. In 1880 Alexander Graham Bell (US) developed the forerunner to the fax machine with his invention, the 'Photophone', based on a selenium cell, which could transmit still images over a telephone line. This led to the development of technology that enabled the news agency wire services to transmit photographs back via a telephone line.

1.6.1 The beginnings of television

In 1884 Paul Nipkow (Germany) took out a patent on his 'electric telescope'. This early conception of television used a mechanical scanning system with motors and large rotating discs, and consisted of a perforated disc with a spiral of holes, spinning in front of a selenium cell, which could produce images with scanning rates of up to 30 times per second. There is no evidence that Nipkow ever demonstrated his idea, as it was impossible with the technology available at the time, particularly to achieve the amplification or synchronization of the receiver to the transmitter.

In 1908, Alan Campbell-Swinton (UK) proposed the concept of electronic television in an article in a British science magazine, describing the principles of scanning, synchronization and display.

As with some other inventions, who can precisely be described as the inventor of television is still a subject of debate, even over seventy-five years later. The three principal claimants to the title are Philo Farnsworth (US), Vladimir Zworykin (US) and John Logie Baird (UK). Other contenders include Jenkins (US), Takayanagi (Japan), Rosing (Russia), Belin (France), Barthelemy (France), Karolus (Germany) and von Mihaly (Hungary). What is certain is that many small inventions and discoveries all contributed to the phenomena of television.

Of the other contenders listed above, Charles Jenkin (US) is of note because he actually produced a television service in the late 1920s and 1930s. In 1925, Jenkins demonstrated his development of the Nipkow system, which he called 'radiovision'. By 1928, Jenkins was transmitting 'radiomovies' on short-wave frequencies, but his audience was restricted to principally radio amateurs. Nevertheless, it kept interest in television alive in the US until NBC (part of RCA) launched an electronic television service in 1939.

Farnsworth conceived his idea of how to produce an electronic television system in 1922, and demonstrated it in 1927 when he transmitted a 60-line image of a dollar sign. Farnsworth patented a complete electronic television system in 1927, based on an 'image dissector' used to scan the image for wireless transmission. At the receiver, an 'oscillite' tube reproduced the picture, and an electron multiplier tube, a 'multipactor', increased the sensitivity of the image dissector. Engaged in time-consuming patent litigation with RCA for over ten years, Farnsworth finally signed a cross-patent agreement with RCA in 1939, opening the way for electronic television in the US.

Zworykin, who worked at Westinghouse with Conrad, invented the 'Iconoscope' (patenting it in 1923) and the 'Kinescope', which were the forerunners of today's television camera and picture tubes. He developed an electronic scanning television system using his inventions, and demonstrated a working system in 1929. Sarnoff tempted him away from Westinghouse in 1929 to develop an electronic television system at RCA.

Baird refined Nipkow's spinning disc system to demonstrate successfully a mechanical television system in 1925. It has been said that Baird only pursued mechanical scanning to get a television system working as quickly as possible, as he lacked the expertise to develop an electronic system. Baird's 'Televisor' used a 16-line scanning disc camera and receiver, with a neon tube as a light source at each end. However, this produced a picture only about 12 mm (1 in) square. By 1926 he had further developed the system to 30 lines of scanning with a larger screen size. In 1928, Baird achieved the first transatlantic television transmission, and shortly afterwards the first demonstration of colour television. The BBC experimented with Baird's mechanical 30-line system via radio transmissions between 1929 and 1935, starting a regular broadcast service in 1932, while Baird tried to further improve his mechanical system. From 1930, with BBC transmissions underway, Baird launched his system onto the market, and by 1936 his system produced 240 lines of scanning. But by 1935, the BBC decided to switch to the 405-line electronic system developed by EMI in the UK, based on a derivative of Zworykin's Iconoscope from RCA. EMI was associated with RCA from the days when the London companies, Victor

Talking Machine Company in Camden and Gramophone Company in Hayes (from which EMI was born), were related.

Throughout the 1920s and 1930s, a number of countries experimented with various mechanical and electronic methods of television, and there are many claims made as to who was first with a true television system or service.

It is acknowledged, however, that in 1935, the Nazi government in Germany inaugurated the world's first public television broadcasting service (using the Farnsworth system) and, on the opening night, broadcast speeches by Hitler and other senior party officials. It also broadcast the 1936 Olympics in Berlin. The BBC officially launched its public electronic television service in 1936, but transmissions were suspended in September 1939 at the outbreak of the Second World War, when it is estimated that 20 000 television receivers had been sold to the public, and the service did not resume until 1946.

In the US, NBC began work on developing a television service in 1935 and launched its television service in 1939. During the World Fair in New York in 1939, Franklin Roosevelt made a speech that was transmitted by NBC on their inaugural service – this transmission was the first use of television by a US president. In 1931, CBS put an experimental TV station on the air in New York and transmitted programmes for more than a year before becoming disillusioned with the commercial aspects of the new medium. Following NBC, CBS began development of a commercial television system in 1937, with a public service beginning in 1941.

1.6.2 Early television news

Local experimental stations in the US began 'newscasts' during the 1930s, though they were restricted to announcers reading news headlines. At least one station is reputed to have begun showing 'newsreels' sometime in the mid-1930s. Newsreels were produced by motion picture studios and shown in cinemas as part of their entertainment schedules, and when the television experimenters wanted to show moving pictures in their newscasts, they used the newsreels.

Covering breaking news was a challenge for television because it was an expensive and awkward proposition to equip vehicles for remote broadcasts with the heavy, bulky cameras and transmitting equipment. In 1939 in New York, NBC developed the ability to undertake 'remotes' (outside broadcasts) with two vehicles – one transporting the transmitter for the link to the studio, the other carrying the camera and production equipment. The BBC did not attempt any remote television news coverage until well after the end of the Second World War.

1.7 Television news after the War

In the UK, the BBC started television transmissions again in 1946, and for the next eight years ran cinema newsreels alone as the sole source of news. In the US, NBC and CBS launched their first regularly scheduled newscasts, but the programmes only went out once or twice a week. As in the UK, the principal diet of moving pictures from outside the studio came from newsreels. These contained feature material rather than hard 'news', compiled by the news agencies and the motion picture studios, and they were the equivalent of a weekly magazine (the most famous newsreels were MGM/Hearst Metrotone, Paramount, Movietone and Pathé). As a result, the newsreels tended to cover planned events and timeless features, edited with titles, music and commentary, so that viewers saw a lot of departures and arrivals of politicians, dignitaries and celebrities, fashion shows, news conferences, and ship arrivals and departures.

In the UK, it was not until 1953, after the sudden rise in popularity of television with the coverage of the Coronation of Queen Elizabeth II, that the BBC established a dedicated television news department. In 1954, the BBC showed their first television news bulletins, but they were visually uninspiring.

On both sides of the Atlantic when breaking news occurred, the audience were offered only a few studio visual aids such as still pictures, charts and maps, or perhaps an on-the-scene report filed by a reporter using the telephone, to accompany stories that were read by the announcer. Because live video from the scene could not be delivered instantly, except in rare situations, television offered little more than radio when immediacy counted, and so these programmes were little more than radio news bulletins combined with still pictures and some newsreel footage.

In the UK, Independent Television News (ITN) was launched in 1955, a dedicated television news company supplying news bulletins to the UK's Independent Television (ITV) commercial channel, and this forced the BBC to further develop its television news output with its own dedicated news film camera crews.

1.7.1 'Film at eleven'

In the US, with newsgathering for television based on film cameras, local stations starting using the 'film at eleven' ruse to encourage viewers to watch a later newscast, by which time the filmed report of a news event would be ready. Having pictures to show in a newscast was highly desirable, as was creating a sense of immediacy. Some stations, as early as the 1950s, began pushing 'eyewitness' formats, based on CBS's idea. For instance, the anchor would introduce the reporter in the studio who had returned after covering a big news story. The reporter would then provide a live 'eyewitness' account of the event and introduce the film report shot at the scene. It was a compromise that worked, as the audience had not come to expect anything more immediate.

1.8 Newsgathering technology before ENG

The technology of newsgathering for television was based on film cameras, and there was the problem of recording the sound and lighting the scene as well. The cameras first used were derivatives of 35 mm cinema film cameras and were, therefore, bulky and heavy, and required considerable electrical power. In the late

1950s, smaller and lighter 16 mm cameras were developed, weighing under 10 kg (22 lb). Sound was recorded onto a magnetic track striped on the film itself and two-man crews were needed to shoot the film and record the sound, with a lighting electrician required where shooting was to occur in poor or difficult lighting conditions. Once the film was shot, there were problems in preparing the film for transmission. It had to be physically transported back to the studio, usually by road by the film crew themselves, or by use of other methods of transportation. Back at the studio, there was at least a 40-minute delay while the film was processed before it could even be edited. The whole process was lengthy, even with the fastest and most skilled technicians.

Television networks did cover news events live from remote locations, but they required considerable notice and pre-planning. Bulky, awkward equipment had to be transported and multiple telephone circuits ordered to set up remote broadcasts. In the US, the audience draw of the political party conventions was a strong influence in establishing the infrastructure required. By 1952, the US networks were linked east to west with coast-to-coast coaxial cable circuits, which provided a nationwide audience for the political parties and their nominees. In the early 1950s, US Senate hearings into organized crime, the Eisenhower presidential inauguration and the McCarthy hearings all offered opportunities for television to deliver news as it happened, and were treated as special coverage. But in each case, arrangements to cover these events were made well in advance.

In the UK in 1953, the BBC in co-operation with the Post Office, had arranged for a coaxial cable ring to be set up around central London (called the 'LoCo') with spurs to various important venues in time for the Coronation. This network was further developed and widely used, only finally disappearing in 1997. The use of cable circuits was not universal – as a result of the radar technology developed during the war, microwave transmitters and receivers had been developed and were used from the 1950s onwards. However, this early transmission equipment was bulky and not particularly reliable, although the development of solid-state electronics (as opposed to the use of vacuum tubes) through the 1960s established terrestrial microwave link technology for coverage of remote events.

Live coverage of breaking news events during regularly scheduled news bulletins was difficult to arrange in those days, because television lacked the technology for speed and mobility. In both the UK and the US, dependence on newsreels ended in the mid-1950s as television stations assembled their own film crews, but delays in showing news pictures were still a problem because of the time required to transport, process and edit film.

In the US, the networks established permanently leased 'wideband' telephone circuits that linked the network centres in New York to their affiliate stations. Edited material could be transmitted over these circuits, making same-day coverage of news events possible, as long as the film crews had enough time to complete their shoot, drive to the nearest affiliate with connectivity to New York, and then process and edit the film before sending it. Only significant stories were sent by this method, because of the very high cost of transmission. Normally the film was flown to New York for processing and editing, and would typically not be broadcast until the next day.

To cover foreign stories during this era, the only reliable way of moving pictures from one part of the world to another was by air, meaning there was a heavy reliance on scheduled airline flights, which were much less frequent than today. This was a very slow way of bringing 'breaking' news to the television screen. To tackle the problem of quickly getting pictures from the US, the BBC used the 'Cablefilm' system, developed during the 1950s. With this process, film pictures had to be obtained (to cope with the use of the different TV standard in the US) which were then converted to electronic pictures and transmitted by a 'slow scan' process while being recorded on film in the UK. The resulting pictures were jerky and of poor quality, and the only transmission of note was the coverage of the arrival of Queen Elizabeth in New York on a state visit in 1958. The process was abandoned shortly afterwards because of the poor results.

In 1962, to try to speed up newsgathering on film, the BBC started experimenting with mobile film processing. The Mobile Film Processing Unit consisted of three vans – nicknamed the 'three ring circus' – which could travel to the scene of a major news story. There was a processing truck, an editing truck and a telecine (film projection-to-television conversion) truck, so the pictures could then be connected to a terrestrial microwave links truck to transmit the pictures back to the studio. However, the system was cumbersome and not particularly successful, and the Unit was dismantled in 1966.

It was clear that an alternative to the use of film was required. Videotape recording was invented during the 1950s, based on 50 mm (2 inch) wide tape, but it was of no use to newsgathering in the field, as the machines were large and designed to be bolted to the floor!

It was not until the advent of electronic newsgathering (ENG) in the 1970s that the solution to the problem of transmitting news as quickly as possible was found. This solution was rapidly developed in the US, where television news was then (and still is) a very aggressive area of the broadcast industry. In a revolutionary development, over approximately a five-year period, the problems of shooting, recording and editing in the field were solved.

1.9 Electronic newsgathering

The rise of the electronics industry in Japan through the 1950s and 1960s, along with the invention of the transistor and the beginning of progress in miniaturization, led to the development of electronic newsgathering. The first problem was the recording medium, and the solution of that problem coincided with the development of compact portable electronic cameras.

As mentioned earlier, 2 inch ('quadruplex') videotape was the standard broadcast television tape format (and remained so until the early 1980s). In Japan, Germany and the US, television video recorder manufacturers worked on producing a smaller tape format for a number of reasons, newsgathering being only one, but the result would be smaller and more portable video recorders. The standard that finally became predominant was the Sony U-matic ¾ inch (19 mm) standard, which remained dominant until it was superseded by the Sony 'Beta' ½ inch (12 mm) format in the mid-1980s. (Currently the battle is between the digital 12 mm and 6 mm formats.)

Soon after the solution to the tape format was becoming clear, the camera manufacturers were working on reducing the size and weight of the television camera. The result was that, by the late 1970s, film was rapidly disappearing as the standard medium for newsgathering, to be replaced by the compact camera and recorder. In the mid-1980s, the first one-piece camcorder appeared, based on the Sony 'Betacam' standard.

Now that the pictures were in electronic format in the field, and the tape recorders and players were smaller, portable editing equipment for field use was now practicable. All the equipment could be transported easily in one small truck that was customized for field use and equipped with a portable microwave transmission system. Television stations now had the technology to record on-site, or broadcast live, electronic pictures from a remote location.

In the US the network-owned stations were the first to be equipped with the new ENG equipment. A Los Angeles station provided one of the first dramatic examples of live ENG reporting. The station broadcast 'live' a police shoot-out with a radical group calling itself the Symbionese Liberation Army. For two hours the station had live coverage of the gun battle, though the live pictures stopped before the shoot-out ended because the camera batteries in operation went flat.

ENG was the biggest change in newsgathering since the invention of wireless. The next big development was the use of satellites.

1.10 Satellites and newsgathering

The development of satellite communications technology was the next leap forward in newsgathering, which ran in parallel to the development of ENG. The technological drive to develop satellite technology came, of course, from the military, but news organizations (and television and radio audiences) ultimately benefited as well. In the 1950s, with the Cold War at its peak, the 'race to space' between the US and Russia was intense. The successful launch of the Sputnik satellite by the former USSR in 1957 was a bitter blow to the US, who were thus spurred on even more.

TELSTAR 1, launched for AT&T in 1962, was the first communications satellite, but it had limited use because tracking stations were necessary to follow its non-geosynchronous orbit. The following year another satellite, SYNCOM II, was put into a geosynchronous orbit at an altitude of 36 000 km (22 500 miles). At this height, the satellite moved in synchrony with the Earth's rotation, thereby remaining stationary relative to the Earth's surface (see Chapter 2). The first commercial communications satellite in geosynchronous orbit, INTELSAT I ('EARLY BIRD') was launched in 1965 (see Chapter 7). As the number of communications satellites increased, more opportunities were available to deliver news from distant locations.

1.10.1 In the US

For the US networks, the Vietnam War helped promote the use of satellite technology as a newsgathering tool. The networks needed a way to obtain video from overseas that was less time-consuming than by air, and satellites provided the means. This was achievable between fixed earth stations, but was not possible in the mobile environment.

In the 1970s, the only transportable satellite earth stations were large and expensive 'C-band' vehicle- or trailer-mounted uplinks, which could be used for major events. However, they could not easily be used at short notice because of the FCC requirement for site surveys before a permit to transmit could be issued, due to the restrictions on the use of C-band (see Chapter 6). Hence, these uplinks were used for events such as sports or for political party conventions in the run-up to presidential elections that were going to be carried by the networks.

Widespread use came with the availability of 'Ku-band' satellite capacity in the early 1980s. Operation on Ku-band capacity was not subject to FCC restrictions, making it more suitable for news, and the uplinks could use cheaper, smaller antennas and systems than those typically used for C band.

In 1984 Stanley Hubbard, the owner of a television station in St Paul, Minneapolis, organized the 'Conus' satellite news co-operative for local television stations to share news material by satellite. At the same time, he encouraged stations to buy vehicles customized for satellite newsgathering, manufactured by another of his companies in Florida, Hubbard Communications. The Conus News Service is still running today and includes local US television stations and as well as international news organizations, who exchange news coverage and collaborate on custom coverage of breaking and planned news events.

The Conus satellite news co-operative spurred many local stations to invest in SNG. Many early systems were constructed by adapting larger, fixed earth station antennas (known as 'cutting') and utilizing other components designed for fixed earth stations. Many stations built their own trucks rather than commissioning construction by bespoke vehicle builders. In these early days there were a few other vehicle manufacturers apart from Hubbard Communications, such as Dalsat and Encom (now long gone).

Nevertheless, through the 1980s a number of companies became established in this area of the broadcast vehicle market, and the current principal companies in the US (and the rest of the world) involved in building SNG systems are listed in Appendix K.

1.10.2 In Europe

The development of SNG in the UK stemmed from the work of the Independent Broadcast Authority (IBA) and GEC McMichael in the late 1970s and early 1980s on trailer-mounted satellite communication uplinks. The IBA was the broadcasting authority for independent television in the UK and, along with the BBC, was very interested in the development of SNG. In 1978, the IBA built a trailer-mounted Ku-band satellite uplink system with a 2.5 m antenna which it was using to experiment with the OTS-2 satellite, with a view to developing a system that could be used for newsgathering.

GEC McMichael was a company with many diverse interests, including fibre-optic technology, television monitors, video-conferencing and television standards conversion, and its strategic position within the GEC group of companies was to act as the conduit for transferring relevant research and development in its defence activities into commercial applications. (These diverse applications may seem at first sight strange bedfellows but, on closer examination, it can be seen that satellite communications, video-conferencing and television standards conversion did have a common linking thread.)

McMichael's position was also unusual within the rigid GEC group in that the philosophy of the McMichael company was to be more liberal, encouraging imaginative thinking from its engineers in order to create new commercial products as a spin-off from military development work.

In 1979, GEC McMichael began to work on developing a transportable satellite system aimed at the news and special events market, with the assistance of engineers from Marconi Space & Defence, another company within the GEC group whose prime business was military satellite communications. One of the biggest challenges was to design an antenna that could meet the rigorous new specification for 2° spacing of satellites required by EUTELSAT, and subsequently adopted by INTELSAT and the US FCC (see Chapters 2 and 6). These new specifications were enforced by EUTELSAT in Europe, but neither INTELSAT nor the FCC actually insisted on these requirements being met for a number of years (partly because there were no satellites in the geostationary arc yet spaced at 2°), much to the relief of US satellite system manufacturers. However, this did prevent most US antenna manufacturers from being able to sell their products into Europe until they could achieve the required performance specification.

By 1983, GEC McMichael had, like the IBA, also developed a Ku-band transportable uplink on a trailer, which could be used for either analogue or digital transmissions. It was used to demonstrate *digital* video-conferencing at the annual defence systems exhibition at Farnborough in 1983 – this was over ten years before Digital Satellite Newsgathering (DSNG) became a reality.

The next objective was to develop the first transportable flyaway SNG uplink that could meet the rigorous 2° satellite spacing requirements. Another company in the GEC group, EEV, was brought in to develop compact high-power amplifiers required for the new flyaway. By 1984 the prototype flyaway was produced and rigorously tested to meet the EUTELSAT specification, and the GEC McMichael 'Newshawk' flyaway was launched at NAB in 1985 – the first SNG flyaway able to meet the 2° spacing specification (Figure 1.2). The US company Comsat General also exhibited their first Ku-band SNG flyaway system, but they did so more as a design concept rather than a commercial product.

The Newshawk flyaway, with its flat elliptical antenna and relatively lightweight, compact construction created a great deal of interest. In 1985, two systems were sold, one to CBS in the US and one to ITN in the UK. Both systems saw their first

Figure 1.2 GEC McMichael Newshawk (ITN) in Kumming, China (for the state visit of Queen Elizabeth II, 1986)

serious outing in 1986 to cover the Reagan-Gorbachev Summit in Reykjavik in Iceland, and they remained in service with both companies for over ten years.

But by the end of 1985, GEC had decided to dissolve the McMichael company and absorb the various activities back into the other companies within the GEC group. Three engineers who had moved to GEC McMichael in 1980 from Marconi Space & Defence to work on the original transportable satellite uplink project, and who were at the heart of the development of the Newshawk, decided to leave GEC and set up their own company to manufacture SNG flyaways. This company became Advent Communications, which had developed its 1.5 m Mantis flyaway system by 1986 and sold its first system to the BBC in 1987, and went on to become the byword globally for SNG flyaways for nearly ten years.

Advent produced their first SNG trucks in 1989, building four 7.5 tonne (16 500 lb) trucks with uplink systems based on their 1.9 m Mantis flyaway product, swiftly followed by a number of 3.5 tonne (7700 lb) vehicles also based on the 1.9 m Mantis flyaway. The company now produces a volume of systems based on a range of vehicle sizes, in addition to their range of flyaway products.

Across the rest of Europe, the uptake of SNG was slow, due both to the very high cost of the equipment and because the use of temporary terrestrial microwave links was well established. The distances to be covered in any one country were not on the same scale as in the US, and therefore there was not the same impetus. However, through the 1990s, other companies in the UK and Europe started producing SNG trucks and flyaways.

1.10.3 Digital SNG

The development of digital SNG in 1994 in Europe heralded the second era in SNG. The cost of the equipment in Europe came very close to matching that of analogue systems and the antenna size reduced, opening up the possibility of using smaller vehicles than the typical 2 m analogue SNG antenna allowed. Most importantly, the lack of ad hoc analogue capacity, combined with the lower cost of digital capacity, could more than offset the slightly higher capital cost of DSNG systems. In Europe, DSNG grew quickly, primarily because of this lack of analogue capacity (particularly ad hoc), as the economic demands on bandwidth meant that more channels could be carried on a satellite, and because a viable digital compression system – the ETSI 8 Mbps derivative (see Chapter 4) – became widely used. The EBU Eurovision News network adopted this compression standard for DSNG in 1995. In fact, the speed of the demise of analogue SNG was surprising, and in Europe, by 1997, DSNG had become the norm rather than the exception.

It was not until 1998 that DSNG started to take a hold in the US. This has been attributed to a number of factors:

- There was not the same pressure on capacity until the loss of the satellites Telstar 401 in early 1997 and Galaxy IV in 1998; for, although these particular satellites carried Direct-to-Home (DTH) traffic, their disappearance created a squeeze as the DTH traffic was shifted to other satellites.

- The cost of MPEG-2 compression equipment began to fall (the US was never interested in pursuing the European ETSI digital compression standard), and through 1999 there was a significant drop in the cost of encoders from some manufacturers.
- The thrust by the CBS network, including its affiliate stations, towards digital domestic contribution.
- The prospect of all television production processes moving into the digital domain with the introduction of HDTV.

This last factor is perhaps not as significant as it might first seem, as no US network had committed to newsgathering in high definition by the end of 1999. But, whatever the reasons for the delay, there is no doubt that the US is now well on its way to embracing DSNG.

1.11 Impact of SNG

As the use of SNG grew through the 1980s, the number of SNG uplinks increased as they were used to cover the major international news events. Such events included: the TWA hostage hijack in the Mediterranean in 1985; the Reagan-Gorbachev Summit in Iceland in 1986; the PanAm airliner disaster in Lockerbie, Scotland in 1988; the Tianamen Square crisis in Beijing, China in 1989; the fall of the Berlin Wall in 1989; and the release of Nelson Mandela in South Africa in 1990.

Without doubt, the event that brought SNG to the fore in newsgathering, as well as having an enormous impact on the public, was the conflict in the Persian Gulf. It began in August 1990 when Iraq invaded Kuwait, culminating in the six weeks of fighting after the Allied attack in early 1991.

All the major newsgathering organizations had a presence across the Middle East, particularly in Baghdad where they were invited by Saddam Hussein from the beginning in 1990 to witness the effects of 'Western aggression'. In the US, the CNN news channel was heralded as the star of news organizations for its early coverage of the Gulf war. CNN was alone in providing live and continuous coverage throughout the first night of the Allied attack on Baghdad on 17 January 1991 (though the coverage was by voice-only, on CNN's 'four-wire' communications channel). The other US networks lost their telephone connections from Iraq during the initial shelling, but their correspondents were able to report live from Saudi Arabia and Israel, which faced possible Iraqi Scud missile attacks. All around the world, on any channel carrying the news, correspondents could be seen donning gas masks to prepare for Scud attacks, repeated night after night. Audiences felt a sense of 'being there' as the correspondents in Saudi Arabia, Baghdad and Israel provided the drama, rather than combat troops, who were seen less often. The correspondents under fire from Iraqi Scud attacks provided an illusion of audience participation in the drama of the war and, for the US, the fact that the TV personalities could actually be hit involved the audience in a new kind of live drama. The combined coverage of ABC, CBS, CNN and NBC attracted 118 million viewers in the US on the first night of the war.

Such was the impact of the Persian Gulf War coverage that virtually no major international crisis or conflict since has escaped being covered by the use of SNG uplinks in some form, either by an SNG flyaway or an INMARSAT satphone. Bosnia, Somalia, Rwanda, Kosovo and East Timor have all featured on news agendas around the world, and it is questionable whether they would have received such prominence if there had not been such instant availability of live or 'near-live' reports.

1.11.1 The 'CNN effect'

Finally, we must mention a phenomenon widely referred to as the 'CNN effect', which is directly attributable to the rapidly growing influence and effect of satellite newsgathering. This phenomenon has evolved in particular from the supposed dramatic impact of CNN news broadcasts on US government decision making in times of war or economic crisis because of the immediacy of the media coverage. For instance, it is widely acknowledged that the US went into Somalia and Rwanda partly because of this influence, and the CNN effect is alleged to heavily influence overall US foreign policy[4]. The operation in Somalia in particular is said to have been set up with coverage by the US networks primarily in mind, and its disastrous consequences ending with the humiliating withdrawal of the US military still haunts US policy makers.

Furthermore, the CNN effect is now said to be exercised by Western television media in general, influencing other governments to take actions that will be widely reported by the global newsgathering organizations. It is said that these organizations have no respect for political boundaries, and pressure is put on governments to act because of their population's emotional responses to televised images. Governments feel compelled by television-generated public opinion to intervene in overseas conflicts, creating situations that would otherwise not occur. Particularly in times of war, there is a debate as to the role of the media, and questions have been raised as to whether media coverage is forcing policy makers to alter their approach.

It is certainly true that in the Gulf conflicts, the Iraqi government has always been keen to see the Western media presence in Baghdad – although always under severe restrictions. On the other hand, in the recent Kosovo conflict, the Serbian government became positively hostile towards newsgathering organizations from NATO-aligned countries remaining in Belgrade, and a number of organizations had staff temporarily imprisoned and SNG uplink equipment confiscated – an action not seen in any other conflict up until that time.

The political influence is mentioned here only because it underlines the dramatic rise in the significance of satellite newsgathering, not only in terms of audience ratings for network television, but also in the way it has extended to government policy formulation. This is a widely debated issue, and although it is probably true to say that there was a period when such political influence undoubtedly did exist, it is questionable whether this is still the case. The extent of the CNN effect is thus a subject of some debate and in recent years its influence has been played down by some observers[5,6].

1.12 Conclusion

Satellite newsgathering has evolved from the technological seed of the electric telegraph in just over a century, and there are several interesting parallels between the telegraph and digital SNG in particular. Both involve a compressed method of coding, both revolutionized newsgathering, and both have had a lasting effect on the impact of news.

If television bulletins of twenty or thirty years ago are viewed today, the difference is startling, and one is left marvelling at where we have got to today. From the stiff and visually boring bulletins of the 1950s, we have progressed to the visually and aurally stimulating style of the 1990s. Over this period, an increasing amount of national and international news has been customized down to local station level. Live interaction between news anchors and on-the-scene correspondents has become an everyday occurrence – expected rather than unusual – and is achieved with ease by the interconnectivity that SNG allows.

In the following chapters we will look at how the technology in the sky and the equipment on the ground are used together, and the processes involved in getting news from out there to the audience at home by the quickest means.

References

1. Gowing, N. (1994) Real-time television coverage of armed conflicts and diplomatic crises: does it pressure or distort foreign policy decisions? Working Paper 94–1, Shorenstein Center on the Press, Politics and Public Policy, John F. Kennedy School of Government, Harvard University.
2. Dunsmore, B. (1996) The next war: live? Discussion Paper D-22, Shorenstein Center on the Press, Politics and Public Policy, John F. Kennedy School of Government, Harvard University.
3. Barnard, P. and Humphrys, J. (1999) *We Interrupt This Programme*. BBC Consumer Publishing.
4. Livingston, S. (1997) Clarifying the CNN effect: an examination of media effects according to type of military intervention. Research Paper R-18, Shorenstein Center on the Press, Politics and Public Policy, John F. Kennedy School of Government, Harvard University.
5. Livingston, S. and Eachus, T. (1995) Humanitarian crises and US foreign policy: Somalia and the CNN effect reconsidered. *George Washington University, Political Communication*, **12**, 413–429.
6. Gowing, N. (1997) *Media Coverage: Help or Hindrance in Conflict Prevention*. Carnegie Commission on Preventing Deadly Conflict.

2

From launch to transmission: satellite communication theory and SNG

2.1 The birth of an idea

In October 1945, the now famous science fiction author Arthur C. Clarke (who was at that time an RAF technical officer) wrote an article that was published in the British magazine *Wireless World*. In this article, he put forward an idea about how the coverage of television transmissions could be improved by using radio 'relays' situated above the ionosphere[1]. Clarke had been working for some time on this concept, which was an astonishing proposal for the time and of such breathtaking prescience that, even now, as one reads the original article, it is uncanny how closely today's satellite communications system mirrors his prediction. Currently, there are over 250 satellites in 'geostationary orbit' and in total over 8500 artificial satellites in various types of orbit around the Earth, not only serving commercial telecommunications needs, but also experimental, scientific observation, meteorological and military purposes.

In his article, Clarke extrapolated the developments that the Germans had made in rocket technology, witnessed in the V1 and V2 rocket attacks on south-east England during the war, and proposed that a rocket of sufficient power could be launched from the Earth's surface into a high orbit. It would be steered on its way by radio control into an orbit that exactly matched the speed of the Earth's rotation – this orbit is approximately 42 000 km from the centre of the Earth. In this way the rocket would become a 'space station', and antennas on the space station could receive a signal from a transmitter on the Earth's surface and re-transmit it back to a wide area. At that time it was not known if signals from the Earth's surface could penetrate the Earth's atmosphere and reach space but, postulating that this would be possible, he calculated the theoretical downlink power from the space station to cover the Earth's surface with enough signal to enable television reception. He also predicted that three satellites placed at three longitudinal positions around the Earth (30°E, 150°E and 90°W) would provide global coverage (Figure 2.1). The relationship between the Earth and the geostationary orbits, relative to the orbit of the Moon around the Earth, is shown in Figure 2.2.

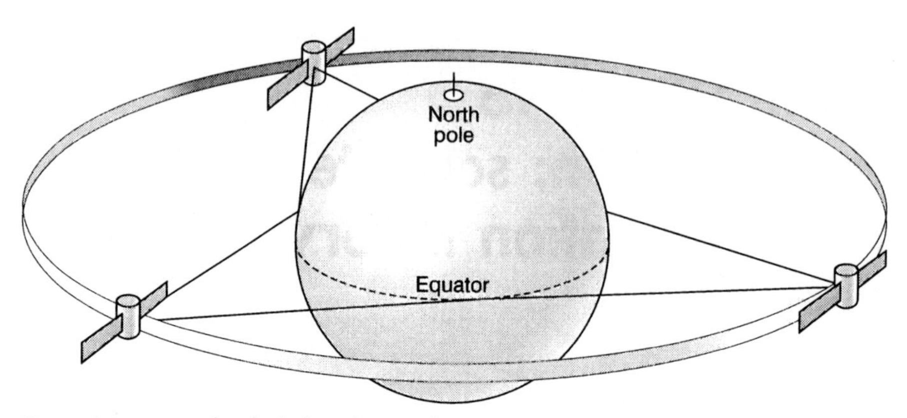

Figure 2.1 Principle of Clarke's theory of covering the Earth

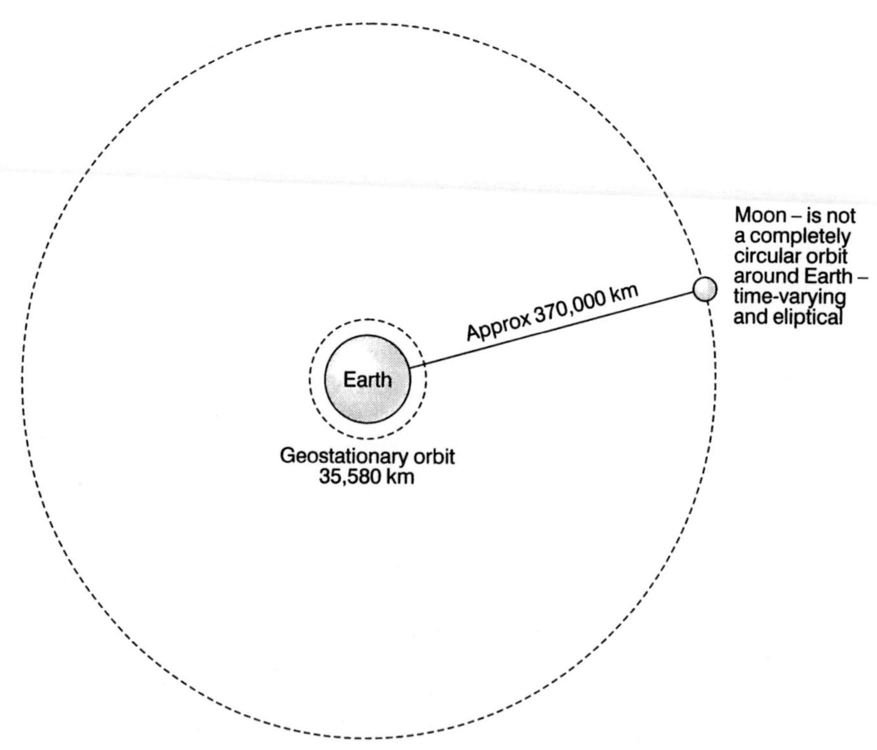

Figure 2.2 Relationship of the orbits of geostationary satellites and the Moon relative to the Earth

In his honour, the orbit that he proposed – which we now know as a 'geostationary Earth orbit' (GEO) – is named the 'Clarke belt' or the 'Clarke orbit'. Clarke never claimed to have invented the geostationary orbit, stating that he simply made a natural deduction from the laws of Newton and Kepler. There are other types of orbit as well (see Chapter 9), but the geostationary orbit is the one currently utilized for satellite newsgathering. Thus, this is the starting point for our discussion in this chapter.

Although it is not necessary to have knowledge of how the satellite arrives in orbit ready for use, and the actions that take place on the satellite during its life, it is useful background for an understanding of the process of satellite newsgathering.

2.2 Geostationary orbit

2.2.1 Principles

The Earth turns about its polar axis one revolution every 24 hours. (Strictly speaking, it actually rotates in 23 hours 56 minutes and 4.1 seconds – this is referred to as the 'sidereal' cycle, which equals one sidereal day.) It can be calculated that,

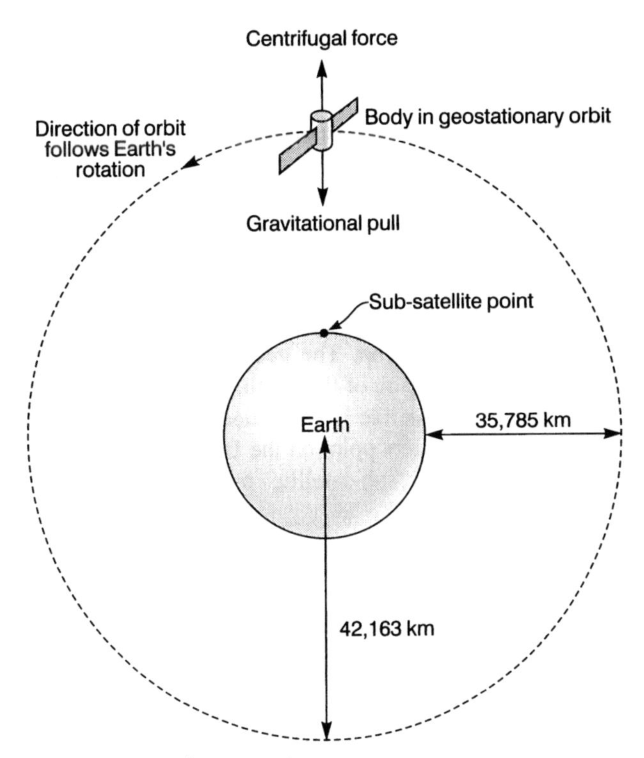

Figure 2.3 Geostationary orbit principle

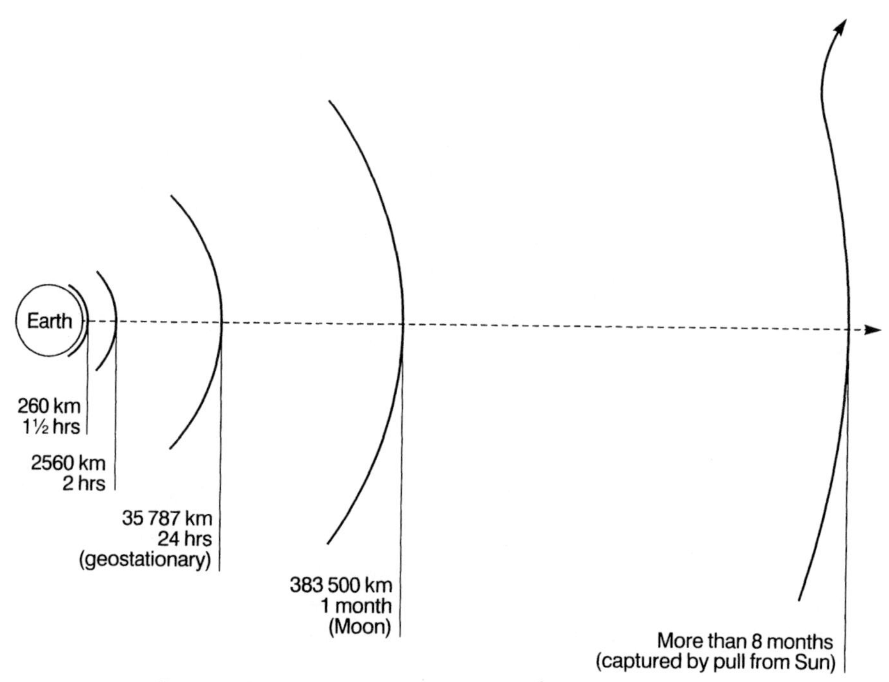

Figure 2.4 Differing orbits in time and distance from the Earth

at a specific distance from the Earth, a satellite in a circular orbit around the Earth will rotate at the same rate as the Earth. This will be essentially in an orbit where the centrifugal force of the satellite's orbit is matched by the gravitational pull of the Earth, as shown in Figure 2.3. The effect of varying the distance from the earth and the resulting effect on the period of the orbit is shown in Figure 2.4.

This satellite orbit is measured from the centre of the Earth, and is called a 'geosynchronous' orbit. If that satellite orbit is above the Equator, then the satellite is said to be in 'geostationary' orbit. The geostationary orbit is at a distance of 42 163 km measured from the centre of the Earth, and the Earth's radius is 6378 km at the Equator. Therefore the satellite is calculated to be 35 785 km (see Appendix B) above the Earth's surface at any point on the Equator. The point on the Equator below the satellite is termed the 'sub-satellite' point.

2.2.2 Longitude and latitude

For those of you who only have a hazy memory of geography, let us briefly look at the principles of longitude and latitude, as this is implicit in understanding how we refer to points on the Earth's surface.

Imaginary symmetrical lines of reference run from the North to the South Poles – these are called lines of longitude ('meridians'), some of which are shown in Figure 2.5. The line that runs through Greenwich in London, UK, is termed the

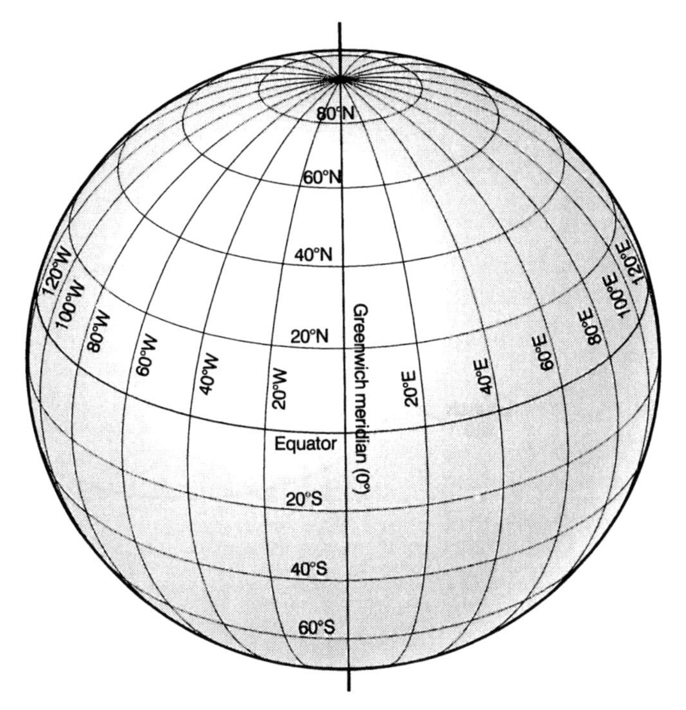

Figure 2.5 Principle of longitude and latitude

Greenwich Meridian, and is the 0° longitude reference point. Lines run North to South around the complete 360° circumference the Earth. Any point can be referred to as being at a certain point °E or °W, depending in which direction around the Earth the point is being referenced. (Satellite orbital positions in the geostationary arc are also referred to in the same way.)

Similarly there are parallel lines that run in rings around the Earth from the North to the South Poles – these are lines of latitude ('parallels'), some of which are also shown in Figure 2.5. The line that runs around the Earth at its middle is the Equator and is the 0° latitude reference point. Lines of latitude run from 0° to 90° running North (Northern Hemisphere), and similarly 0° to 90° from the Equator to the South (Southern Hemisphere). Any point is referred to as being at a certain point °N or °S, depending on whether it is in the Northern or Southern Hemispheres respectively.

2.3 From launch to orbital position

2.3.1 Into orbit

The satellite is placed into a geostationary orbit by use of a rocket launch vehicle. Ideally, the launch vehicle would lift the satellite straight from the Earth's surface to the destination orbital position. However, the amount of force and hence power

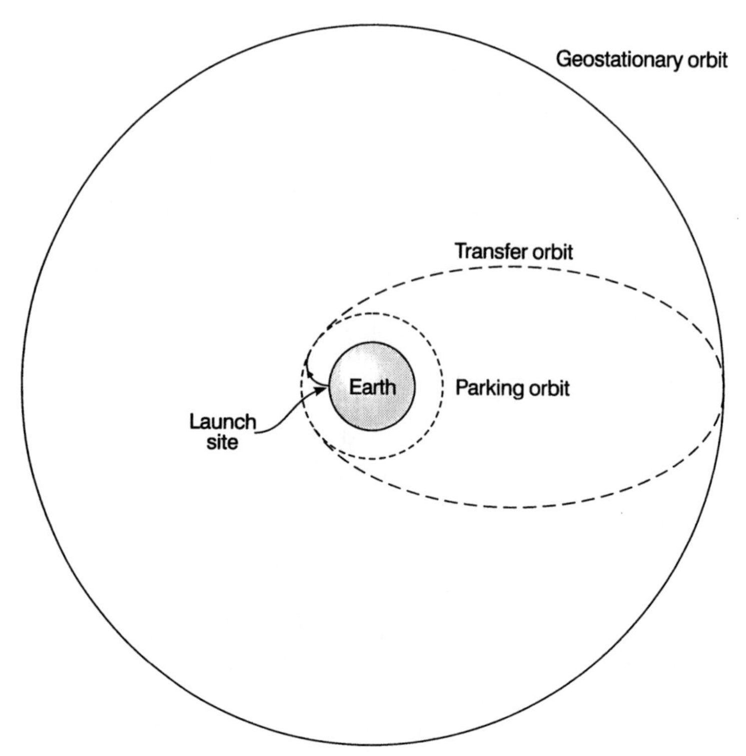

Figure 2.6 Phases in placing satellite in geostationary orbit

required are far beyond current capabilities, and the centrifugal force of the Earth's rotation has to be used in combination with the rocket's motors to catapult the satellite, in stages, into the correct orbit. The ideal launch site is at or near the Equator, to be as close to the final orbital plane as possible, and also to harness the 'sling effect' of the Earth's rotation, which is greatest at the Equator.

In a geostationary orbit mission, there are essentially four phases: the launch, the parking orbit, the transfer orbit and the final geostationary orbit positioning, as shown in Figure 2.6. There are a number of different types of launch vehicle used for the launch of commercial satellites, varying in size and range depending on customer choice and the payload, and types include Delta (US), Atlas (US), Ariane (European Space Agency or ESA), Proton (USSR) and Long March (China). Commercial satellites are made by companies such as Hughes Space & Communications (US), Space Systems Loral (US), Lockheed Martin (US) and Alcatel (France).

Typical costs to place a satellite into a geostationary orbit are US$100 million for construction of the satellite, and a further US$100 million for the launch, and US$40 million for insurance. On average, the failure rate of launches is 10%; hence the high cost of insurance. The satellite, carried in the nose cone of the rocket launch vehicle,

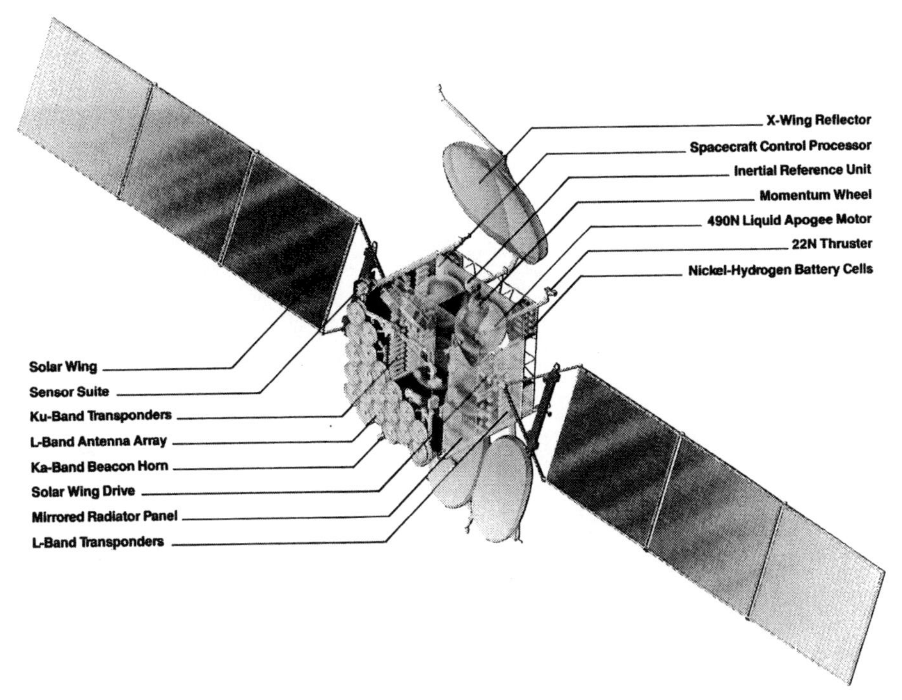

Figure 2.7 Hughes HS-601 satellite. *(Photo courtesy of Hughes Space & Communications)*

is very small relative to the size of the rocket. Primary launch sites for commercial satellites are near to the Equator and include, for example: Cape Canaveral, Florida, US; Kourou, French Guiana; Baikonur, Kazakhstan; and XiChang, China. The launch vehicle is composed of a 'payload' module, containing either one or a number of satellites depending on the size of the satellites and the launch vehicle, and a multi-stage set of motors. Typically, there are three stages, which detach from the vehicle in sequence as each one burns out during the launch phase.

To describe the whole process, we can consider a launch from Kourou, French Guiana, which was similar to that used to put PanAmSat's PAS-6B satellite into orbit in late December 1998. PAS-6B is an HS-601 series satellite built by the Hughes Space & Communications Corporation. Its body dimensions are 3.4 × 2.8 m and it was 6 m tall at launch, reaching a span of 26 m with the solar panels deployed (Figure 2.7). It weighed 3600 kg at launch, but its mass on reaching the beginning of its life in geostationary orbit reduces due to the burning of fuel by its thruster motors – most of the satellite's fuel is burnt in the initial phase of its life. It has a predicted life of fifteen years, and its orbital position is at 43°W serving South America – it is actually in orbit to provide DTH services, but the launch process described applies in just the same way to any geostationary satellite.

Figure 2.8 (a) Launch of Ariane Flight 115 (PAS-6B) on 21 December 1998. *(Photo courtesy of ESA-CNES-ARIANESPACE)*

Figure 2.8 (b) PAS-6B (HS-601 satellite) prepared ready for shipment to Kourou for launch on Ariane 42L rocket. *(Photo courtesy of PanAmSat/Hughes Space & Communications)*

This launch used an Ariane-42L launch vehicle which was 51 m high and weighed approximately 365 tonnes at launch, which included 85 tonnes for the launch vehicle itself and 275 tonnes of fuel, as well as the 3.6 tonnes for the satellite (Figure 2.8). The launch example is simplified for the sake of brevity (and to avoid complex mathematics).

The timing of the launch and the weather conditions – particularly wind speed at both ground level and at high altitude – are critical. At launch, the rocket is on a near-vertical trajectory, the angle of which will be close to the latitude angle of the launch site due to the curvature of the Earth's surface. In the case of Kourou, French Guiana, the trajectory angle is at 7° to the Equator, as shown in Figure 2.9. The angle of trajectory then changes within the first minute to 40°, and the trajectory

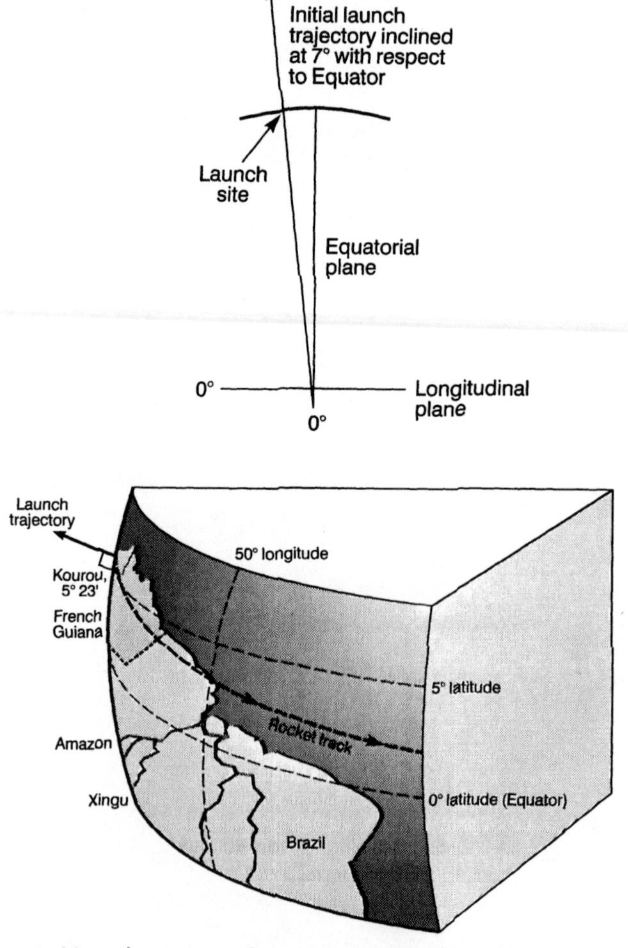

Figure 2.9 Typical launch trajectory from Kourou, French Guiana

angle will be 90° on reaching entry into the parking orbit. For safety reasons, for a launch to be initiated, wind speed is critical and has to be below 17 m/s at ground level, with both the wind speeds on the ground and at high altitude (between 10 and 20 km) being taken into account.

After launch, the rocket leaves the ground and carries the satellite to a position typically 200 km above the Earth's surface, which it would normally reach approximately 20 minutes after launch. Five seconds after the first stage engines have ignited, the launch vehicle leaves the ground, and in the first 3½ minutes the first stage will burn 230 tonnes of fuel to reach 75 km altitude, reaching a velocity of 3 km/s. The first stage is then jettisoned and the second stage fires for 2 minutes using 34 tonnes of fuel to reach 150 km altitude at 5 km/s. The second stage then falls away and the third stage is fired for 13 minutes using 11 tonnes of fuel to reach 200 km altitude at a velocity 10 km/s. Thus, 275 tonnes of fuel is burnt in just 20 minutes.

At that point, the final third motor stage of the launch vehicle will detach and the satellite is then in a circular but inclined parking orbit at a defined velocity. It will remain in this orbit for approximately 1 hour, while various checks are carried out and the status of all systems can be verified, including the partial deployment of its solar panel array to provide some on-board power and begin to recharge the batteries.

At a point when the satellite is crossing the line of the Equator, it will then be fired using its own motors into the elliptical transfer orbit at a further defined velocity of typically 10 km/s – equivalent to 36 000 kph (22 500 mph). A commercial telecommunications satellite typically weighs 3500 kg at launch, and this reduces to approximately 2100 kg on final positioning in geostationary orbit due to burning of fuel.

The transfer orbit has two points that define the degree of ellipticity in relation to the Earth: the 'perigee' and the 'apogee'. The perigee is the point of the orbit when the satellite is closest to the Earth's surface and is usually the same as the height of the parking orbit at this time. The apogee is the point of the orbit when the satellite is furthest away from the Earth. Typically the apogee of the transfer orbit is equivalent to the geostationary orbit distance, i.e. 35 785 km, although there are super-synchronous transfer orbits where the apogee is 40 000–120 000 km (120 000 km is approximately one-third of the distance to the Moon). The use of super-synchronous transfer orbits is used on some missions as these orbits have the advantage that, since the apogee velocity is lower, inclination changes to achieve final geostationary positioning use less fuel. The overall key to the transfer, positioning and 'station-keeping' of the satellite is ensuring all the manoeuvres in the life of a satellite can be achieved with the minimum expenditure of fuel. Essentially, fuel capacity defines the life of a satellite, as this obviously cannot be replenished despite the systems on the spacecraft potentially having a much longer life. The on-board fuel is used for both early orbit manoeuvres and station-keeping for the rest of its operational life.

Both the perigee and the apogee have to be above the line of the Equator, and this is the reason for the need for a parking orbit where the launch is from a location not on the Equator. It has to be remembered that the Earth is not static during orbital manoeuvres, and its movement has to be taken into consideration. For instance, while the satellite is in the parking orbit the Earth rotates 15° for each orbit of the satellite. This, combined with the movement of the satellite, means that there are extremely complex calculations and subsequent manoeuvres to ensure the satellite is on the right course, at the correct inclination and at the correct orbital position by the end of the final phase. It is analogous to a soccer player kicking a soccer ball from the middle of the pitch towards the goal, but as soon as the soccer ball leaves the boot of the player, the pitch begins to rotate and the ball has to be steered towards the changing position of the goal to score – literally 'moving the goalposts'!

The process of correcting the inclination to 0° with respect to the Equator is incrementally progressed throughout the process of the transfer orbit and the final positioning of the satellite. Assuming the typical transfer orbit, where the apogee is at or close to 35 785 km (the nominal geostationary orbit distance), then the period of this orbit is initially approximately 10 hours. The Earth rotates 150° in every 10-hour orbit. After four orbits, the first apogee motor firing will occur for about 40 minutes, increasing the perigee to 5500 km and the orbital period to 12 hours. The Earth rotates a further 180° in every 12-hour orbit.

After another two orbits, the second apogee motor burn occurs, lasting $1\frac{1}{4}$ hours, which brings the satellite into a practically circular orbit, raising the perigee to approximately 30 000 km. The orbital period will now be over 22 hours, and the Earth rotates a further 330° in every 22-hour orbit.

After a further two orbits, the final firing of the apogee motor occurs at the apogee to move the satellite into geostationary orbit – sometimes referred to as the 'injection orbit' or 'geostationary transfer orbit' (GTO) – where it will orbit the Earth almost in synchronism. The satellite will not yet be at its precise longitude, but 4–5° away. This third burn also ensures that the satellite will move extremely slowly to its final location. The natural forces of centrifugal force and gravity are used as much as possible, and the satellite is 'drifted' into its final orbital position within five to six days, using a series of precise East–West manoeuvres to decrease its advance and finally halt its drift. Under control from the ground, the deployment of the satellite antennas is completed with the antennas pointed towards Earth, and the solar array is fully unfolded and the solar panels tracked to follow the path of the Sun.

This whole process, from launch to final positioning, takes approximately 10 days. There is constant monitoring from a number of ground-based Tracking, Telemetry and Control (or command) (TT&C) stations, and the various sub-systems are methodically activated, tested and monitored. The time taken to bring the satellite online usually needs to be kept to a minimum, as the customer – the satellite system operator – will want to commence telecommunications and control system testing, leading to commercial operations, as soon as possible. A compromise must be reached between the time required for a full test programme and the lowest consumption of fuel.

2.4 In orbit

2.4.1 Station-keeping

The satellite, once in orbit, has to be maintained in a constant position to meet the geostationary requirement. This process is termed 'station-keeping'. It could be inferred from what has been said so far that the orbital position will be maintained purely naturally. However, there are a number of forces that can cause disturbances in the orbital path. If these are uncorrected, the satellite position will not be under the ITU definition of 'geostationary'. (For the role of the International Telecommunications Union, see Chapter 6.)

The orbit is generally considered circular, but it is in fact slightly elliptical as the Equator 'bulges' at 15°W and 165°E. This causes a slight circular oscillation of the satellite orbit and, more significantly, a shift in the gravitational pull that appears to be slightly off-centre from the centre of the Earth. If uncorrected, the force from the uneven gravitational field would cause satellites to move slowly to positions 90° away from each of these points (105°W and 75°E) where the gravitational pull is more stable. (These positions correspond to the eastern Pacific and India respectively.)

As these are the positions to which a satellite will move to and remain if there is no control, these are used as 'graveyard' orbits, though satellites are moved approximately 5000 km further out into space from the geostationary arc at the end of their operational life (termed 'de-orbiting'). The advantage of collecting what is effectively 'space junk' at these positions is that these orbits can be monitored to ensure that there is minimal risk of this space debris damaging any active spacecraft. Space surveillance, by the use of radar and optical telescopes, provides information on what is orbiting Earth, and this information includes each object's orbital parameters, size and shape, and other data useful for determining its purpose. Only a small fraction represents defunct satellites; the remainder is the consequence of almost forty years of payload and rocket fragmentation. These objects are a hazard for future space missions due to the increasing probability of an impact with satellites or spacecraft. Active and inactive satellites, along with space debris such as boosters, shrouds and other objects are tracked; the smallest object that can be tracked is about 10 cm in diameter. Over 10 000 objects larger than 10 cm are known to exist, and the population of particles between 1 and 10 cm in diameter is estimated at between 100 000 and 150 000. The number of particles smaller than 1 cm probably exceeds tens of millions. Untrackable objects within the 1–10 cm size range could seriously damage an operational spacecraft. At geostationary orbits, the ability to detect orbital debris is limited, but studies indicate that the orbital debris population at geostationary orbits is probably less severe than in low Earth orbit.

There are further effects on the satellite's orbit. The Earth is not completely circular as the Poles are slightly 'flattened' (termed 'oblate'), and this causes a distortion in the Earth's gravitational field. Add to this the disturbances from both lunar and solar forces, and all this accounts for the distorting effects on the ideal geostationary orbit. If uncorrected, the satellite would oscillate both on its 'North–South' axis and its 'East–West' axis over a 24-hour period (one rotation of the

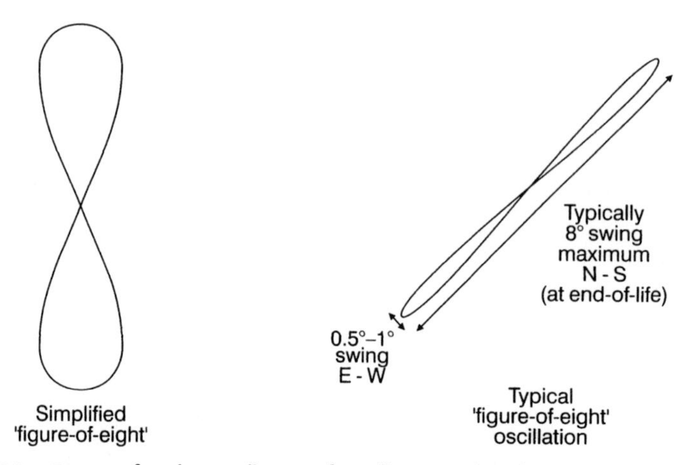

Figure 2.10 'Figure-of-eight' oscillation of satellite in inclined orbit

Earth) and the orbit would thus be described as 'inclined'. This is sometimes described as having a 'figure-of-eight' orbit. However, this description of the shape is exaggerated, as the true shape of the oscillation of a satellite in inclined orbit is very much more elongated on the North–South axis than the East–West axis, resulting in a very stretched 'figure-of-eight' shape, as shown in Figure 2.10. If uncorrected, this would cause an increase in the inclination of the satellite orbit by an average of 0.8° per annum, and cause increasing difficulties on the ground in maintaining tracking of the satellite by earth station antennas. This inclination leads to the apparent North/South and East/West oscillation as viewed from the Earth.

The distorting effects on the satellite's orbit create the need for station-keeping manoeuvres on the satellite. The nominal spacing for satellites in the geostationary arc is 2°, as defined by the agreement of the principal satellite authorities and some national administrations (see Chapter 6). If the 0.8° average drift per annum were allowed to go unchecked, there would be a serious risk of multiple illumination of adjacent satellites from uplink signals where there is common frequency band usage. It would also cause problems for the large antenna of a fixed land earth station (LES or 'ground station') in trying to track the satellite. It should also be noted that there are often a number of satellites co-located at the same orbital position – this is particularly used for DTH services so that consumers need only have an antenna pointed at one orbital position for multiple services. For example, in Europe, there are eight DTH satellites in the Astra series co-located at 19.2°E. This particularly increases pressure to maintain accurate station-keeping on Societé Europeanne de Satellites (SES), the Luxembourg-based operator of the Astra series.

The relationships between all the parameters of the geostationary orbit are shown in Figure 2.11a. The area in which the satellite has to be maintained while in station-kept geostationary orbit is termed the 'box' (Figure 2.11b), and the satellite operator has to maintain the satellite in position to within 0.1° principally in the vertical and

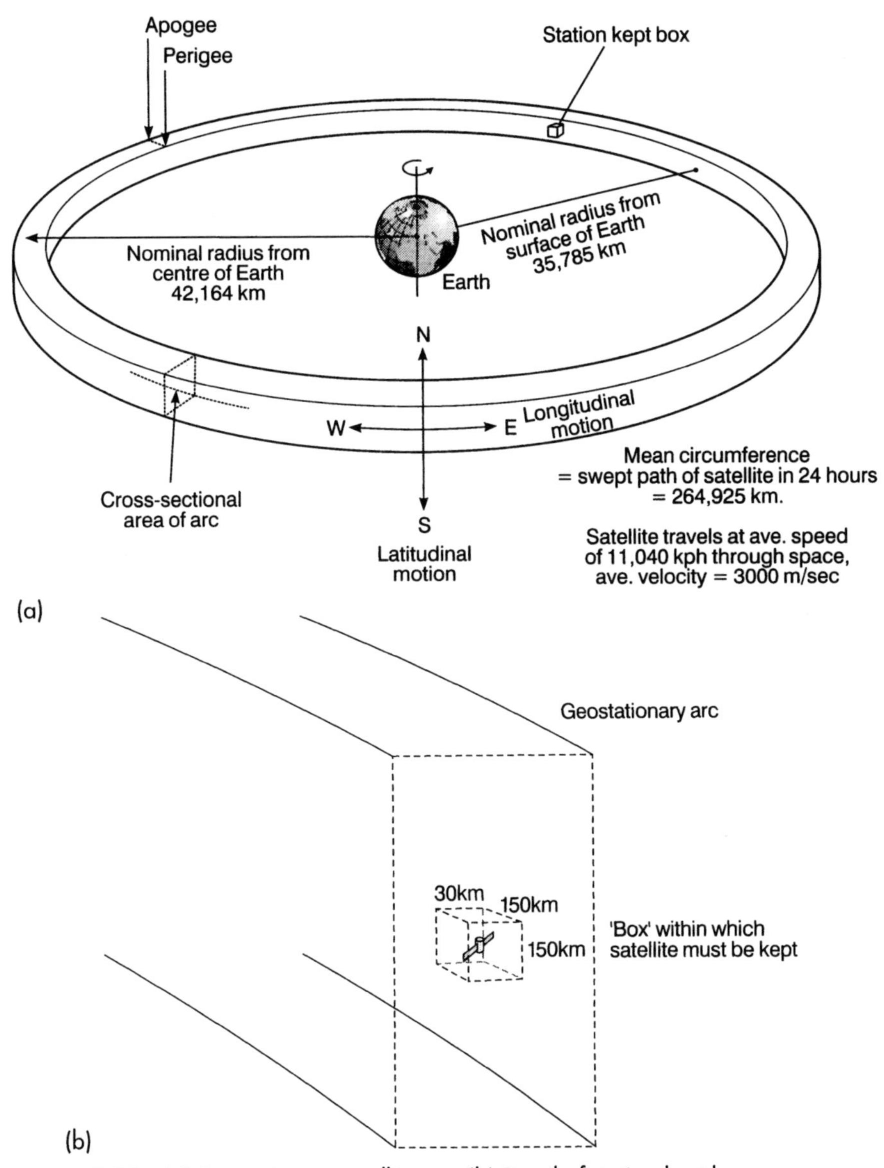

Figure 2.11 (a) Geostationary satellite arc. (b) Detail of station-kept box

horizontal planes. This translates to a cube of space, the dimensions of which are typically 150 km high × 150 km wide × 30 km deep. To give an idea of the width of a beam from an uplink antenna, a Ku-band signal from a 1.5 m SNG uplink antenna will have spread to approximately 600 km wide by the time it reaches the satellite at the geostationary arc. The centres of the 'boxes' spaced at 2° are approximately 1500 km apart.

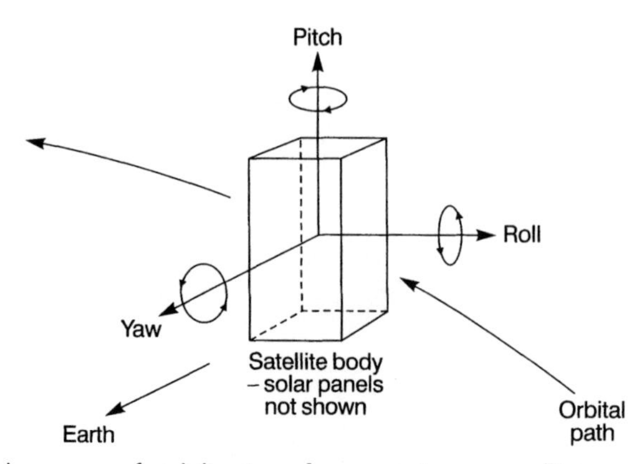

Figure 2.12 Three axes of stabilization of a geostationary satellite

There are two principal types of station-keeping – East–West and North–South – and the process is controlled by an integral attitude control system on the satellite. There are typically two types of attitude control on satellite, spin-stabilized and three-axis stabilization. Most modern geostationary satellites used for commercial telecommunications traffic use three-axis stabilization control, so we will concentrate on describing this system.

The aim of the attitude control system is to keep the satellite in position with the antennas pointing towards the Earth, and the solar panel array pointing towards the Sun. In a three-axis stabilization system, a gyroscope in each of the three planes is used to control the satellite attitude (see Figure 2.12).

Station-keeping is achieved by use of small rocket booster motors called 'thrusters'. A three-axis stabilized satellite typically has five sets of thrusters to correct the movements in orbit (roll, pitch, yaw, East–West and North–South). These are regularly fired under control from the main TT&C station that is receiving information on the satellite's attitude back from the satellite's on-board sensors.

As described in Chapter 7, satellites are maintained in a station-kept state for as long as is possible. At some point near the end of the operational life of the satellite, the North–South station-keeping manoeuvres are virtually abandoned to extend the operational life of the satellite (though there are some occasional minor adjustments to support the East–West station-keeping). This makes the satellite only usable for certain services and this can include SNG operations. When the North–South inclination exceeds a swing of 7–8° daily, the decision is usually taken to end the operational life of the satellite, and it is moved to a 'graveyard' orbit.

2.4.2 Locating the satellite from the Earth

The satellite has to be 'seen' from the Earth by any transmitter or receiver that is going to work through it. The transmitter is referred to as the 'uplink' and the receiver as the 'downlink'.

Both the uplink and the downlink have to be accurately pointed at the satellite, and the 'azimuth' and 'elevation' of alignment to the satellite have to be calculated. The azimuth is the compass bearing of the satellite in the horizontal plane from a point on the Earth's surface and elevation is the angle of the satellite above the horizon. Both of these parameters vary according to the longitude and latitude of the location on the Earth's surface and define the 'look angle' of the uplink or the downlink towards the satellite (Figure 2.13).

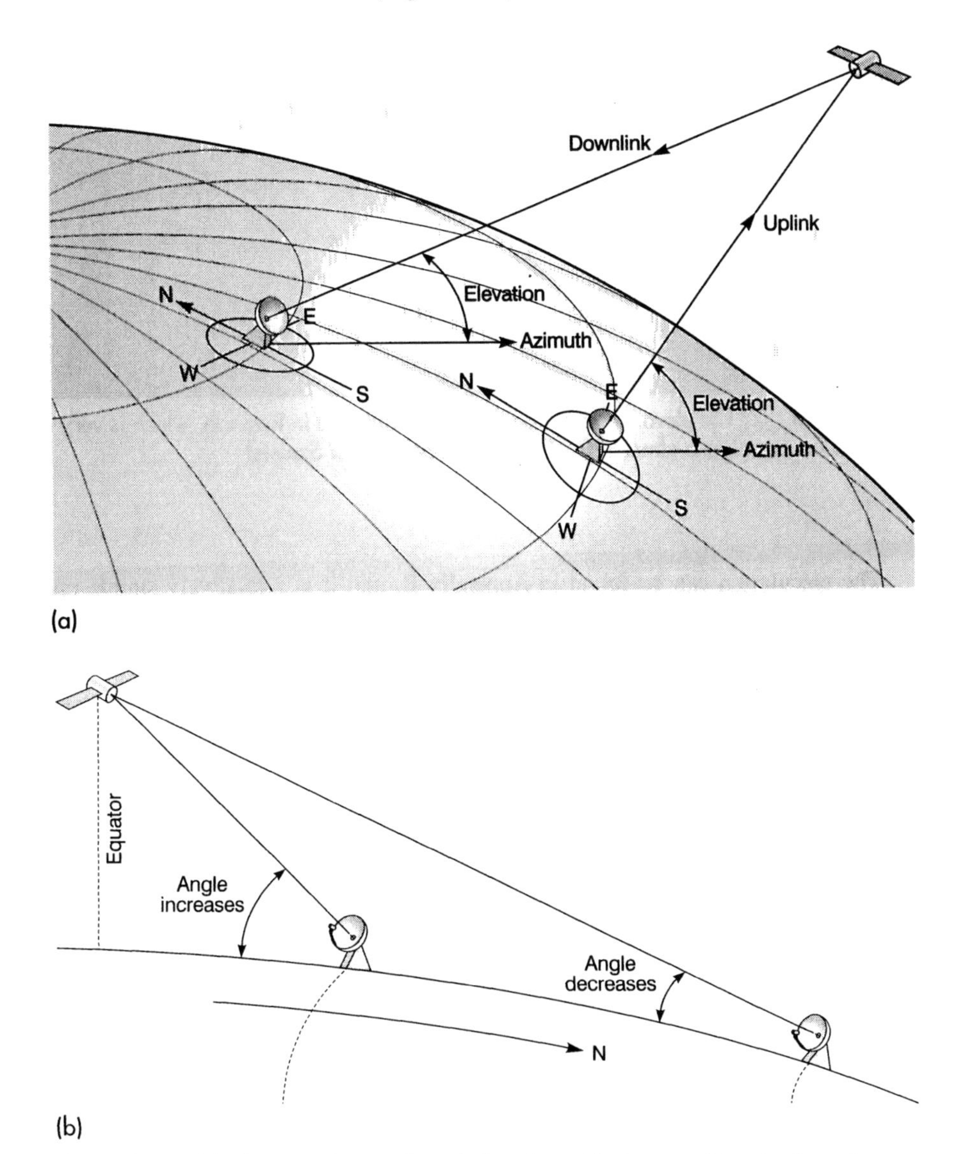

Figure 2.13 (a) Change of azimuth and elevation with change of longitude and latitude. (b) Elevation angle change with change of latitude.

Figure 2.13 (c) The elevation angle on this uplink antenna in Rwanda, which is very close to the Equator, is almost 90°. *(Photo courtesy of Paul Szeless)*

The calculation can be found in Appendix B, and it is a relatively simple one to program into a spreadsheet on a palmtop computer for use in the field for SNG.

The theoretical calculation produces an azimuth value that relates to the True North Pole, expressed in degrees East of True North (°ETN), which is a geographical designation at 90°N latitude, 0° longitude. The True value is the theoretical position of the North Pole, where all the lines of longitude meet at the top of the Earth and the axis about which the Earth rotates – a point some 750 km north of Greenland. However, it is the magnetic bearing in degrees East of Magnetic North that is actually required (°EMN) as it is the Magnetic North Pole towards which a compass needle will point, and is actually further south than the True North Pole. (It is currently about 500 km NW of Resolute Bay in Canada in the Arctic Sea and moves northwards on average 10 km per year – it can vary up to 80 km each day from its average position.)

There are maps that provide a correction value for any point on the Earth's surface which when added to the True value will give the Magnetic value, which can then be used with a compass to find the azimuth on which the satellite lies. A magnetic variation of 5–10° from True North is not unusual, and this variation is significant when trying to locate a satellite and must therefore be factored into the compass bearings used.

2.4.3 Telemetry, Tracking and Control (TT&C)

It is briefly worth describing the principal functions of Telemetry, Tracking and Control (TT&C), as this is a background management which operates 'behind the scenes' so far as users of satellites are concerned.

The satellite contains a complex system of sensors and control mechanisms, which are all remotely controlled from TT&C ground stations around the globe operated by the satellite operator. These play a role in both the launch and throughout the whole life of the satellite, and these ground stations are the crucial control centres that sustain the operation of the satellite. The following primary functions are carried out:

- Orbit and attitude control, including solar panel pointing for optimum power.
- Configuration of satellite and back-up switching (such as which transponder, or channel, is connected to which satellite 'beam', 'footprint' steering and adjustment of transponder amplification).
- Monitoring of external and internal parameters.

Typically, the control and monitoring channels between the satellite and the TT&C ground station operate on frequencies well away from the telecommunications channels. There is usually a basic TT&C system used in the parking and transfer orbit phases, operating in the L-band (see Appendix A), while the main TT&C system operates during the rest of the operational life of the satellite in the C-band or Ku-band. Many of the primary satellite functions are transmitted as part of the 'beacon' signal, which is normally in the same frequency band as the telecommunications channels as it also serves as a unique identifier for the satellite users on the ground.

The satellite management activity is, for obvious reasons, regarded as a security issue for satellite operators, and there is no specific information officially put into the public domain. A satellite is an extremely costly and valuable asset, and a satellite operator must use highly complex and secure control systems to ensure that the satellite remains protected from accidental or malicious interference. In many satellite organizations, there is almost a 'wall' between the TT&C operation and the commercial operations of the satellite. It is useful, however, to be aware of the existence of the TT&C operation.

2.4.4 Solar eclipses and outages

Before moving on to describe the telecommunication principles of satellites, we need briefly to discuss the phenomena of solar eclipses of satellites and the effects of solar outages. Both these types of event can cause disruption to satellite transmission and reception respectively, and can therefore impact on SNG operations.

The Earth revolves around the sun once every 365 days while rotating once every 24 hours on its own axis. The Earth spins on its axis at a tilt of 23.5° from a line perpendicular to its own orbital path around the Sun, resulting in the North Pole

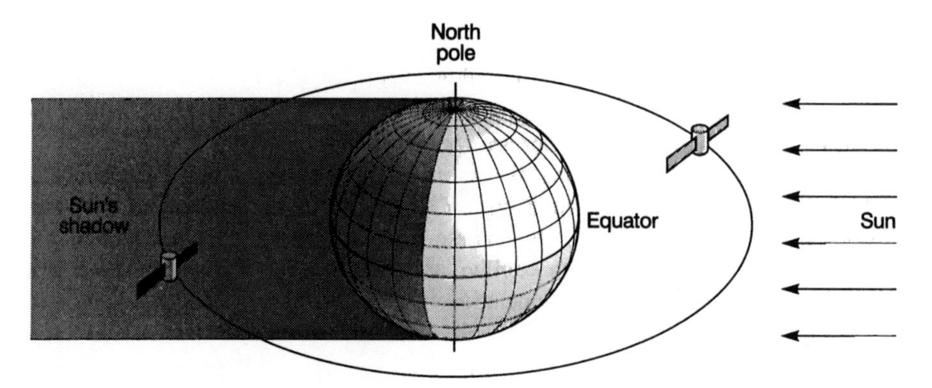

Figure 2.14 Solar eclipse of satellite

always pointing in the same direction in space, and it is this tilting of the Earth's axis that causes the seasons. Thus in June the Northern Hemisphere is tipped slightly toward the Sun increasing the length of daylight (summer), and the Southern Hemisphere slightly away from the Sun decreasing the length of daylight (winter). In December the opposite is true. In March and September (the spring and autumnal equinoxes) both hemispheres are equally exposed to the sun.

For most of the year, the orbit of the satellite keeps the satellite in view of the Sun. However, the satellite is out of view of the Sun for a few days around midnight at the time of the spring and autumn equinoxes at the sub-equatorial position, when the Earth moves between the satellite and the Sun. This causes a solar eclipse of the satellite, as shown in Figure 2.14. These do not occur during the summer and winter equinoxes, as the satellite is in view of the Sun at all times.

During the eclipse, the Sun's energy is obscured from the satellite solar panels that provide the primary electrical power on the satellite. The satellite has batteries on board which are charged by the solar panels, and during the period of an eclipse the satellite is switched to use the stored energy in these batteries to maintain operation. Satellite operation should therefore normally be unaffected despite the temporary loss of solar power.

Solar outages (or 'sun-outs') occur during the equinoxes when the satellite is on that part of its orbit where it comes between the Sun and the Earth. The result of this is that a downlink antenna receiving the satellite signals is 'blinded' as the Sun passes behind the satellite, as shown in Figure 2.15. This is not due to the visible light from the Sun, but to the invisible electromagnetic 'noise' that is also radiated from the Sun, which swamps the signal from the satellite. The Sun has a surface temperature typically in excess of 6000 C, and this level of radiation is produced across the rest of the electromagnetic spectrum.

The photograph in Figure 2.15b shows the effect of an outage viewed at an antenna in Jerusalem, where the Sun's rays are directly focused from the parabolic antenna onto the 'feedhorn'. The reception of signals from the satellite towards which the antenna is pointing is impossible under these conditions.

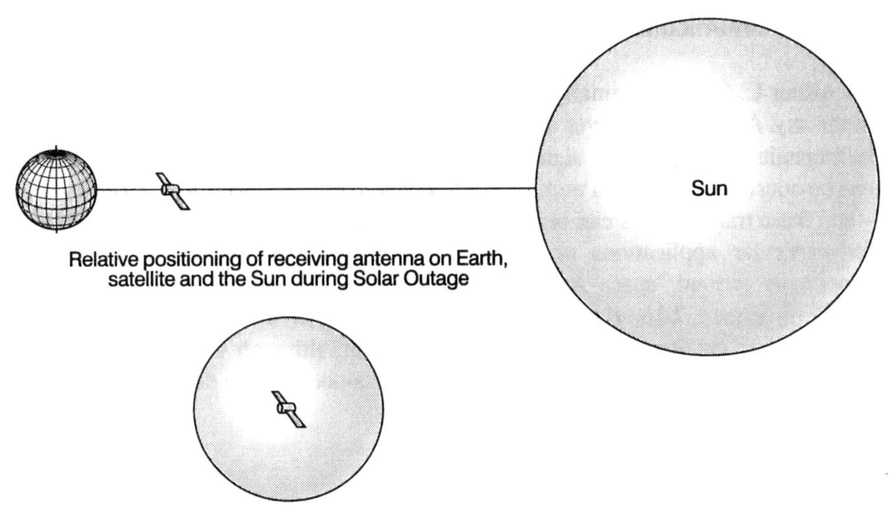

(b)

Figure 2.15 (a) Solar outage. (b) Solar outage of satellite in Jerusalem *(Photo courtesy of Paul Szeless)*

2.5 Communication theory

As Arthur C. Clarke originally envisioned, satellites are effectively 'radio mirrors' in the sky. A satellite receives a signal from a ground transmitter – the uplink – and re-transmits ('reflects') the signal to a ground receiver – the downlink. A satellite can be considered to have a number of these 'radio mirrors' – termed 'transponders' – and these transponders can be dynamically configured by ground control (TT&C) for particular applications or services. The part of the system in the sky is generically termed 'space segment'. The concept of the transmission system is shown in Figure 2.16. Because the uplink is of most interest to us we will be concentrating on describing this part of the system, although we will briefly cover the other elements. However, there are few basic concepts to be dealt with before we examine the uplink system.

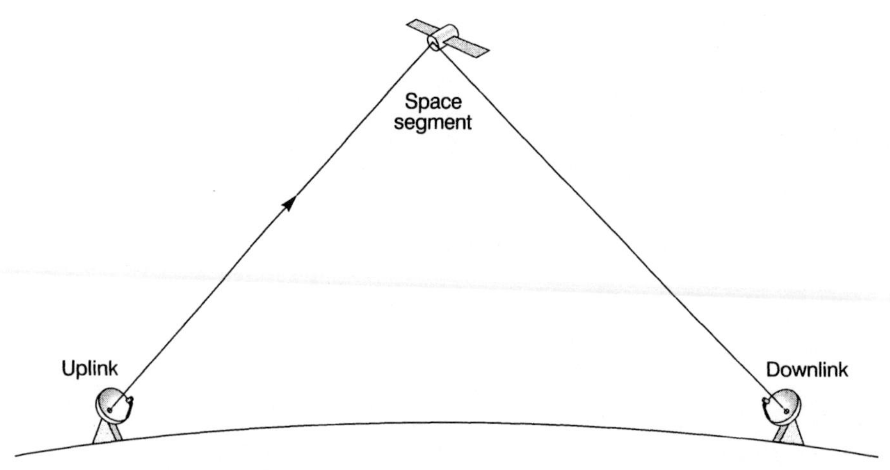

Figure 2.16 Overview of uplink–downlink system

2.5.1 Analogue and digital signals

The words 'analogue' and 'digital' are in everyday parlance, and before we can begin to cover the principles of satellite transmissions, we need to cover the concept of analogue and digital – with the minimum of maths. Those readers who already understand these principles can skip to Section 2.6.

In communications engineering there are two types of transmission, analogue and digital, which describe the nature of the signal used to convey the video and audio programme information. It is not necessary for you to have a deep understanding, but the basic concepts need to be appreciated.

2.5.2 Analogue

An analogue signal is a 'wave' signal that carries information by the continuous varying of its amplitude (size) and frequency (period), as shown in Figure 2.17. The concept is analogous to two people, each holding the end of a rope, and standing some distance apart. One person acts as the 'transmitter' and one as the 'receiver'. The 'transmitter' flicks the end of the rope up and down, creating a wave shape that passes down the rope to the 'receiver'. The harder the rope is flicked, the greater the amplitude of the wave, and the more frequently it is flicked, the faster the waves are created and the faster the rope oscillates. This demonstrates the two fundamental properties of an analogue signal – amplitude and frequency. One complete wave (as shown in Figure 2.17) is equal to one cycle or 1 hertz (Hz). The essential difference between an analogue signal and a digital signal, as we shall see shortly, is that the analogue signal amplitude can take *any* value between pre-defined limits.

This is how many electrical signals are conveyed, such as those reproduced by hi-fi systems. In terms of audio, the amplitude is the 'loudness', and the period is the

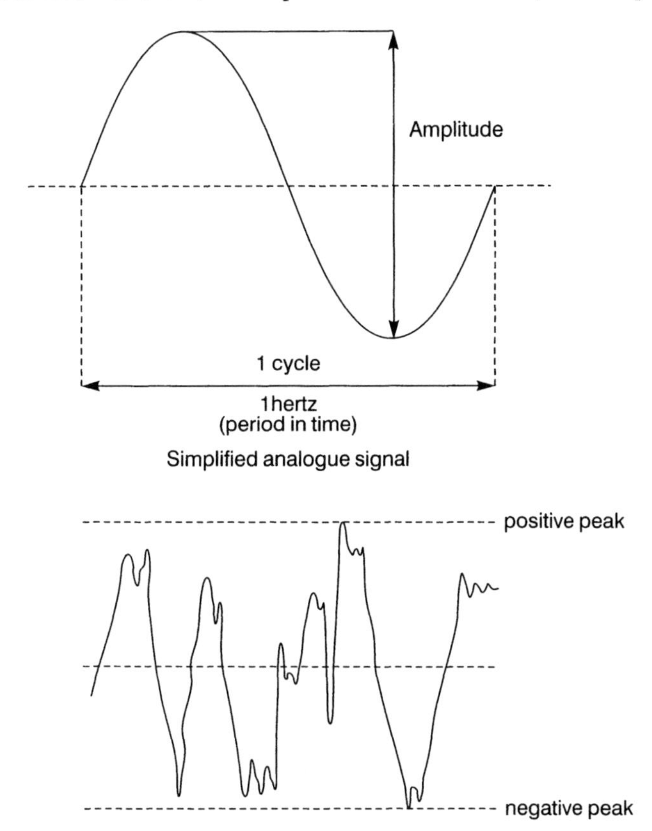

Figure 2.17 Analogue signals

frequency or 'pitch' of the programme signal. The human ear has a range of approximately 50–20 000 Hz. Loud, high-pitched sound signals correspond to large and closely spaced waves, and quiet, low-pitched signals to small and widely spaced waves. Obviously, these two components vary independently to create a wide spectrum of sounds.

A radio transmitter is a device that converts this type of 'message' signal (often termed the 'baseband' signal) to a signal at a much higher frequency and power that can travel over a distance (often termed the 'carrier' signal.) Although it is at a much higher frequency, the amplitude and the frequency of the original signal can be received at the distant point. The process of converting this signal to a much higher frequency to become a radio signal is called 'modulation', and the inverse process of recreating the original baseband message signal, 'demodulation'.

There are several ways of modulating a carrier with a baseband analogue signal, but the two most common methods are amplitude modulation (AM) and frequency modulation (FM). Amplitude modulation is the process where a baseband message signal modulates (alters) the amplitude and frequency of a high frequency carrier signal, which is at a nominally fixed frequency, so that the carrier signal varies in amplitude. If this signal is analysed, it can now be seen to contain the original carrier signal plus lower and upper 'sidebands' of frequencies. The entire modulated carrier now occupies a frequency range termed the 'RF bandwidth' (where RF is an abbreviation for radio frequency). AM is commonly used in radio broadcasts as amplitude modulation has the advantage of carrying great distances with a relatively moderate amount of power from the transmitter.

Frequency modulation is of more interest as it is used for analogue SNG transmissions. The principle of frequency modulation is based on a carrier that shifts up and down in frequency ('deviates') from its 'at rest' (centre) frequency in direct relationship to the amplitude of the baseband signal, though its amplitude remains the same, as shown in Figure 2.18. The maximum deviation (shift) of the frequency is equal above and below the centre frequency, and is usually referred to as the 'peak-to-peak' deviation.

In an FM signal, the instantaneous frequency of the baseband signal is represented by the instantaneous rate of change (speed of change) of the carrier

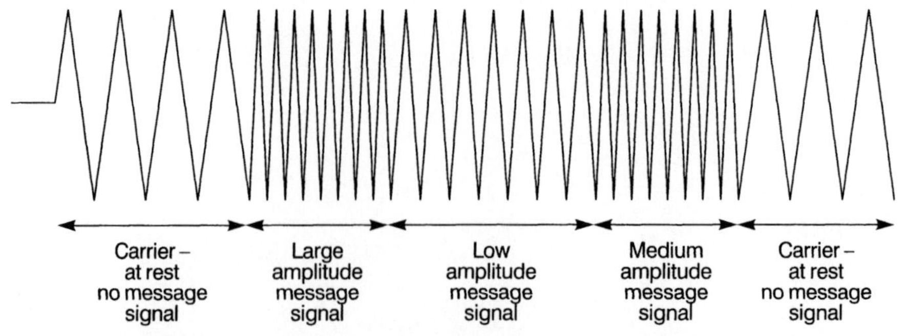

Figure 2.18 Frequency-modulated signal (simplified): constant input frequency, but varying amplitude

frequency. If the modulated signal 'waveform' is viewed on an electronic measurement instrument such as an oscilloscope or a spectrum analyser, the relationship appears much more complex than in amplitude modulation, and it is difficult to see exactly what is occurring.

However, the advantage of frequency modulation is that the signal has significantly greater immunity from random effects ('noise') than AM, and is much more effective than AM at the very high carrier frequencies used for satellite transmissions. FM is spectrally very inefficient, i.e. the amount of bandwidth required to transmit a message signal is greater, but if the right conditions are met, it produces a much higher quality signal compared to AM. Satellite communications have traditionally been power limited rather than bandwidth limited, so the spectral inefficiency has not been an issue.

The principles of modulation, and in particular frequency modulation, are difficult concepts to fully grasp, so you need not be unduly concerned if you don't fully understand. There are innumerable communications engineering textbooks that cover the subject in considerable depth (and which require a good grounding in mathematics!), but let this description suffice, alongside you accepting that frequency modulation is the preferred method for analogue SNG transmissions.

2.5.3 Digital

A digital signal can convey information as a signal containing a series of 'on' and 'off' states, which can be thought of as corresponding to numbers 1 and 0 based on the binary number system (Figure 2.19). In a stream of 1s and 0s, each 1 and each 0 is termed a 'bit', and a digital signal is defined by the parameter of 'bits'. For instance, the signal is transmitted at a certain speed or 'bit-rate', measured in bits per second (bps).

The analogue signal can be converted to a digital signal (using the processes of 'sampling', 'quantizing' and 'coding', covered in Chapter 4), and the digital signal can then be transmitted. Alternatively, some video equipment can directly produce a digital signal. The advantage of a digital signal, particularly in relation to satellite transmissions, is that the amount of power required to transmit it successfully can be significantly lower than the power required for analogue transmission. This is because, in general, the signals are more easily recovered from the background 'noise' and interference.

If digital signals were used in direct replacement for the same analogue information, the resultant bit-rate at which the information would have to be

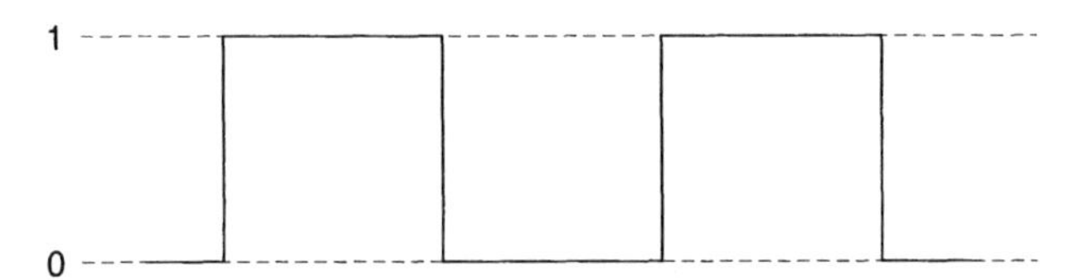

Figure 2.19 Digital signal

transmitted would be very high. This would use much more bandwidth (frequency spectrum) than an equivalent analogue transmission, and is therefore not very efficient (e.g. 36 MHz for analogue FM video, and perhaps 270 MHz for digital video). There is a need to strip out the redundant and insignificant information in the signal, and this process of data rate reduction, or 'compression' as it is more commonly referred to, is dealt with in more detail in Chapter 4. The important point to grasp is that digital signals are increasingly the norm for SNG transmissions, and understanding the fundamental difference between analogue and digital transmissions is crucial to the appreciation of the advantages and disadvantages of each type of system.

Analogue transmissions require significantly greater power and bandwidth than compressed digital transmissions, but the quality of the signal recovered can be very high. This greater power can only be delivered by a larger transmitting amplifier and/or a larger antenna than is required for a digital transmission.

Digital transmissions are always compressed for SNG, thus achieving savings in power and bandwidth. Although the quality of the signal can look very good at first glance, there are particular picture degradations (termed 'artifacts'; see Chapter 4) that can be objectionable for certain types of picture content. However, as the cost of satellite capacity and the size of the SNG equipment are often primary considerations, these artifacts may have to be tolerated as the compromise for the financial savings. The compression process also increases the overall delay of the signal, as the processing involved in compressing and decompressing the signal takes an appreciable time. This is in addition to the delay that the signal suffers in travelling from the Earth to the satellite and back to Earth, as we shall see later. Any delay can be problematical in conducting 'live' interviews.

The issues of gain and bandwidth crop up frequently in discussions on satellite communications. Having identified the two principal types of transmission, we can now move on to look at the transmission system. Because we will be dealing with specific components in the system, we need to look briefly at the frequencies used for the transmission and reception of signals.

2.6 Frequency bands

There are two principal frequency bands used for satellite transmissions for television, including those used for SNG (Appendix A). The band in the frequency range of approximately 4–6 GHz is called the C-band, and the frequency band between approximately 11 and 14 GHz is called the Ku-band (1 GHz is equal to 1000 000 000 Hz, or 1×10^9 Hz). The use of terms such as 'L', 'C' and 'Ku' for frequency bands dates from the Second World War.

2.6.1 C-band

The C-band is the frequency band that has been used for telecommunication transmissions since the 1960s (see Chapter 3). The 'transmit' frequency is typically in the range 5.8–6.5 GHz and the 'receive' frequency is 3.4–4.8 GHz; and, as far as SNG is concerned, C-band is used predominantly for digital transmissions. In 1994, INTELSAT began promoting the use of C-band for digital SNG as there was an increasing amount of capacity becoming available. However, the use of C-band DSNG does have some limitations from a regulatory aspect (see Chapter 6).

2.6.2 Ku-band

The Ku-band is the frequency band that is now, in many parts of the world, the predominant band for video (particularly DTH services) and has been the dominant frequency band for SNG since the early 1980s. The 'transmit' frequency is typically in the range 14.0–14.50 GHz and the 'receive' frequency is split into two bands: the lower band of 10.7–11.7 GHz and the upper band of 11.7–12.75 GHz.

2.6.3 Ka-band

The Ka-band extends from around 17 GHz upwards, but there are different interpretations as to exactly what constitutes Ka-band depending on the application. However, at present the Ka-band is not used for SNG, though there is discussion of its use in the future. For the time being, we will limit our discussion to the C- and Ku-bands.

2.6.4 Polarization

The transmission signals to and from satellites in any frequency band have a property termed 'polarization', and this relates to the geometric plane in which the electromagnetic waves are transmitted or received. There are two types of polarization, circular and linear, and the signal is transmitted and received via the 'feedhorn' – the assembly on the end of the arm extending out from the antenna. This either delivers the signal into the 'focus' of the transmit antenna on an uplink or satellite, or receives the signal via the focus of the receive antenna on a downlink or satellite (remembering that the satellite both receives and transmits signals).

As with many other aspects of satellite transmission, you need not be overly concerned with the technical detail, but for the purpose of discussion here, you need simply be aware of the property of polarization, as it is one of the defining parameters for a satellite transmission. Polarization is described in more detail later in this chapter in the discussion of uplink antennas. Suffice to say that, in general, circular polarization is used in transmissions in C-band and linear polarization is used in the Ku-band.

What is the reason for having different polarizations? Principally, the use of polarization allows the maximum utilization of frequency bandwidth on the satellite, as it allows the re-use of a frequency on a different polarization on the satellite. An antenna switched to one particular polarization will not detect or be subject to interference from signals on the opposite polarization, even though they may be at an identical frequency.

In general, a signal transmitted (uplinked) on a particular polarization is received (downlinked) on the opposite polarization. The uplink and the downlink antennas have operationally adjustable polarization, but the receive and transmit antennas on the satellite are fixed.

2.7 The transmission system

As we look at the functions within a complete satellite transmission system, there are certain elements that are part of the uplink and the downlink which will be seen in principle to be common, as the signal is modulated or demodulated, amplified, and converted up or down in frequency. There are also similarities in some of the processes on board the satellite. Although the actual components vary in construction, as they are built for the environment in which they will be operating, they essentially fulfil the same functions.

Within an uplink system there are a number of processes, some of which are typically mirrored at the downlink. The uplink is primarily composed of a 'modulator', an 'upconverter' and a 'high power amplifier' (HPA), which is connected to an antenna ('dish'). This is shown in overview in Figure 2.20. Some form of monitoring of the uplinked signal would also be implemented. Additionally, for a digital system, there would be an 'encoder' – the compression device.

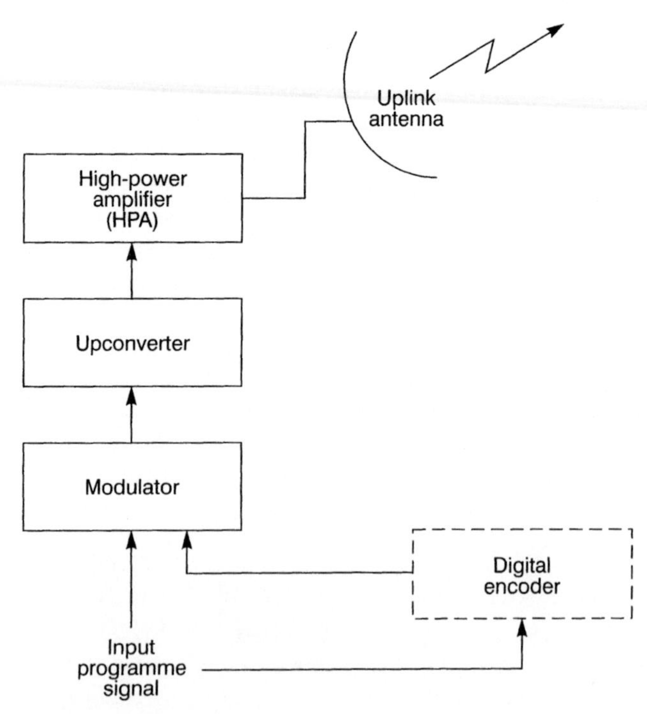

Figure 2.20 Uplink overview

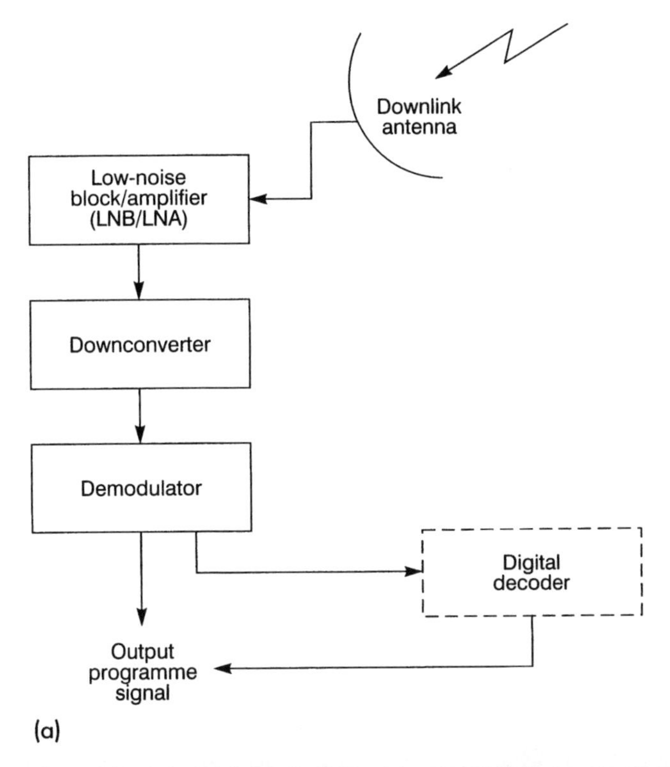

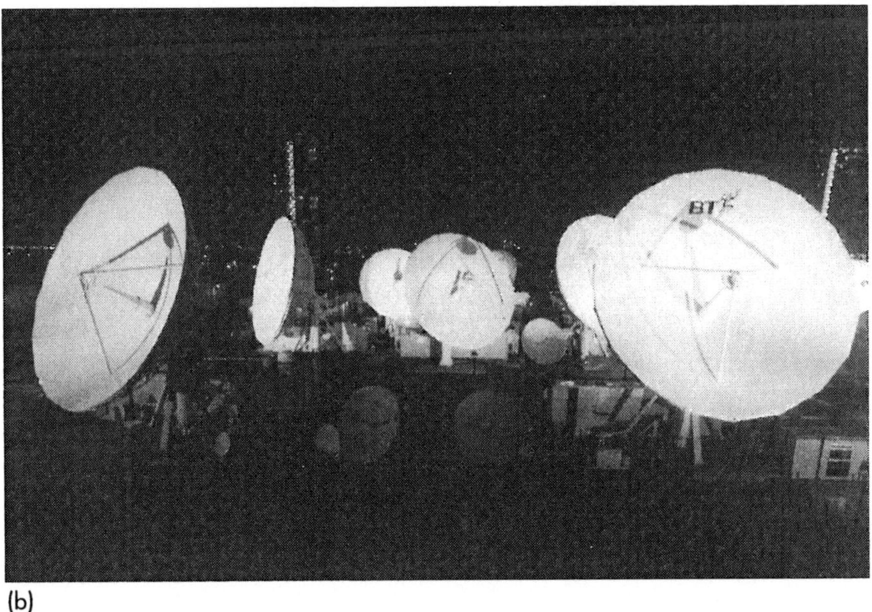

(b)

Figure 2.21 (a) Simplified downlink overview. (b) BT London teleport. *(Photo courtesy of BT Broadcast Services)*

(c)

(d)

Figure 2.21 (c) PanAmSat teleport at Homestead, Miami. *(Photo courtesy of PanAmSat)* (d) PanAmSat teleport master control room at Homestead, Miami. *(Photo courtesy of PanAmSat)*

The downlink has an antenna, a 'downconverter' and a 'demodulator', and as with the uplink, additionally a 'decoder' for the digitally compressed signal. This is shown in overview in Figure 2.21a. There would also be a sophisticated monitoring system able to measure all the parameters of the received signal. Typically the downlink antenna will be significantly larger than the SNG uplink antenna (from 4–30 m in diameter) and a cluster of antennas at one location is termed a 'teleport'. National PTTs, PTOs, satellite operators, individual broadcasters and private enterprises operate teleports.

2.7.1 Delay

As we have seen earlier in the chapter, the signal path from the uplink to the satellite is approximately 35 785 km above the Earth. The signal therefore travels a 'round-trip' distance of 71 570 km, and as radio signals travel at the speed of light, this is calculated to take 238 ms if both the uplink and the downlink are on the Equator (the sub-satellite point). If either (or both) are moved away from the Equator, this delay will increase as the path length increases. The maximum delay would be 277 ms, and so the delay is generally approximated to 250 ms.

Each earth-satellite-earth path is termed a 'hop', and care has to be taken to allow for the multi-hop delay if a signal has to traverse several satellite links from the point of origination to the final destination. This delay causes unnatural pauses or hesitations if a 'live' two-way conversation is attempted, and this delay can be further exacerbated by digital coding and decoding caused by the compression process on a digital link (see Chapter 4).

2.8 The analogue uplink

As uplinks are clearly the focus of attention in this book, this section will be the most descriptive. In terms of the equipment, there is a degree of commonality between analogue and digital systems, so we will describe the typical analogue system first (and it is still widely used). Where possible, we will give some idea of the size of each of the pieces of equipment in the uplink, expressed in terms of how much space is occupied in equipment racks.

The uplink has a baseband video and audio signal as its input, and from the antenna it produces a high-power radio signal directed towards the satellite. For the purposes of describing the process, we will use a Ku-band signal as it is the more common band used for SNG. Figure 2.22 shows the analogue uplink 'chain' in more detail – note the uplink transmission chain is often referred to as a 'thread'.

2.8.1 Modulator

In an analogue signal chain, the input video and audio baseband signal is processed and then applied to a modulator. The RF carrier frequency of the modulator output is usually either at nominally 70 or 140 MHz (these are standard carrier frequencies

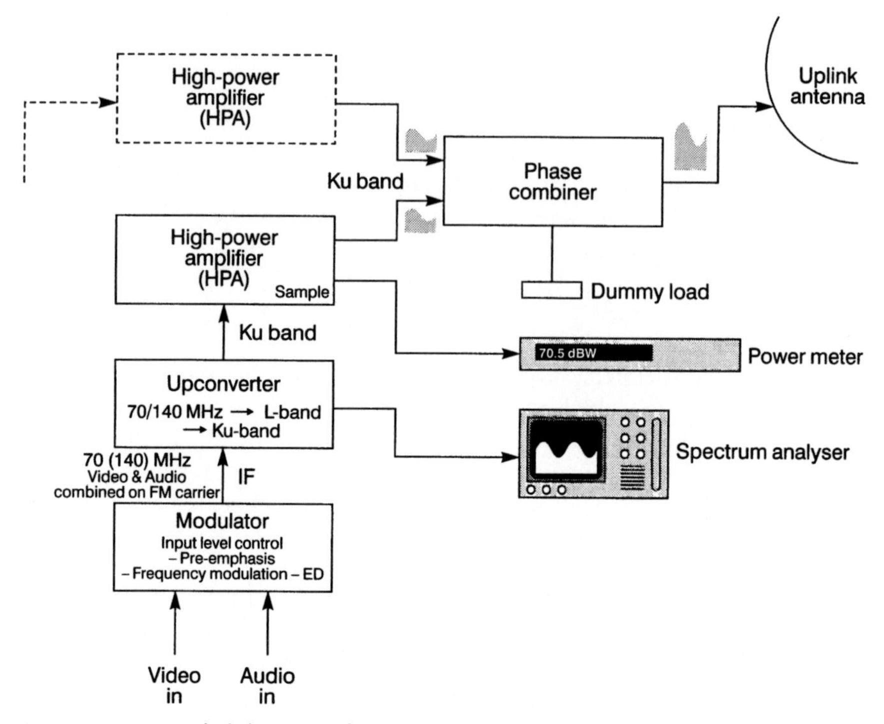

Figure 2.22 Detailed diagram of an analogue uplink chain

for telecommunications equipment) or, depending on the manufacturer, sometimes in the L-band. This is referred to as an intermediate frequency (IF) signal. The term 'intermediate' is used as it lies between the baseband frequency range and the final transmission frequency.

The modulator, which is typically 1 or 2 RU in height (19 in rack-height units; see Table 3.1), has a number of adjustments available as front-panel controls, such as input video 'gain' (boost), input audio gain and RF carrier controls (e.g. deviation). An example is shown in Figure 2.23, which shows a modulator with the upconverter on top. The modulator also has a control for 'energy dispersal', which we will describe later. Before modulation, however, the video and audio has to be processed.

2.8.2 *Video processing*

The primary video processing is called 'pre-emphasis' (a type of frequency-dependent level boosting/cutting or 'filtering'). This overcomes problems introduced in the rest of the transmission process (primarily an aberration called 'group delay' affecting the colour element of the video signal) that would be apparent on the demodulated video output of the downlink. The parameters of the pre-emphasis

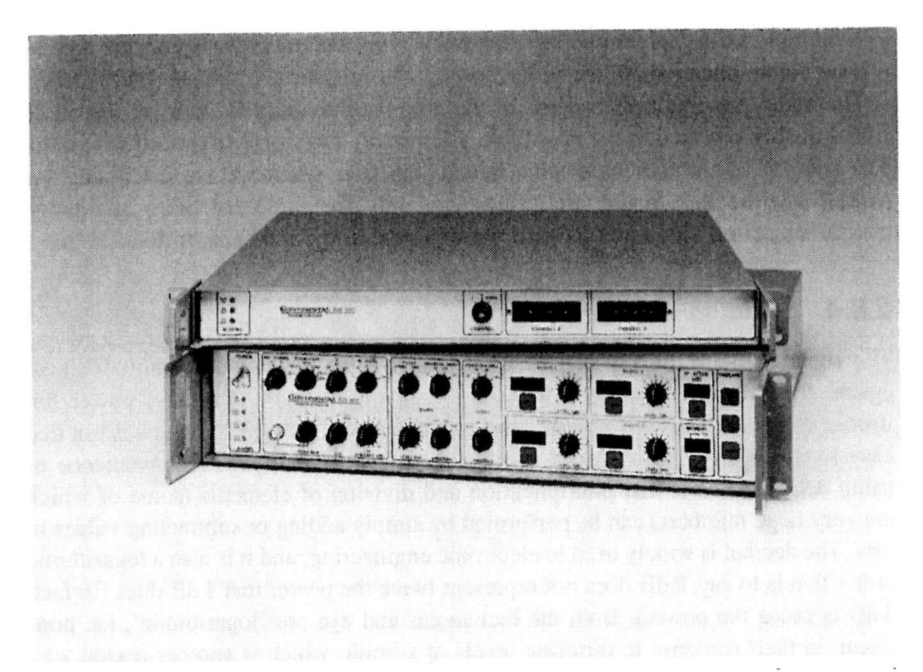

Figure 2.23 Modulator (bottom) and upconverter (top). *(Photo courtesy of Continental Microwave Ltd)*

can be varied by controls on the modulator, along with the video input level. The demodulator at the downlink must have the same 'de-emphasis' standard selected as the pre-emphasis at the uplink, otherwise the demodulated video will suffer distortion. The precise video pre- and de-emphasis filtering chosen depends on the TV standard in use; e.g. 525-line filtering is different from 625-line filtering.

2.8.3 Audio processing

The audio processing stages of the modulator (which can handle at least two analogue audio channels) have controls allowing the audio 'sub-carrier' frequency for the audio channels to be selected, as well as the type of audio pre-emphasis that is applied. The audio sub-carrier is an additional FM signal that is applied at a frequency just above the highest video frequency (which is usually taken to be either 5.0 or 5.5 MHz, depending on the TV standard). Typical audio sub-carrier frequencies are 6.6 and 7.2 MHz (these are the centre frequencies of the FM audio sub-carrier).

Alternatively, there is a European system where the audio can be carried as a digitized signal in the 'sync' (line timing) pulses of the TV video signal, and this is called 'sound-in-sync' (SIS). This then frees up the audio sub-carriers to carry other information, such as remote-to-studio communications, which can even be carried as a multiplexed signal on one sub-carrier carrying several channels of this type of private communication.

The audio pre-emphasis applied in the modulator is used to improve the audio signal quality received at the downlink, specifically the signal-to-noise (S/N) ratio. The concept of the S/N ratio will occur again later when we are discussing the overall satellite link quality (it is used not only for audio but other qualitative measurements on video and RF), so we will take a few moments to look at this.

2.8.4 Signal-to-noise ratio

The signal-to-noise ratio is exactly what it says – the ratio of the (wanted) signal against the (unwanted) background noise. It is a measurement of relative power, but instead of being expressed as a straightforward ratio (e.g. 1000:1), which is not that easy to use in calculations, it is expressed in decibels (dB). The convenience of using decibel units is that multiplication and division of elements (some of which are very large numbers) can be performed by simply adding or subtracting values in dBs. The decibel is widely used in electronic engineering, and it is also a logarithmic unit – that is to say, 2 dB does not represent twice the power that 1 dB does (in fact, 3 dB is twice the power). Both the human ear and eye are 'logarithmic', i.e. non-linear, in their response to differing levels of stimuli, which is another reason why we use a logarithmic measurement unit.

Mathematically, a ratio expressed as decibels is:

$$X\ (\text{dB}) = 10 \times \log_{10}(Y)$$

where X is the value in dB and Y is the original numerical value. The '$\log_{10}$' may terrify some readers as it may bring back memories of grappling with logarithms at school, but don't worry, we will limit reference to them only to here, so that you may appreciate why we use the decibel.

So, for instance, if we wanted to know how many dB represents an increase by 1000 (or a ratio of 1000:1), the calculation looks like this:

$$X = 10 \times \log_{10}(1000)$$

The logarithm (to base 10) of 1000 is 3, and thus:

$$= 10 \times 3$$

$$= 30\,\text{dB}$$

Returning to S/N, which is a ratio, then it is expressed as:

$$\text{S/N} = \frac{P^{\text{signal}}}{P^{\text{noise}}}\ \text{dB}$$

where P = power.

So a S/N ratio of 30 dB is where the signal power (level) is 1000 times greater than the power (level) of the noise.

In terms of audio, a S/N ratio of 30 dB is not very good. Hi-fi audio quality is generally considered to be in excess of 60 dB (1000 000:1), and an audio CD is generally reckoned to have a S/N ratio in excess of 90 dB (1000 000 000:1). A good analogue SNG link audio S/N ratio should be in the region of 55–60 dB.

2.8.5 Audio pre-emphasis

There are three basic audio pre-emphasis standards, quoted here for reference only:

- 50 μs time constant (μs meaning microseconds)
- 75 μs time constant
- J17 – an ITU standard

Alternatively, an audio companding standard may be selected. There are several of these, the most common being the Wegener Panda standard. Audio companding (COMpressing/exPANDING) is a system of dynamically varying the amount of pre-emphasis depending on the frequency being processed to optimize the signal-to-noise performance (similar to the use of Dolby in hi-fi). As with the video pre-emphasis, the downlink must have the same de-emphasis/companding standard selected to avoid distortion and obtain the benefits of applying the process in the first place.

2.8.6 Carrier controls

The modulator will have controls to enable the carrier to be switched on or off, the carrier level to be varied, and the modulation to be on or off, with further selections for either one of a number of pre-set test signals to be modulated onto the carrier or the programme video and audio.

There will also be a selection for the deviation of the modulated carrier, and that is usually selectable in steps from 17 to 36 MHz/V (peak-to-peak). The analogue TV signal has a peak-to-peak value of 1 V and, typically, an analogue SNG transmission will be set to an FM deviation of 20 MHz/V.

The value depends on the bandwidth allocated – the higher the deviation, the more bandwidth required – and this is dictated by the transponder channel bandwidth allocated by the satellite operator, who will decide on the amount of 'guard-band' required between transmissions in the 'frequency plan' for the satellite transponder. A frequency deviation of 25 MHz/V with a TV signal of 5.5 MHz bandwidth allows the modulated signal to fit into a transponder channel of 36 MHz (which may be either half-transponder or full-transponder occupancy depending on the transponder configuration) with a 'guard-band'. From the point of view of the uplink operator, more bandwidth will allow for a better quality signal, but that increased bandwidth will have a price. It is therefore an issue of playing off quality against cost.

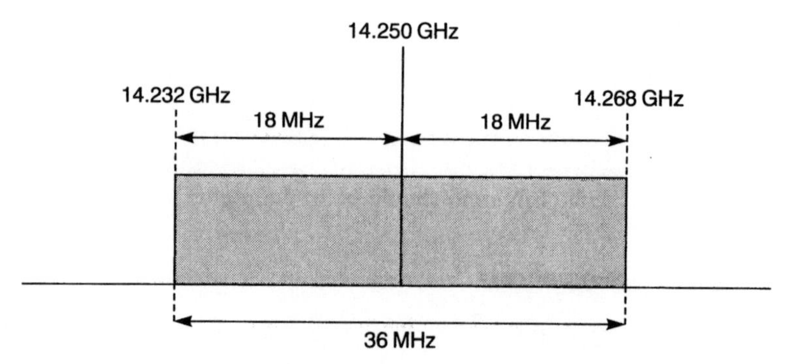

Figure 2.24 Spectrum of typical analogue TV Ku-band channel/transponder

Let us say that that the satellite operator has allocated a transponder channel with a centre frequency of 14.25 GHz. This is typically expressed by the satellite operator in MHz and so, in this case, it would be 14250.000 MHz. With a deviation of 25 MHz/V selected, and assuming the TV signal has a bandwidth of 5 MHz, the minimum peak frequency deviation would be 14232.000 MHz (14250.000 – 12.500 – 5.500) and the maximum peak frequency deviation would be 14268.000 MHz (14250.000 + 12.500 + 5.500); this is shown in Figure 2.24.

As mentioned earlier, there will also be an energy dispersal control. Energy dispersal (ED) is necessary in an analogue uplink because of some fundamental characteristics of the TV signal. The analogue TV signal is made up of lines and fields, which form the scanned elements of the pictures. The number of lines varies according to the TV standard, and is either 625 lines per picture in the European colour TV standard (PAL), or 525 lines in the US and Japanese standard (NTSC). A picture is referred to as a 'frame', and there are 2 'fields' in every frame. The field rate also varies, and is either 50 fields per second (25 frames per second) or 60 fields per second (30 frames per second). Although these rates are not necessarily tied exclusively to either PAL or NTSC colour standards, the line/field rates are generally 625/50 for PAL and 525/60 for NTSC.

2.8.7 Energy dispersal

The analogue TV signal has timing pulses that occur at the start of every line and every field. Because these signals are repetitive, compared to the picture information that is usually constantly changing, constant fixed 'nodes' of energy at spot frequencies are created. These could cause interference to adjacent signals in the same frequency band, and this is particularly of concern in relation to many fixed point-to-point terrestrial (ground) microwave links that also operate in the Ku-band.

It is always desirable from the viewpoint of interference that energy be spread as evenly and randomly as possible across the 'occupied' bandwidth. Under most conditions, with a constantly changing picture content, this would not necessarily cause a problem. However, if static test signals were being transmitted, the picture energy would have a constant pattern, which could cause interference.

Energy dispersal is an additional signal added to the video signal, usually at frame rate (25 or 30 Hz) with a nominal frequency deviation of 4 MHz (i.e. ± 2 MHz) peak-to-peak. This is with no video signal, i.e. a 'clean' unmodulated carrier; with video applied, it automatically drops to 2 MHz peak-to-peak. This effectively disperses the energy more evenly over a wider band of frequencies, so avoiding the spot frequency nodes of energy. The peaks of this signal are timed to occur in the interval between successive frames of the picture to avoid any disturbance to the picture information.

The ED signal is removed at the downlink in the receiver stage (if it were not removed, a flickering picture would result). It is a requirement with most satellite operators that analogue transmissions have ED added. However, a control on the modulator is necessary as the ED is often switched off in the early stages of the line-up for a satellite transmission. A satellite operations control centre may request to see a 'clean' (unmodulated) carrier with no ED before modulation is applied and ED is switched on (see Chapter 7).

2.8.8 The upconverter

The function of the upconverter is to transform the modulated IF signal from the modulator up to the desired Ku-band frequency by a process of frequency shifting or 'conversion'. This is sometimes performed in two discrete steps: from 70/140 MHz to L-band, then L-band to Ku-band (obviously only the latter step is necessary if the modulator output is in the L-band). However, any process of upconversion has a number of smaller steps of frequency shifting within it.

The reason for using 70 or 140 MHz standard IF frequencies is that, although 70 MHz has been the 'traditional' IF standard, the 140 MHz standard has the advantage that its output frequency can be varied over a range of 90 MHz (compared to around 45 MHz with a 70 MHz modulator). This matches the typical bandwidth of a wideband transponder (72 MHz), and therefore the upconverter frequency control can be used to select the transponder, and the modulator frequency control then used to select the channel within the transponder.

An obvious question might be 'why is the signal not directly modulated onto a 14 GHz carrier?' One reason is the order of magnitude between the baseband signal and the Ku-band frequency. If we assume that the baseband video signal ranges from 0 to 5 MHz, and the Ku-band frequency is 14 000 MHz (which is another way of expressing 14 GHz), then the transmission frequency is 2800 times greater than the highest baseband frequency. The transmission frequency stability has to be maintained within very precise limits, and a direct upconversion that meets the required tolerance at this very high frequency is not achievable in a single simple process.

Therefore, a number of steps are required to arrive at the final modulated transmission frequency. We have already derived a 70 MHz modulated frequency. This is now upconverted to the next IF, the frequency of which depends on the

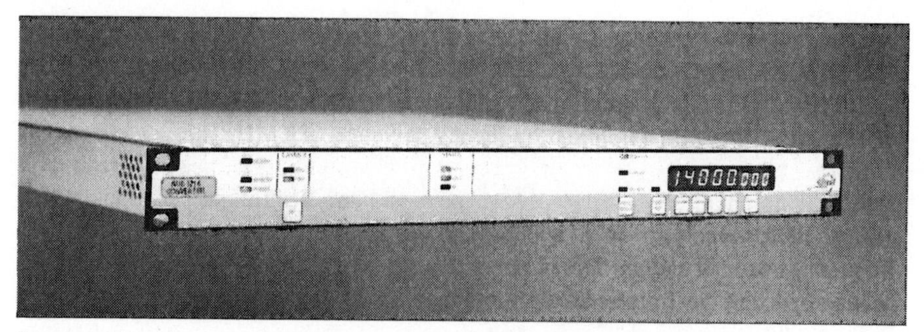

Figure 2.25 Upconverter. *(Photo courtesy of Advent Communications Ltd)*

manufacturer of the upconverter. One particular manufacturer has an adjustable IF of 950–1450 MHz in their upconverter, which allows for the adjustment of the IF so that the exact transmission frequency is derived at the output of the next fixed stage of upconversion.

The upconverter, typically 1 RU high (see Figure 2.25), usually has few controls on it, the most significant of which is the upconverter output frequency. Assuming the next stage of upconversion is a fixed frequency transition, then this frequency control is usually calibrated in the final Ku-band frequency. Therefore this is effectively the equipment on which the operator 'dials-up' the actual transmit frequency.

The output of this first stage of upconversion then passes to the fixed upconverter. This unit may either be rack-mounted or, if mounted as near as possible to the HPA to minimize losses, is normally contained within a small weather-proofed box. The final upconverter produces a low level Ku-band 'drive' signal that is applied to the HPA, which then amplifies this to the desired transmit level.

It should be noted that in an SNG RF transmission chain, the upconverter is one of the two most common components to fail – the other is the HPA PSU (power supply unit). This is because the upconverter has to carry out a number of stages of frequency multiplication, maintaining high accuracy in frequency, generally to within a few tens of kHz in 14 GHz.

Incidentally, the modulator and upconverter stages are often generically referred to together as the 'exciter'.

2.8.9 The high power amplifier (HPA)

High power amplifiers for analogue SNG use range in size from 300 to 800 W, and they are generally relatively large and heavy pieces of equipment. The HPA is typically one of the two most expensive components in the uplink, the other being the antenna and mount.

The function of the HPA is to amplify the low-level signal from the upconverter to a very large power that is then fed to the antenna. At the heart of the HPA is typically a device called a travelling wave tube (TWT). It is beyond the scope of this book to describe the operation of a TWT, but essentially it is a very powerful amplifying vacuum tube. It is relatively fragile and requires sophisticated control electronics and a power supply unit – some of the voltages are extremely high, in the range of several thousand volts.

Although the output power of the HPA is measured in watts, it is often expressed in decibel watts, or dBW. The expression dBW means decibels greater or less than 1 W (0 dBW) – indicated by a '+' or '–' sign. Earlier in the chapter we referred to the measurement unit of a dB, and the convenience of using decibels in measurements is that power calculations can be achieved by simply adding values in dBs together. The power used in satellite transmissions produces large numbers resulting in unwieldy calculations, so by using decibels the calculations are more manageable. As we shall see later in calculating 'link budgets', the standardization of all the factors into a single unit of measurement – the dB – eases overall calculations. So, for instance, a typical HPA used for SNG transmissions might have an output power of 300 W, or +25 dBW in dB notation: the '+' shows that the signal is 25 dB greater than 1 W (in ratio terms, this is an amplification of over 300:1).

The TWT device is generally non-linear in amplifying a signal; in other words, an increase in input signal level is not matched by a directly proportional increase in output power. The input-to-output relationship – termed the 'power transfer characteristic' – is shown in Figure 2.26. At first sight this non-linearity might not seem to matter, but in fact operation of the TWT in the non-linear part of the

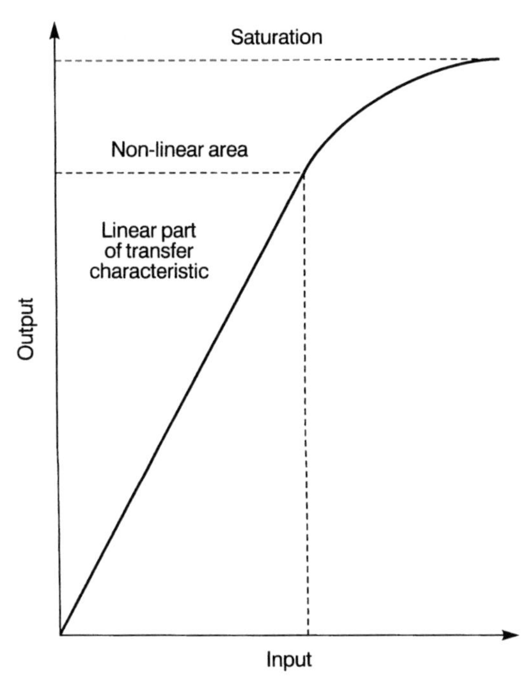

Figure 2.26 HPA power transfer characteristic

'transfer curve' results in spurious frequencies being produced in addition to the desired signal at the output of the HPA. These spurious signals, called 'intermodulation products' (IPs), are a form of distortion, causing interference to other adjacent signals, and are therefore undesirable. But in the lower part of the power transfer curve, the relationship is linear. Therefore, in operation, the input signal to the HPA is 'backed-off' so that the TWT is operating on the linear part of the transfer curve. This is often called 'input back-off' (IBO) and is defined in dB. For a given value in dB of input back-off, a number of dBs of 'output back-off' (OBO) results, and as TWTs are also used in satellite transponders, this term is often heard in reference to the operation of a transponder.

The HPA can be operated at maximum power (termed 'saturation') where a single analogue signal is being uplinked. Where a single programme signal is being uplinked, this mode of operation is termed 'single channel per carrier' (SCPC), and applies to both analogue and digital operation. If there are multiple carriers being combined and uplinked, this is termed 'multiple carriers per channel' (MCPC). A larger amount of input back-off is always required in MCPC operation.

Ideally, the HPA for the system should be selected so that it is generally operating at between 60–80% of its maximum rated (saturated) output power. The determination of the size of the HPA (in combination with the size of the antenna) has to be made considering the power of the satellites it will be used with. As a rule of thumb, the older the satellite, the greater the uplink power required.

The cost of the TWT makes up a significant amount of the overall price of an HPA, and if the TWT fails the cost of replacement can be in excess of 50% of the total cost of the HPA. As mentioned previously, the PSU is one of the two most common elements to fail because of the number and range of voltages it has to produce. On powering up, these voltage supplies also have to be switched automatically to supply the HPA in a particular timed sequence, and if one voltage is applied at the wrong time, the HPA may not power up.

The output power of an uplink is defined by two parameters: the output power of the HPA and the gain of the antenna. The output power of the HPA is often referred to as the 'flange' power, which is the power delivered to the output port flange (connector) of the amplifier. The connections throughout the system so far have been with coaxial cable, but the connection from the HPA to the antenna uses a waveguide (rigid, semi-rigid or flexible). This is a specially formed hollow metal tube, rectangular in cross-section (often referred to as 'plumbing' for obvious reasons), and is manufactured with differing degrees of flexibility. Rigid waveguide is the most efficient with the lowest loss, but semi-rigid or flexible waveguide is required to allow for movement of the waveguide (e.g. between the HPA and the antenna on a flyaway, or through the roof of an SNG truck).

At frequencies above several GHz it is usual to use waveguides, as these conduct microwave signals in the most efficient manner. It is vital that the connections of waveguide joints are as effective as possible, as microwave signals leak very easily from an incorrectly made joint. In particular it must be noted that the power levels are hazardous, and reference to Chapter 8 is recommended to appreciate fully the hazards and the necessary safety measures.

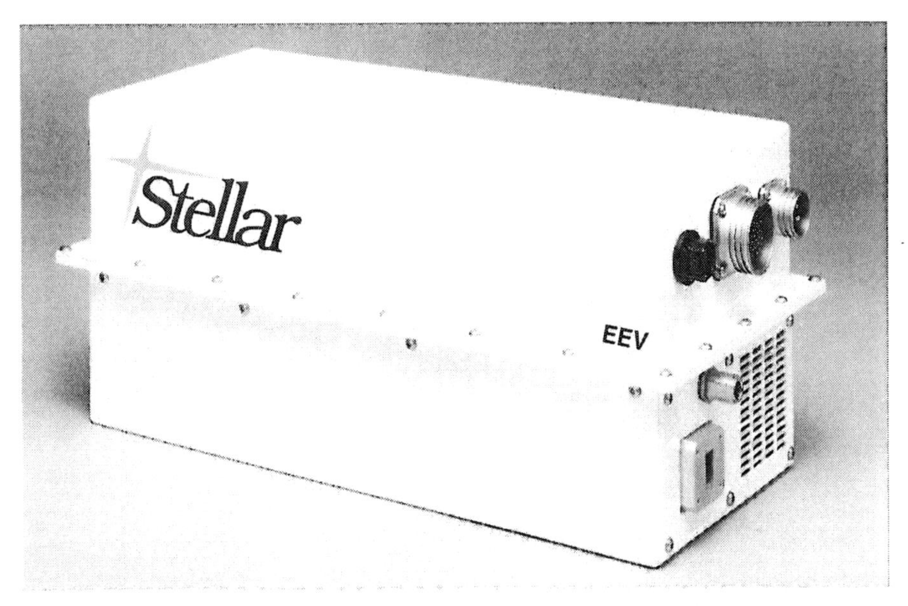

(a)

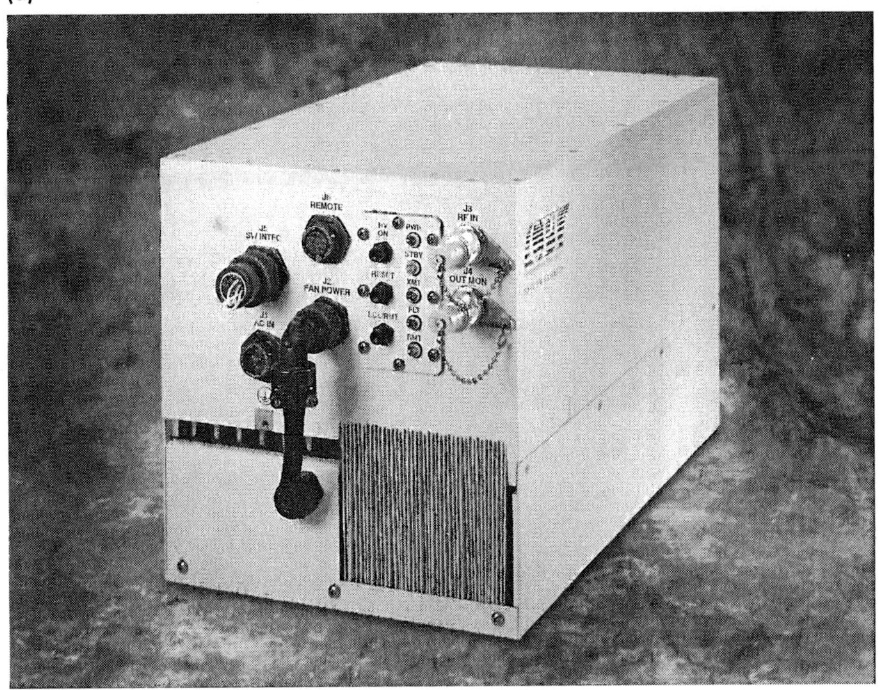

(b)

Figure 2.27 (a) 180 W Ku-band hub-mounted HPA. *(Photo courtesy of Marconi Applied Technologies)* (b) 160 W Ku-band hub-mounted HPA. *(Photo courtesy of CPI Satcom Division)*

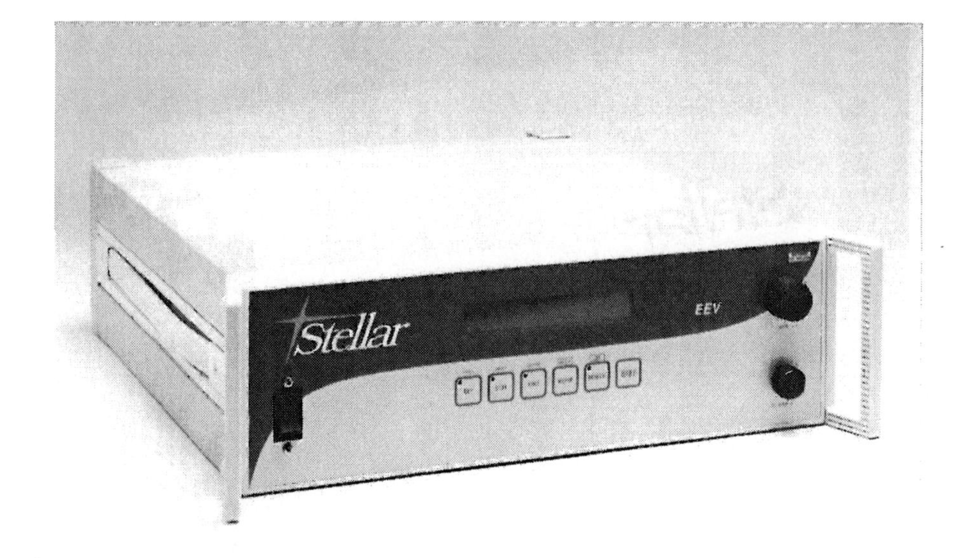

(a)

(b)

Figure 2.28 (a) 500 W Ku-band rack-mounted HPA. *(Photo courtesy of Marconi Applied Technologies)* (b) 400 W Ku-band rack-mounted HPA. *(Photo courtesy of CPI Satcom Division)*

HPAs are generally 'packaged' in one of two different ways, either as an antenna-mount ('hub-mount') or as a rack-mount, as shown in Figures 2.27 and 2.28. Hub-mount HPAs are usually in the lower power range (50–200 W) and are typically used in SNG for digital operation because of the lower level of power, and are particularly suited for compact DSNG flyaway systems. The hub-mounted HPA is mounted very close to or on the dish assembly (hence the name) on a fixed installation as might be found on an SNG truck (although, less typically, it can be used in this configuration on a flyaway). The advantage of mounting it in such a fashion is that the power losses in the waveguide connection between the HPA output flange and the antenna feedhorn are minimized because of the short physical distance involved. The rack-mounted HPA is, as the name suggests, mounted more remotely from the antenna in an equipment rack, in either a flyaway or vehicle installation, and is offered in a wider range of output powers.

2.8.10 Phase combiner

A 'phase combiner' is a device that combines two HPA outputs together. It is essentially an 'adding' device that operates at microwave frequencies, located between the HPAs and the antenna, and is often referred to as a 'variable phase combiner' (VPC). This is used either:

- to produce higher power in the antenna to increase the overall uplink power;
- to provide immediate redundancy in the event of failure of one of the HPAs;
- to combine two programme signals from two uplink chains into a single MCPC signal for transmission;
- or provide flexibility to achieve any of the above as required.

The system typically consists of a VPC unit with a remote control panel, which provides manual or automatic switching and combining of two HPAs used in any satellite earth station uplink, including SNG uplinks (see Figure 2.29). The remote control unit monitors both HPAs and controls the switching or combining in either 'single thread' or phase combined modes through the VPC. In the 'hot standby' mode, if a fault is detected from the 'online' HPA, the VPC switches in the backup HPA to maintain the transmit signal to the satellite. Alternatively, the two outputs from the HPAs can be combined to provide increased power. The phase-combiner is designed to ensure that the individual outputs from each of the HPAs are combined together so that they are added 'in-phase' – the concept of phase is described more fully later. The 'variable' in the name VPC is because the output of one HPA has to be adjusted to match the phase of the other HPA, as an 'out-of-phase' addition would result in a reduction in power and wasted energy. The phase combiner will usually allow the output of one or both HPAs to be fed to the antenna, or one or both HPAs to be switched into 'dummy load'. Note that if both HPAs are switched to the dummy load, there will still be a small degree of leakage into the antenna, and the antenna will, therefore, still be radiating.

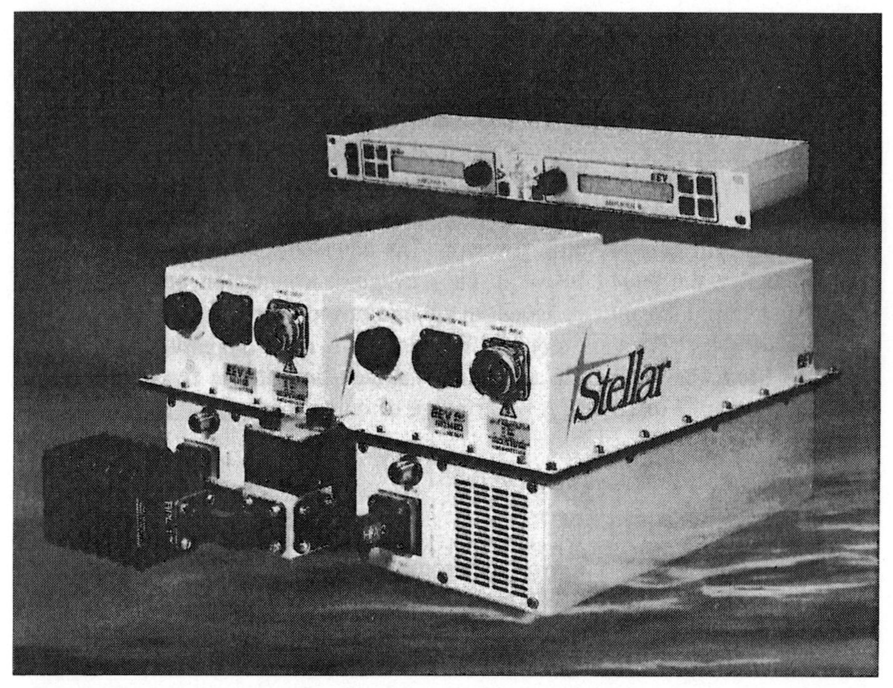

Figure 2.29 Two 180 W HPAs with phase combiner and dummy load, and remote control unit. *(Photo courtesy of Marconi Applied Technologies)*

A dummy load is a device that will absorb most of the power from the HPAs safely, and enables the uplink to be tested at full power as if it was transmitting to the satellite. The huge amounts of energy are dissipated as heat, and it is a routine procedure to test the uplink by running it up to full transmit power into the dummy load. Because of the small amount of leakage of the antenna, it should be 'skyed' (pointed in the opposite azimuth to the geostationary arc) with no risk of directing the small leakage power towards people.

The phase combiner does attenuate the combined signal to a small degree; it has a 'through' loss of around 1 dB and this loss has to be subtracted from the gain of adding the outputs of two HPAs. For example, if two 300 W HPAs are combined, the power output from the phase combiner would be equivalent to a single 500 W HPA (and not 600 W). However, when combining a number of signals, careful consideration has to be taken in calculating and achieving the correct power balance. If all the signals are driven at as high a level as possible out of the HPA, IPs will result from interference with the adjacent signal(s) from the other HPA(s). This may also cause interference to other adjacent signals in the transponder (and degrade the signals being uplinked if they are digital, as interference from 'in-band' IPs will create distortion in the digital domain). The output levels from the HPAs therefore have to be adjusted ('balanced') to achieve the maximum power output while minimizing the IPs. This is particularly important in digital operation, as we shall see shortly.

The power from the HPA, either directly or via a phase combiner, is now delivered into the antenna. But before we deal with that, let us look at the differences in a digital SNG uplink.

2.9 The digital uplink

The processes used in digital SNG (DSNG) for adaptation of the source signal to the transmission medium includes the compression, error correction and modulation processes, as well as various kinds of format conversion and filtering. The principal difference between an analogue and a digital uplink lies in the modulator, which is a digital rather than an analogue type, and the addition of a digital encoder. The upconverter is also slightly different but only in terms of its specification. Figure 2.30 shows a block diagram of a digital uplink.

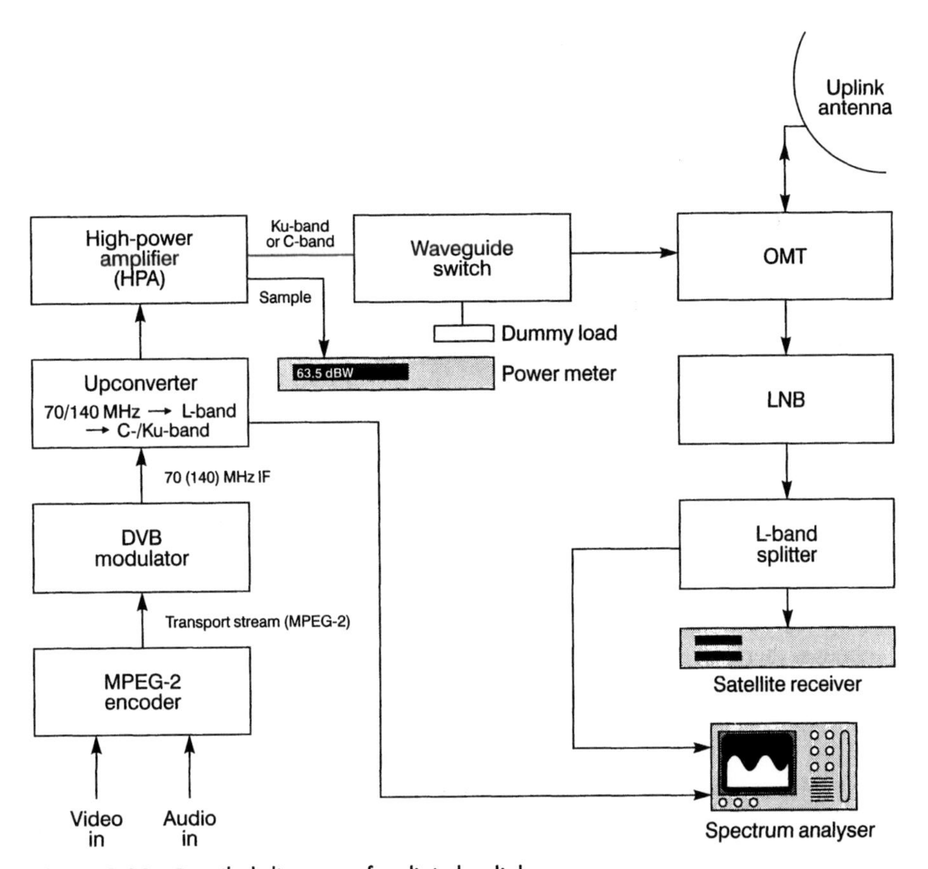

Figure 2.30 Detailed diagram of a digital uplink

2.9.1 Encoder

Since around 1994, digital SNG uplinks have been made possible by the development of low bit-rate digital compression encoders. Digital links had been possible for some time, but the development of a new generation of digital compression encoders that could run at bit-rates as low as 8 Mbps created an opportunity for use in SNG. The advantage offered was lower power uplinks and narrower bandwidth channels on the satellite, offering the possibility of smaller, lower cost uplinks and lower satellite charges.

A digital encoder essentially converts full bandwidth (uncompressed) video and audio to a compressed digital signal, and then presents it in a form suitable for the process of upconversion on the output of the encoder. The process of digital compression is covered in detail in Chapter 4.

The most common type of digital compression used for DSNG is MPEG-2, so let us assume that an MPEG-2 encoder is being used for an uplink. The encoder, which is typically 2 or 3 RU in height (although some are as much as 6 RU), has analogue (or serial digital) video and audio inputs, with some degree of input level control; see Figure 2.31. There are also front-panel controls for setting a number of digital

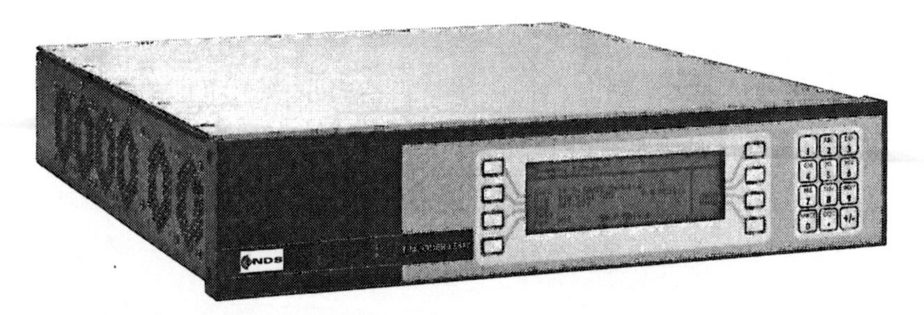

Figure 2.31 MPEG-2 encoder. *(Photo courtesy of Tandberg Television)*

parameters such as bit-rate, symbol-rate, FEC, horizontal and vertical resolution, and delay mode – some of these terms are dealt with later or in Chapter 4. The typical bit-rate for DSNG is 8 Mbps (this is known as the 'information rate'), although there is a move to lower information rates as quality improves with advances in technology.

The output is produced as a multiplexed video and audio 'DVB-compliant' signal that can then be fed to the modulator. This is also referred to as ASI (asynchronous serial interface) standard. DVB refers to the digital video broadcasting standard (see Chapter 4). Rather confusingly, recent advertising by some MPEG-2 equipment manufacturers refers to an encoder used for SNG as a 'DSNG' – as if that is all that is required for a DSNG uplink!

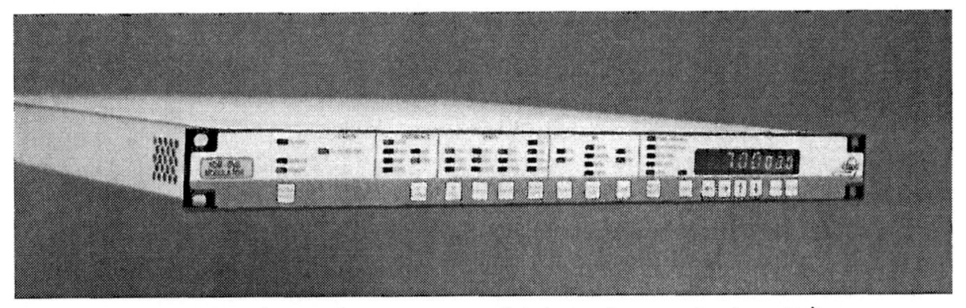

Figure 2.32 DVB modem. *(Photo courtesy of Advent Communications Ltd)*

2.9.2 Modulator (modem)

The modulator in a digital SNG system is often referred to as a modem, and is a different device to the modulator used in an analogue system. In computer terminology a modem is a modulator/demodulator (hence MOdulator/DEModulator), which implies a bi-directional process. However, this is not necessarily the case in a DSNG system, so do not necessarily assume that the 'modem' is capable of demodulating, for instance, for 'check' purposes. To differentiate between a digital modulator used for satellite transmissions and any other type of modulator/ modem, it is often referred to as a 'satellite' modulator or modem, and the unit is typically 1 or 2 RU in height (see Figure 2.32).

The modulator is often referred to as a DVB modulator, as it requires a DVB ASI standard input, with the incoming signal containing the compressed multiplexed video and audio programme information. The output is produced as a signal modulated onto an IF carrier signal as described previously, although some digital satellite modulators offer C- or Ku-band outputs for direct input to the HPA, without any further upconversion being necessary.

Although not as likely a problem as in the analogue domain, it is considered desirable to add an energy dispersal signal to the DVB stream from the compression encoder. This is based on a pseudo-random 'noise' signal to distribute any constant peaks of energy in the spectrum of the digital bit-stream, and this process is carried out in the DVB modulator.

The difference between an analogue modulator and a digital modulator is the method of modulating the input signal, by the use of phase modulation, and the provision for error correction. Error correction is applied before the modulation but, for clarity, we will look at these processes in reverse order.

2.9.3 Phase modulation

We have already dealt with frequency modulation, but a further derivative of frequency modulation, namely digital phase modulation, is used for digital signals. Historically, digital phase modulation has been referred to as 'phase shift key' (PSK) modulation, and there are a number of different forms.

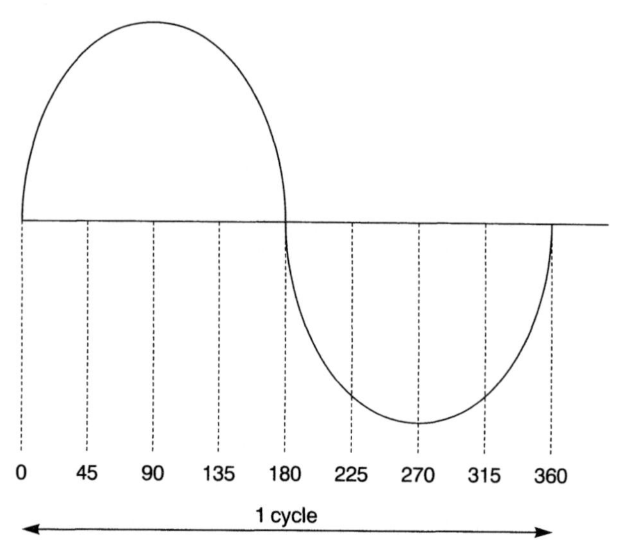

Figure 2.33 Properties of phase

Before we can describe the process of PSK modulation, we must first explain the concept of phase. Any signal can be described as having parameters of amplitude, frequency and, in addition, 'phase'. In Figure 2.17 we saw a simplified analogue signal, which was in fact a sine wave. If we consider that the point where the sine wave begins is 0°, and that by the time it has gone through one cycle it has reached a value of 360°, we have given the sine wave the properties of a circle, i.e. it describes 360° (Figure 2.33). If we consider this as our 'reference' signal then, if we compare it to another sine wave that is running slightly later, the difference can be measured as a 'phase shift' in degrees. In Figure 2.34 we can see two signals, where the bottom signal is 90° phase-shifted with respect to the top signal. By using the property of phase, we can modulate a signal in different phases in relation to a reference signal, and so convey information.

If we use the same phase property with a digital signal we can send a signal composed of bits that have one of two phases, 0° and 180°. This is called a two-state or binary phase shift key (BPSK) signal (Figure 2.35a), but because it is not efficient in its use of bandwidth, a further derivation is more commonly used.

If we generate two streams of BPSK signals, with one stream phase-shifted by 90° with respect to the other, and then add these together, we can have any one of four phases of signal, and have created four-state or quadrature phase shift key (QPSK) modulation. One BPSK stream has bits with a phase value of either 0° or 180°; the other BPSK stream has bits with a phase value of either 90° or 270° (Figure 2.35b). Therefore, the combined signal effectively has one of four phases, i.e. 0°, 90°, 180° and 270°, and so two bits can be transmitted at the same time (Figure 2.35c). Therefore, twice the amount of information can be communicated in the same bandwidth for the same data rate as BPSK requires.

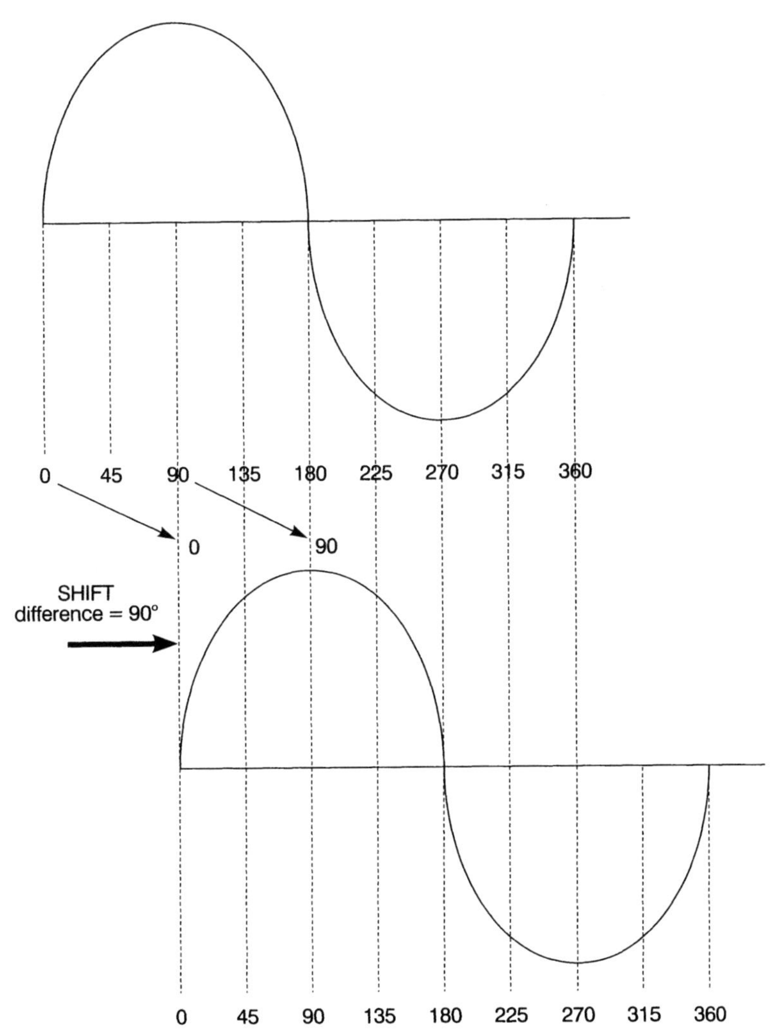

Figure 2.34 Comparison of phase of two signals

The term 'symbol' is often used in connection with the transmitted bits, and all this means is that each bit in a BPSK signal represents a 'symbol', and also that each pair of bits in a QPSK signal also represents a 'symbol'. Therefore, a symbol in a QPSK signal carries twice the information that a BPSK symbol does for the same bandwidth. One symbol per second is termed a 'baud'.

The process of combining bit-streams with phase differences can be further extended, and there are satellite modems on the market that offer the facility of 8-PSK (8-phase shift key), where each symbol now contains 3 bits of information. The total number of symbols is now $2 \times 2 \times 2 = 8$ ($= 2^3$); hence 8-PSK. By doing this we seem to be getting 'something for nothing', but the problem is that the quality of the received signal has to increase significantly with each step up in the number of phases

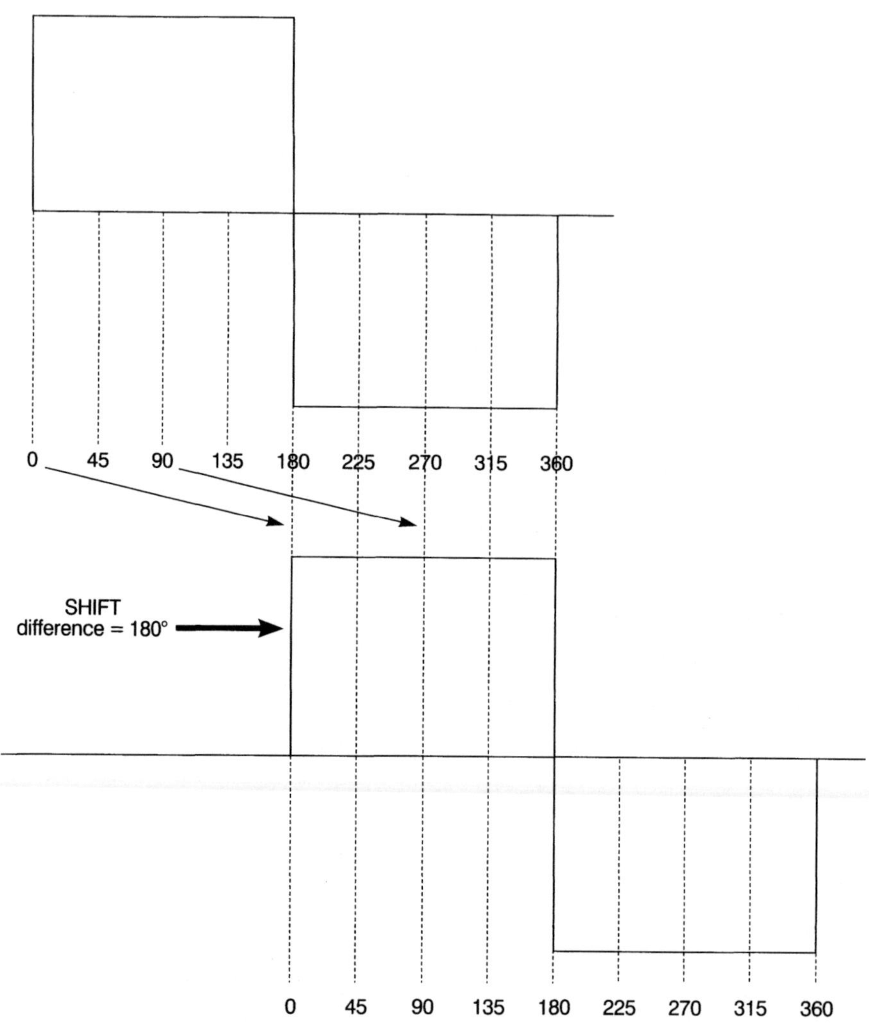

Figure 2.35 (a) Digital BPSK signal

modulated. This can typically only be achieved by increasing the amount of power to improve the S/N ratio – a concept we covered earlier in the chapter. Increasing power means a larger HPA and/or antenna, and this increases cost.

It should be noted that there is actually no reference signal transmitted – the demodulator at the downlink has to generate its own reference signal by examining the incoming bit stream. If the reference signal cannot be correctly generated or it becomes 'out-of-sync' with the incoming bit-stream, the incoming signal cannot be decoded and a period of time will pass while the demodulator tries to recreate the correct reference signal. In short, to generate the reference signal (termed the 'clock') and decode the incoming data stream, the incoming signal needs to be as 'clean' as possible – the enemy here is interference and noise. Later, when we look

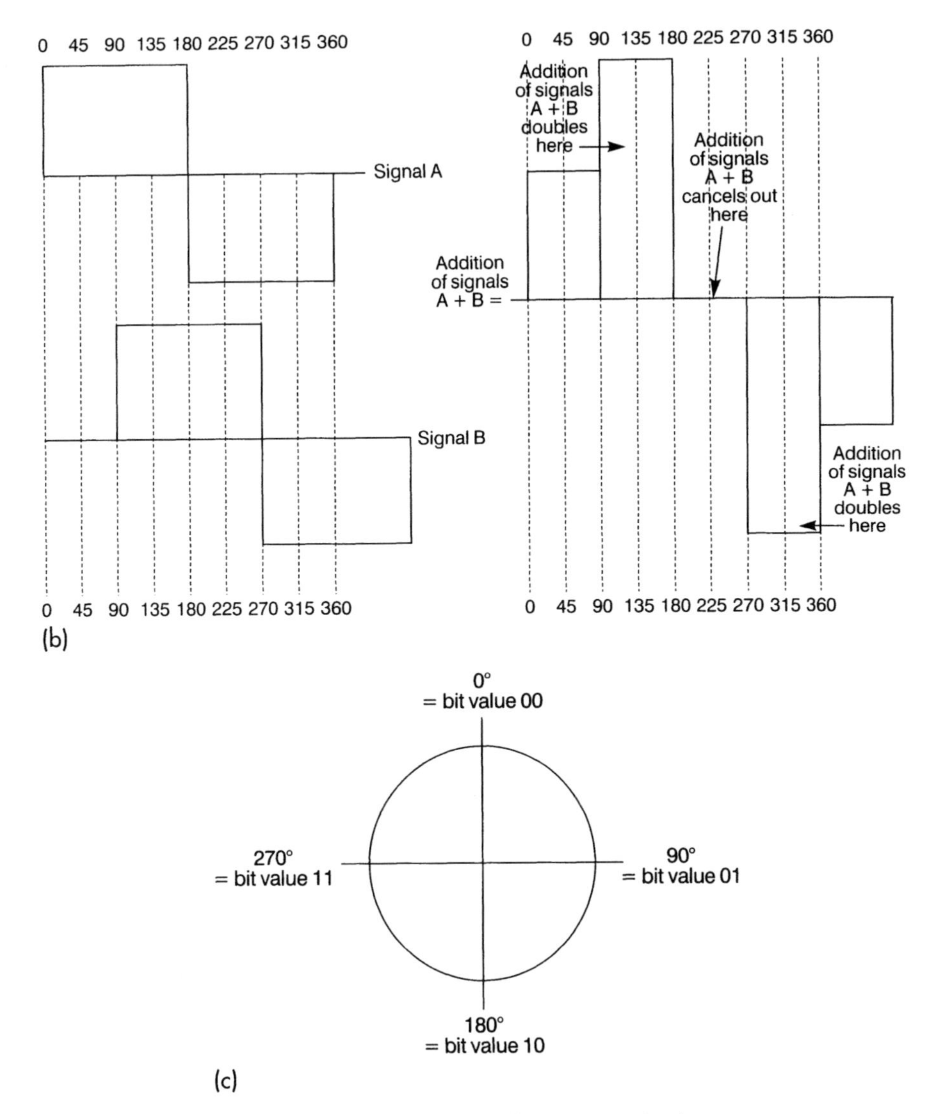

Figure 2.35 (b) digital QPSK signal; (c) digital QPSK signal values

at link budgets, we will see the effect of noise and how we measure the quality of the received signal.

QPSK is currently the typical modulation scheme used for DSNG, but the other types of modulation, as well as 8-PSK, include 16-QAM (16-state quadrature amplitude modulation) and 64-QAM (64-state quadrature amplitude modulation). Quadrature amplitude modulation is a scheme where both the amplitude and the phase are used to convey information, but it is far more difficult to distinguish between the small differences in phase and amplitude; as has already been said, more power is required to improve the S/N ratio. It is beyond the scope of this book to further describe these processes, and as they are not currently widely used for DSNG, we shall assume QPSK is the modulation scheme in any further discussion of DSNG.

2.9.4 Error correction

The processes involved in digital encoding and modulation are not as clearly separated as in the analogue process. We have talked of encoding in terms of compression and MPEG-2, but there is also an element of digital encoding within the digital modulation process. This is because bits are added in the modulation of a digital signal to correct for errors in transmission – termed 'error correction'. The signal is inevitably degraded in transmission due to the effects of noise and interference, and to compensate for this inevitable consequence, 'check' bits are added to the bit-stream in the modulator to enable errors to be detected at the downlink. Although error correction coding reduces the transmission power required, as the error correction can tolerate higher levels of noise, the demand for bandwidth increases because of the increased overall amount of data being transmitted.

There are two levels of error correction coding – an 'outer' and an 'inner' code – applied to a DSNG signal before (typically QPSK) modulation. There are a number of different types of outer code, but a typical type used in DSNG is 'Reed-Solomon' (RS) code, named after the mathematicians who devised it. The outer code adds a number of parity (check) bits to blocks of data, while the inner code – typically a 'convolutional' code – is applied after the outer code. The inner convolutional coding used in DSNG is typically referred to as forward error correction (FEC), although strictly speaking both the outer RS coding and the inner convolutional codes are types of FEC.

Outer code: RS

In Reed-Solomon code, for each given number of symbols forming a block, an additional parity check block of data is added to make up a complete block of data (termed a 'codeword'). This additional parity block is to compensate for any bursts of errors that the signal may suffer on its passage from the transmitter to the receiver.

So, for instance, in a typical data stream used in a DSNG uplink, the Reed-Solomon code is defined as (204,188), meaning that a codeword is 204 symbols long but actually has 188 information symbols, and a parity block of 16 symbols is added (188 + 16 = 204). Therefore the size of the data block has grown from 188 to 204 – an increase of some 8.5%. The numbers involved in this calculation are derived from the mathematics of the Reed-Solomon theory.

Inner code: FEC

The inner FEC is a 'convolutional' type of code that adds bits in a pre-determined pattern. This is decoded at the receiver (using a decoding algorithm called 'Viterbi')

Table 2.1 Effect of varying FEC on occupied bandwidth [8.448 Mbps, QPSK, RS (204,188)]

FEC	*Occupied BW (MHz)*
1/2	12.4
2/3	9.3
3/4	8.3
5/6	7.4
7/8	7
15/16	6.6

to detect any loss of information bits and attempt to reconstruct the missing ones. The number of bits added by this process defines the FEC ratio for the signal, and is typically 3/4; in other words, for each 3 bits, 1 extra bit has been added. Other FEC rates that can be used are 1/2, 2/3, 7/8, 5/6, 15/16, but the current standard DSNG rate is generally 3/4. As mentioned above, however, the bandwidth demand is increased, so that a 1/2 FEC QPSK-coded signal requires the same amount of bandwidth as a BPSK-coded signal with no FEC. Table 2.1 shows the effect of varying the FEC of a 8.448 Mbps QPSK-coded signal on occupied bandwidth. The greater the degree of FEC applied, the more rugged the signal will become, but the occupied bandwidth will need to increase to cope with the error correction overhead. So why use QPSK with FEC rather than BPSK? Because it is still the case that more information can be carried in the same bandwidth.

In between the two processes the signals are 'interleaved' – this is where consecutive bits leaving the outer FEC coding stage are spread over a certain period of time before the resulting bit-stream is then further protected by the inner FEC coding stage. This further decreases the risk of damaging corruption of the data stream in the transmission process.

2.9.5 Overhead framing

Although not now often used in SNG applications, it is worth mentioning 'overhead framing' (OHF). This is an additional data signal added to the incoming data stream from the encoder, which is often at a rate of 96 kbps but may depend on the standard being used. It is defined in an INTELSAT Earth Station Standard (IESS), IESS-306[2], describing how auxiliary information data are carried in wider tele-communications applications. Often generically referred to as IDR framing (IDR being the abbreviation for intermediate data rate, which refers to INTELSAT's general data services), it can be largely ignored for SNG applications; for example, it is not part of the DVB specification. However, it should be noted that if OHF is present on a signal being downlinked, the demodulator must also have OHF decoding switched on, otherwise the signal will not be demodulated correctly.

2.9.6 Transmission and symbol rates

The aggregate data rate, including RS, convolutional encoding and the information rate (and, strictly speaking, overhead framing as well), is termed the transmission rate. So now we can calculate the actual transmitted symbol rate – this is an important defining parameter for a digital signal.

If we assume that a typical DSNG signal is an 8 Mbps information rate signal, then this is an actual bit-rate of 8.448 Mbps – a standard data rate in the digital 'hierarchy'.

The data stream is QPSK modulated, so the information symbol rate is 4.224 Mbaud (2 bits per symbol, therefore 8.448/2 gives 4.224).

The information symbol rate (S^{info}) is now multiplied by 4/3, to account for the inner convolutional code:

$$S^{info} = 4.224 \times 1.333 = 5.632\,\text{Mbaud}$$

This now gives us the symbol rate including the FEC rate (S^{FEC}), but excluding Reed-Solomon coding.

The actual transmitted rate (S^t) is the FEC symbol rate (S^{FEC}) with the Reed-Solomon code added:

$$\begin{aligned} S^t &= S^{FEC} \times 204/188 \\ &= 5.632 \times \frac{204}{188} \\ &= 6.1113\,\text{Mbaud} \end{aligned}$$

Therefore, the 8 Mbps DSNG signal is expressed as a 8.448 Mbps signal transmitted with QPSK modulation at 3/4 FEC inner rate, and 204,188 Reed-Solomon outer coding, giving a modulated symbol rate (modulation rate) of 6.1113 Mbaud (or it can be described as 6.1113 Msps). Sometimes the transmitted symbol rate is expressed as before RS coding i.e. 5.632 Msps. There is a specification for the encoding of the audio in such a signal, but as it is part of the compression process, it is described in Chapter 4.

The RF bandwidth required is approximately the modulation rate multiplied by a factor – a value between 1.2 and 1.5, but typically 1.35 is used. This signal will fit within a 9 MHz channel; it will actually occupy just over 8 MHz, but there is a 'guard-band' allowed, minimizing any interference with signals in adjacent channels on the satellite. The 9 MHz channel has become a nominal standard for DSNG signals in most satellite operators' systems. However, when in future the common bit-rate and/or modulation schemes change, this nominal channel size will also alter.

2.9.7 Controls

The modulator will typically have the following front-panel controls.

- Modulation scheme: QPSK, 8PSK, 16QAM.
- FEC.
- Data rate.
- Output carrier frequency: usually in the range 50–90 MHz or 100–180 MHz.
- Carrier on/off.
- Output level control.

The Reed-Solomon inner code settings are set (204,188) and are not user-configurable.

2.9.8 The upconverter

It may seem that, having digitized, compressed and modulated the baseband signal, everything is the same in the rest of the equipment in the transmission chain. However, although the upconverter performs the same function, it does differ in one particular respect – its frequency and phase stability. The upconverter has to be very stable in terms of its frequency performance for analogue operation, but because digital transmission involves phase modulation, the frequency stability has to be even greater and the phase stability of the upconverter has to be maintained within very close tolerances.

Upconverters suitable for digital transmissions do not necessarily look different from those only suitable for analogue use, and the key parameter that defines the difference is the INTELSAT specification.

INTELSAT have issued two reference documents as INTELSAT Earth Station Standards that define minimum performance specifications for modulation in satellite systems: IESS-306[2] and IESS-308[3]. IESS-306 covers the modulation standard for analogue TV frequency-modulated carriers and IESS-308 that for QPSK modulation. Although both of these documents define standards for users of the INTELSAT system, they have become the general *de facto* standards for each type of modulation for the whole of the satellite industry.

An upconverter to the IESS-308 specification is essential for digital SNG transmissions; and, as the specification exceeds that required for IESS-306, an upconverter to the IESS-308 standard can also be used for analogue transmissions. The upconverter has few front-panel controls – some have none. Typically, there is a control to set the absolute output centre frequency and possibly a level control, i.e. for output amplitude.

As previously noted, the upconverter is one of the two most common elements to fail (the other is the HPA PSU). The upconversion process is even more critical in a DSNG transmission chain. This is because the upconverter has to carry out a number of stages of frequency multiplication, maintaining not only a high accuracy in frequency but, more critically, phase stability and accuracy. Bearing in mind the rigours of the environment in which this equipment has to operate, it is perhaps not surprising that this delicate electronic process is more liable to fail.

2.9.9 The HPA

Generally, the only difference in an HPA used in a digital uplink is that it can be lower in maximum power output, as this is one of the intrinsic benefits of digital operation. As previously said, analogue HPAs range in size from 300 to 800 W, but in a digital SNG uplink system the maximum output power range is 50–350 W, with typically a 125 to 175 W hub-mount HPA being used in conjunction with a 1.2–1.8 m antenna. Because of the nature of the digital signal, it is even more critical that the HPA is operated in the linear part of its transfer characteristic, to avoid increased bit errors. These errors are introduced from IPs generated in-band, as mentioned previously. Therefore the HPA is not usually operated at anywhere near its maximum rated output power, and on a Ku-band uplink it is 'backed off' typically by at least 4 dB from saturated power output. As we discussed earlier, if there are a number of digital carriers being combined, care has to be taken in achieving the correct power balance. Therefore, the power output back-off has to be increased to approximately 7 dB for two carriers, and by 9–10 dB for three carriers. A device called a 'linearizer' can be included in the HPA design (at significant financial cost), which effectively compensates for the non-linearity near the top of the power transfer characteristic – but this is not offered by all HPA manufacturers.

It is possible to use a solid state power amplifier (SSPA), which has solid-state electronics as opposed to a vacuum tube device as the amplifying element, where the output power demand is not very large. SSPAs come in a range of sizes from 20 to 100 W, but the significant drawback in their use is the amount of primary (line input) power that is required. An SSPA is typically 50–70% less efficient in input to output power than a TWT-based HPA and, historically, SSPAs have had a reputation for lower reliability than TWT-based HPAs.

2.10 The antenna

2.10.1 Background

We now come to the focal point of the uplink – the antenna. There is a wide range of antennas available for satellite transmission systems, but we will concentrate on the type of antenna used for SNG uplinks.

The function of the antenna is to take the signal from the HPA and both further amplify the signal and focus the beam towards the satellite, to 'launch the signal into free space'. Although there is a wide variety of antennas used in satellite communications, SNG uplink systems always use a 'parabolic' type, which has a reflector shaped in a parabola. (By including INMARSAT systems as a type of SNG system, we should note that a parabolic antenna is used on the older Standard-A systems and a 'flat-plate' type on the Standard-M, Mini-M and -B.) Downlink systems for reception of geostationary telecommunication satellite transmissions also tend to use parabolic antennas.

2.10.2 Essential characteristics of parabolic antennas

The parabolic antenna (Figure 2.36) is essentially an electromagnetic wave lens which, like a magnifying glass can focus sunlight energy into a narrow beam, focuses the radio frequency energy into a narrow beam. By doing so, it also has a gain characteristic as it amplifies the signal. A parabolic surface produces a parallel beam of energy if the surface is illuminated at its 'focus', and the parabolic dish is commonly referred to as the 'reflector'. The signal is transmitted via the feedhorn assembly on the antenna, and this is often loosely referred to as the 'waveguide', 'feed' or 'launcher' (as it 'launches' the signal).

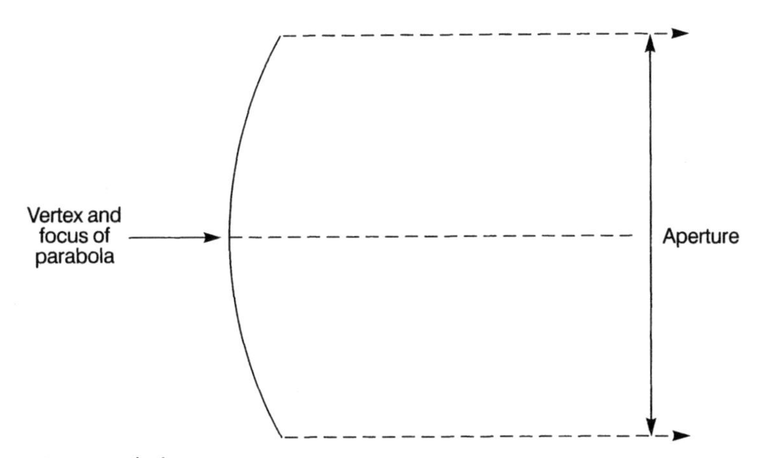

Figure 2.36 Parabolic antenna

The parabolic antenna family has a number of variants, but the types used for SNG systems are primarily the 'prime focus' and 'offset prime focus' types. In the offset prime focus, the focus of the antenna is displaced and, instead of being at the centre of the parabola (the 'vertex'), is shifted up. This is shown in Figure 2.37, where it can be seen that the lowest part of the parabola is not used. Figure 2.38 shows an antenna of this design.

A parabolic antenna is fundamentally described by two parameters. The physical diameter of the antenna is expressed in metres (or centimetres) and the amplification factor ('gain') of the antenna is measured in 'dBi'. The two are interrelated, as the gain of the antenna is a function of its diameter and frequency of operation.

The measurement 'dBi' means dBs with reference to the gain of an 'isotropic' antenna (the 'i' in dBi). An isotropic antenna is a theoretical ideal antenna that radiates equal electromagnetic power in all directions (the Sun is an optical isotropic radiator because it emits the same amount of light in all directions), and therefore

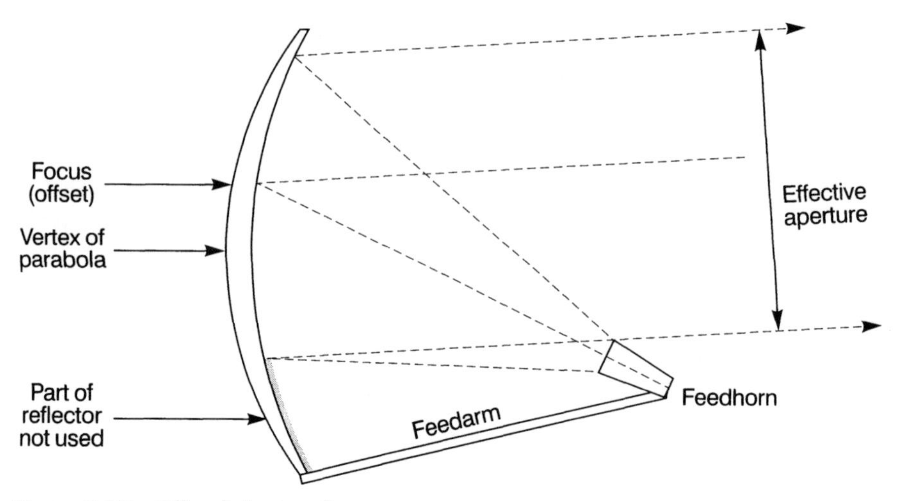

Figure 2.37 Offset fed prime focus antenna

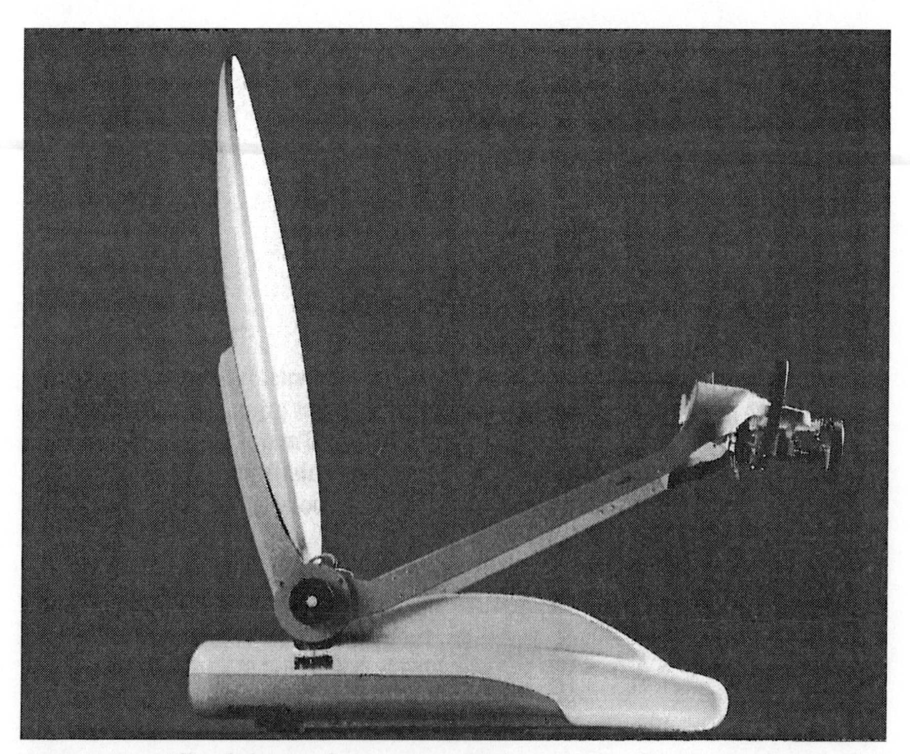

Figure 2.38 Offset fed prime focus antenna (Advent 'Newswift'). *(Photo courtesy of Advent Communications Ltd)*

the gain of an isotropic antenna is 0 dBi. The gain figure of a parabolic antenna is a measure of the increase of power over an isotropic antenna radiating the same power, in a particular direction defined by the focus of the antenna. The direction of maximum gain is often referred to as the 'boresight'.

The construction of the antenna varies according to the application, depending on whether it is to be used in a flyaway or is mounted on a vehicle (see Chapter 3). Irrespective of what type of application it is designed for, most antennas now have to meet the requirement of being able to operate to satellites spaced only 2° apart in the geostationary arc (see Chapter 6). The reason for the demand for 2° spacing is that in the 1980s it was anticipated that the number of satellites was going to increase steadily, with a resulting need to move satellites closer together to increase capacity in the geostationary arc.

This means not interfering with adjacent satellites to the intended one, and therefore the radiation pattern from the antenna has to be accurately defined. The antenna does not produce a completely perfect radiation pattern, which would be a single focused beam, but has a main 'lobe' (centred on the boresight) and a number of 'sidelobes' radiating out from the antenna, as shown in Figure 2.39. The sidelobes

Figure 2.39 Illustration of main lobes and sidelobes of a parabolic antenna

can potentially interfere with adjacent signals on the satellite, and one of the aims of good antenna design is to seek to minimize the sidelobes while maximizing the main lobe. Typically, up to 70% of the signal energy will be on boresight – this is a measure of the efficiency of the antenna. Too much energy in the sidelobes will reduce energy in the main lobe signal and will interfere with signals on adjacent satellites. Hence, the requirement to meet a performance target based on 2° satellite spacing.

The characteristic to meet the 2° spacing requirement is defined by the mathematical expression of gain (dBi) = $29 - 25 \log_{10}\theta$, where the sidelobes must not (in theory) exceed this mathematical 'mask' or 'envelope'. The Greek letter theta (θ) is the angle in degrees away from the boresight. In the absence of any

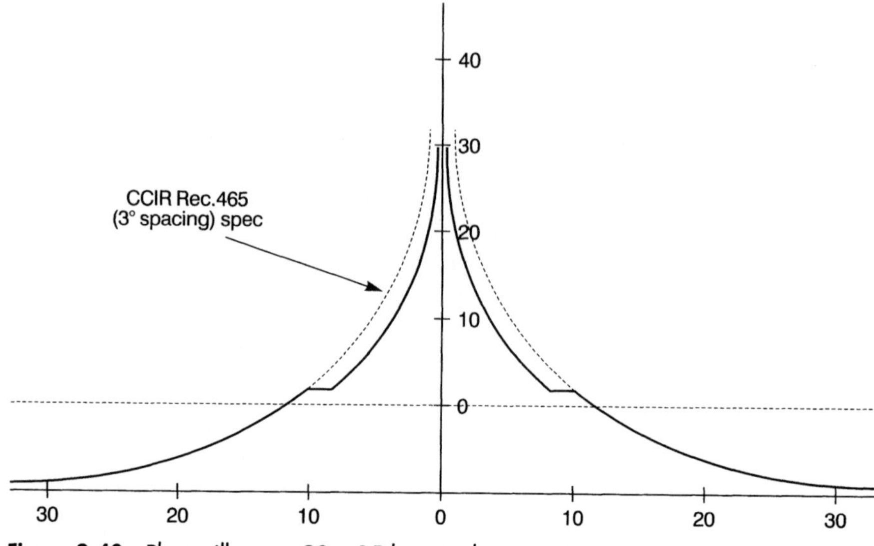

Figure 2.40 Plot to illustrate $29 - 25 \log_{10}\theta$ characteristic

requirement to meet 2° spacing, the ITU-R Recommendation 465 applies[4], requiring $32 - 25 \log_{10}\theta$ for 3° spacing.

The $29 - 25 \log_{10}\theta$ characteristic is shown in Figure 2.40, where the distance in degrees away from the boresight in each direction is shown along the horizontal axis, and the relative signal level is shown on the vertical axis. This pattern will vary with frequency and, as part of proving the performance of an antenna (see Chapter 6), the manufacturer will have to be produce a number of plots of this type across the range of frequencies the antenna is intended to operate. In practice, the specification is generally relaxed slightly as it is very difficult to produce antennas commercially that can exactly meet the specification, due to the variance that can occur in production.

As we have said, no antenna is 100% efficient, i.e. it does not radiate all the power on boresight. This is due deliberately to reducing the illumination towards the edges of the antenna (to reduce sidelobes) and blockage of the beam. This is particularly true of in-line fed prime focus dishes, where the feedhorn is directly in front of the antenna (see Figures 2.41 and 2.42). This type of feed is typically used on larger SNG antennas where the amount of RF 'shadowing' is a small proportion of the whole antenna surface area. Even with a perfectly smooth parabolic surface, the beam spreads out due to diffraction, and so it is not perfectly parallel in any case. Efficiency therefore has to be taken into account when calculating the gain of an antenna. Most high-grade parabolic antennas have efficiencies ranging from 55 to 70%. A typical SNG antenna of 1.5 m diameter and 60% efficiency has a gain at 14.25 GHz of around 45 dBi.

The gain figure is the single most important descriptor of the antenna, as together with the power output rating of the HPA, the total system power is defined in the

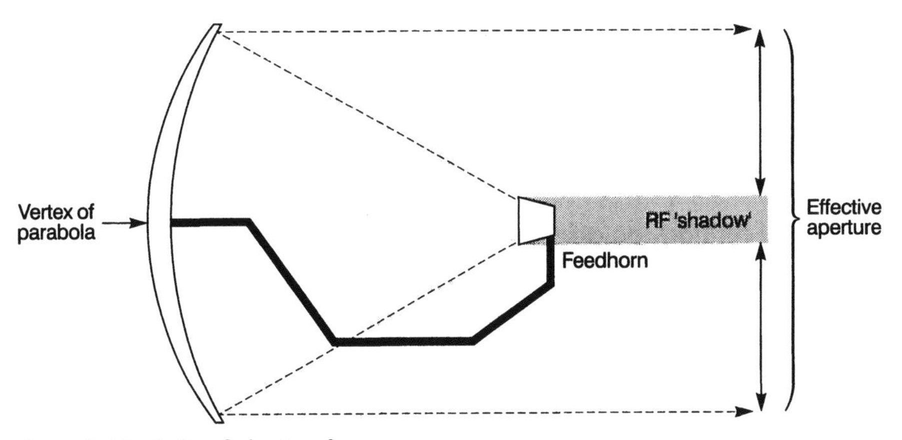

Figure 2.41 In-line fed prime focus antenna

Figure 2.42 In-line fed prime focus antenna (Advent 'Newswift'). *(Photo courtesy of Advent Communications Ltd)*

'effective isotropic radiated power' (EIRP). This figure is calculated by adding the HPA dBW figure to the dBi figure from the antenna. Hence, we have quoted 25 dBW for a 300 W HPA and 45 dBi for a 1.5 m antenna, and by adding these two figures together we have a system EIRP of 70 dBW – a typical minimum uplink power requirement for a Ku-band analogue SNG system.

2.10.3 Feedhorn polarization

Earlier in the chapter, we discussed the property of polarization. The two types of polarization – circular and linear – are each further subdivided and determine how the signal is transmitted from the feedhorn.

Circular polarization can either be clockwise (left-hand circular polarization or LHCP) and counter-clockwise (right-hand circular polarization or RHCP). The C-band polarization depends on whether you are talking about transmitting to or receiving from a satellite, and is often referred to by the INTELSAT terms of 'A-polarization' or 'B-polarization'. You can transmit to the satellite in LHCP, A-polarization, and you would receive the signal in RHCP, A-pol. Alternatively, you can transmit to the satellite in RHCP, B-pol, and you would receive the signal in LHCP, B-pol. Thus, the satellite converts or changes the 'direction' of the signal polarization from left-hand to right-hand, or from right-hand to left-hand, so as not to interfere with transponder inbound and outbound signals.

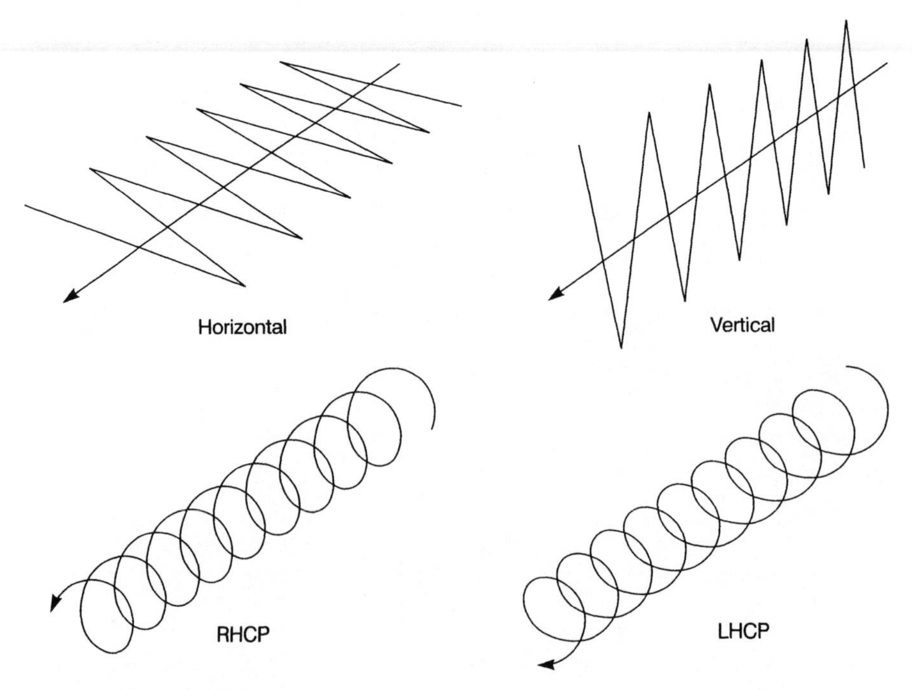

Figure 2.43 Polarization

Linear polarization is sub-divided into vertical (X) and horizontal (Y) polarization, and a signal uplinked on one particular polarization is typically downlinked on the opposite polarization. Both circular and linear polarization are shown in Figure 2.43, noting that C-band transmissions are always circularly polarized on INTELSAT, but are often linearly polarized on other operators' satellites (e.g. US domestic capacity and PanAmSat). Ku-band signals are generally linearly polarized (exceptions are some DTH services and services for Russia).

Although the polarization configuration of the waveguide determines the polarization of the transmitted or received signal, on both the uplink and downlink the waveguide is additionally rotated to compensate for the angular difference between the antenna position on the Earth's surface and the satellite position. This is referred to as 'polarization skew', and the degree of skew can be calculated from the latitude and longitude of the uplink (or downlink). Particularly for linear polarization, it is critical in achieving a good transmission or reception of a signal that the polarization skew is correctly applied to the antenna orientation. It is typical for a satellite operator[5] to require polarization to be set to within 1° accuracy (sometimes relaxed to 2° for digital carriers).

It is also important that any receive antenna is as 'blind' as possible to signals on the opposite polarization – this is termed 'cross-polar discrimination' (XPD) – so that any potential interference from a signal on the same frequency but opposite polarization is minimized. The ratio of 'good' cross-polar discrimination is at least 35 dB (and nearer to 40 dB), and professional-grade precision uplink antennas must at least match this to comply with most satellite operators' technical specifications for using their satellites.

2.11 Monitoring

In addition to any baseband monitoring, an uplink has to have some form of monitoring for the radio frequency signal processing, and this is achieved by the use of a 'spectrum analyser'.

A spectrum analyser is a very complex piece of equipment. Fundamentally, it is a radio receiver with the ability to repeatedly and automatically tune across a band of electromagnetic spectrum, displaying the amplitude of signals present in that band on a display screen. A signal is fed into the analyser, and as the analyser rapidly and repeatedly tunes across the radio spectrum, the signal sweeps from left to right across the display screen and individual sources of energy are displayed.

On the HPA there is a connection called the sample port, from which the analyser can measure the output signal. To be able to view the signal coming back from the satellite, an 'ortho-mode transducer' (OMT) needs to be fitted to the back of the feedhorn. An OMT is a small, multi-port microwave device that allows the transmission of signals on one polarization to be combined with the reception of signals on the opposite polarization (remembering that the downlinked signal is normally on the opposite polarization).

The OMT is also connected to the satellite receiver, via a combined amplifier and downconverter called a low noise block converter (LNB), so that the uplink operator is able to view signals received from the satellite. The LNB amplifies the weak signal from the satellite and frequency shifts it down from Ku-band (or C-band) to L-band, as required at the input to the satellite receiver.

The satellite receiver is also required to enable identification of traffic on a satellite. Particularly for analogue transmissions, this may allow the uplink operator to directly monitor the uplink signal directly back from the satellite, subject to the uplink being in the downlink footprint as well as the uplink footprint of the satellite (which is not necessarily always the case). On a digital uplink it is not always the case that the antenna has enough gain adequately to receive and decode the uplink signal, but the satellite receiver is still used as an aid to identify satellites. Nevertheless, even if the digital signal cannot be demodulated or decoded at the uplink, it can still be monitored on the spectrum analyser.

The array of monitoring equipment typically required is shown in Figure 2.44. A spectrum analyser can directly measure the high-frequency signals that are transmitted, but a good quality spectrum analyser typically costs US$15 000–25 000. Alternatively, a simple spectrum monitor can be used, which can monitor the signal at the intermediate L-band frequency; but if a problem occurs with the uplink and the operator is expected to venture beyond simple signal monitoring to fault detection and analysis, then a spectrum analyser is required.

There is one additional piece of measurement equipment that may be used by some uplink operators. This is a power meter, which can be connected directly to the sample port of the HPA to give a constant readout of the power being produced from the HPA. Power meters can either have an analogue or digital display. Measurement of absolute power output is possible with a spectrum analyser, but power control is much easier and sometimes more accurate with a digital power meter, as often the uplink (particularly for digital operation) may be operating at low to medium power.

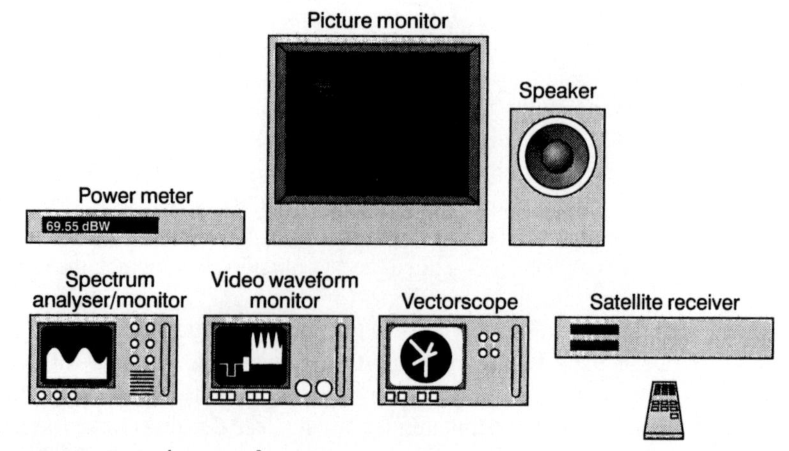

Figure 2.44 Typical array of monitoring equipment

The uplink operator can make both absolute and relative measurement of HPA power in either watts or decibels. However, it is another expensive piece of equipment, and another box to carry – though it should be borne in mind that some satellite operators[6] require the uplink EIRP to be measurable by the uplink operator and set to an accuracy of ±0.5 dB.

2.12 The studio-to-remote communications system

Before we leave looking at the uplink, we need briefly to look at the issue of studio-remote communications. Communications with the studio is a vital component of an SNG uplink, and with the development of the digital uplink came the possibility of easily integrating a digital transmit and receive carrier for the 'comms'. The comms system involves principally audio circuits that provide both 'talkback' and a feed of the studio audio output to the remote site. This is so that the reporter can hear instructions from the studio and is able to conduct a two-way dialogue with the studio presenter, which requires the reporter to hear the questions from the studio. There is also a talkback circuit from the remote location to the studio, and as the talkback circuit is a bi-directional path, the uplink system has to have both a transmit and a receive 'side-chain' to handle the bi-directional signals.

The feed of the studio output is known by a number of names, but the most common is IFB – the various definitions are dealt with in more detail in Chapter 3. Other communication circuits will typically include at least one telephone-line circuit (usually directly interfaced into the telephone switch at the broadcast centre) and a data connection with the newsroom computer system; and these can all be provided via an additional transmit/receive side-chain on the satellite uplink. The comms is uplinked as a separate carrier in the same frequency band as the main programme uplink channel, and is typically at a data rate of 64 kbps – hence it is common to talk of comms 'channels' in terms of 64 kbps slots. The combining of the comms transmit signal with the main programme signal can be carried out at any one of three points:

- At the lower 70/140 MHz IF.
- At the upper IF in the L-band.
- At C-/Ku-band before or after the HPA (but this is not common in the case of a comms carrier).

Mixing the comms carrier with the main programme signal at the 70/140 MHz IF has the advantage that it can be done at relatively low cost. The disadvantage is that the degree of frequency 'agility' of the comms carrier is limited to the maximum offset from the 70/140 MHz available in the comms modem. The uplink operator also has to calculate this offset and adjust the comms modem each time a new main programme and comms channel allocation is given by the satellite operator. For the purposes of illustration, we will assume that we have a system with a 70 MHz IF comms modem. If the modem has an output carrier frequency in the range

50–90 MHz, the maximum that the comms carrier can be offset from the main programme signal is 20 MHz (as that is the maximum offset up or down from 70 MHz). It may not be possible for the satellite operator to allocate a comms carrier slot this close to the main programme, and therefore it can be operationally limiting to configure the system in this way.

The advantage of carrying out the combining at L-band is that:

- There is full frequency agility by the addition of a separate comms up/down converter.
- The C-/Ku-band transmit and receive frequencies can be 'dialled up' directly on the up/down converter (which is performing a fixed frequency shift), making it easier for the uplink operator.

However, the disadvantage is that:

- An L-band upconverter is required, which is expensive (it is common to purchase a combined up/down converter).
- The uplink system has to be configured to allow insertion of a comms carrier at L-band.

Combining at C-/Ku-band before the HPA can be achieved using a simple combiner, but the disadvantage is that the overall signal level will be reduced by 3 dB because of the loss through the combiner. Combining at high power after the HPAs is rarely seen on an SNG uplink, as the power levels are so high that large components are required which can dissipate large amounts of heat. However, combining at this level does minimize the generation of IPs and is carried out on large fixed earth stations. In general, on an SNG uplink the addition of a comms carrier is carried out at either 70/140 MHz or at L-band.

The comms carrier does not need to be transmitted at as high a power level as the main programme signal, because of the much smaller bandwidth requirement of the data carrier of the comms. At whichever level the two carriers are combined (assuming it is before the HPA), the HPA will need to be backed-off to prevent the generation of IPs and to achieve satisfactory power balancing of the two signals. One also has to be wary where the satellite transponder is being operated at or near saturation of an effect called 'small signal suppression', when mixing the relatively low-power comms carrier with the high-power programme carrier. Observed on the downlink, as the level of the higher power carrier increases, the low-power carrier is reduced in level. This is caused by non-linear effects in the amplification stages (TWT) on the satellite. It can also be seen in a transponder where there is a mixture of high-power analogue signals with low-power digital signals – usually satellite operators try to avoid this scenario because of these effects.

Figure 2.45 shows a comms system with a digital uplink where the combining is carried out at L-band, and Figure 2.46 shows the detail of combining at 70 MHz IF. If it is a simple comms system, with just a single audio 'go' and 'return' for IFB, then the signals can be sent and received via a comms modem with an integral

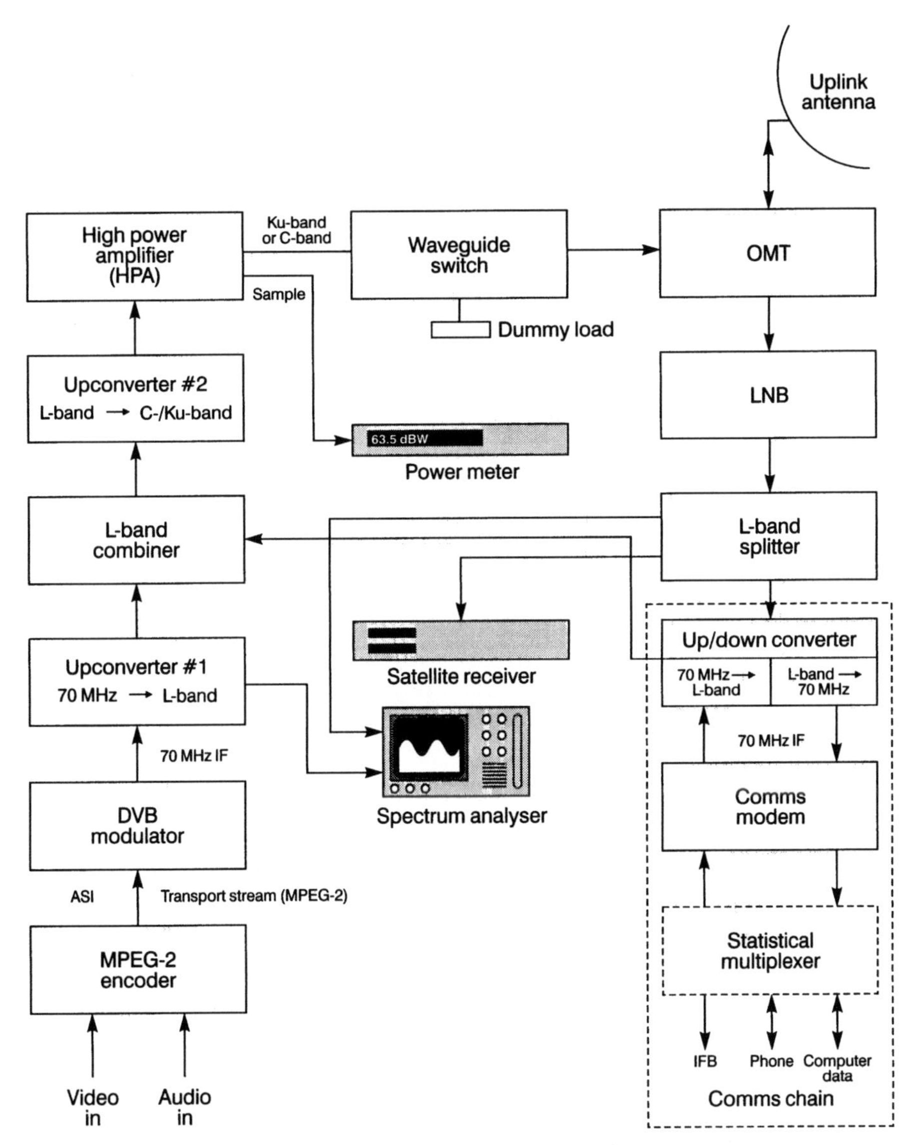

Figure 2.45 Digital SNG uplink with L-band comms system

ITU-T G.722 codec fitted. For a more complex system a 'statistical' data multiplexer is used, to maximize bandwidth, by varying the allocation of the relatively narrow data bandwidth of each of the audio channels according to the wider simultaneous bandwidth demand of a number of circuits. Channels on the statistical multiplexer can be programmed with different priorities, and it is also common to be able to reconfigure each end from the other if necessary.

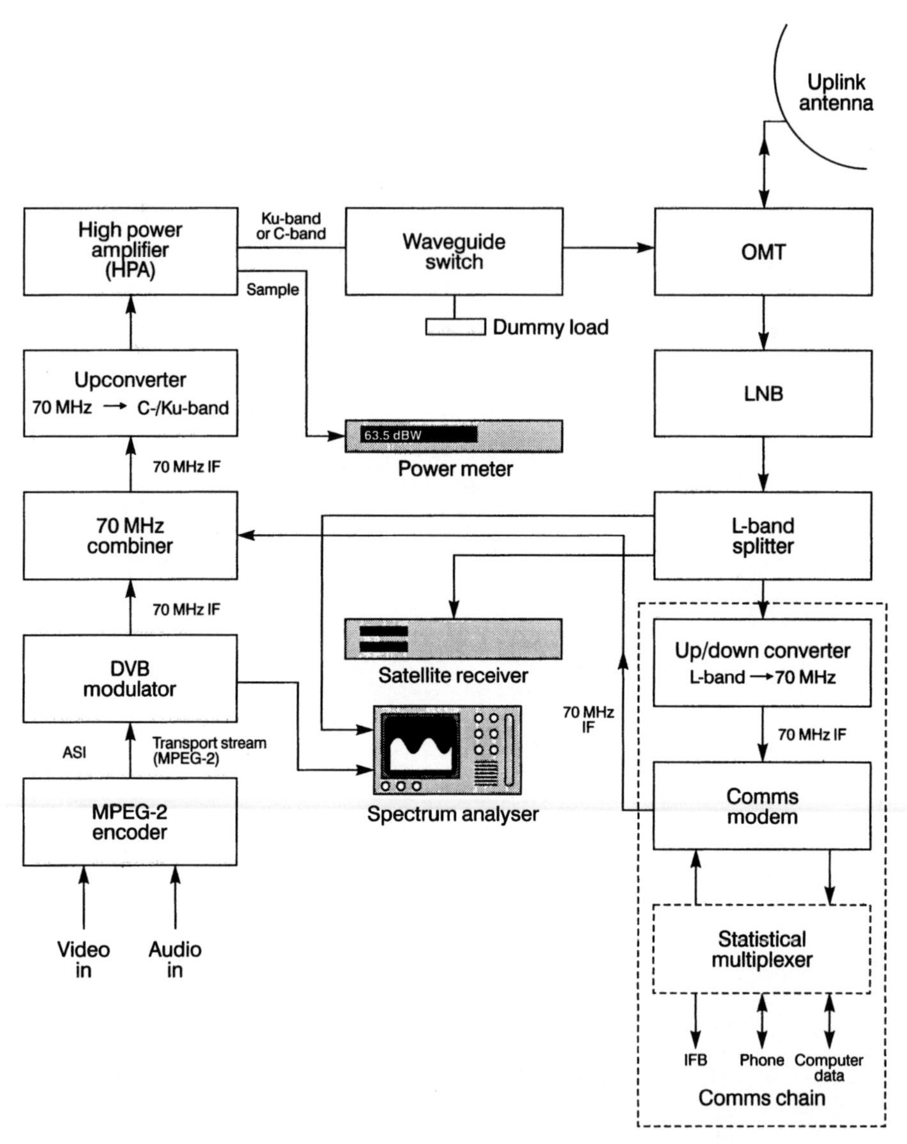

Figure 2.46 Digital SNG uplink with 70 MHz IF comms system

The use of an integrated communications system with the uplink is highly advantageous, from both a cost and convenience point of view, as the bandwidth and power requirements of the comms carriers on the transponder are much lower than for the main programme channel. Hence, satellite operators, even on ad hoc occasional capacity, will either make only a modest charge for the use of this bandwidth or none at all; often the comms carriers can be squeezed into the main programme channel 'guard-band', with no noticeable degradation to either signal. There is further discussion of the studio-remote communications system in Chapter 3.

2.13 The satellite

We will briefly look at the processes on board the satellite itself. The satellite has a receive antenna and, for each transponder, a frequency translator (converter), an input multiplexer, a switching matrix, an HPA, an output multiplexer and a transmit antenna. This is shown in Figure 2.47. A switching matrix is a router which, under remote control from the ground, can connect different source inputs to different destination outputs. A multiplexer combines a number of separate signals into one signal. As we are primarily concerned with the processes on Earth, it is not proposed to go into any further detail about the signal processing on board a satellite. Figure 2.48 shows a EUTELSAT Series 2 satellite deployed.

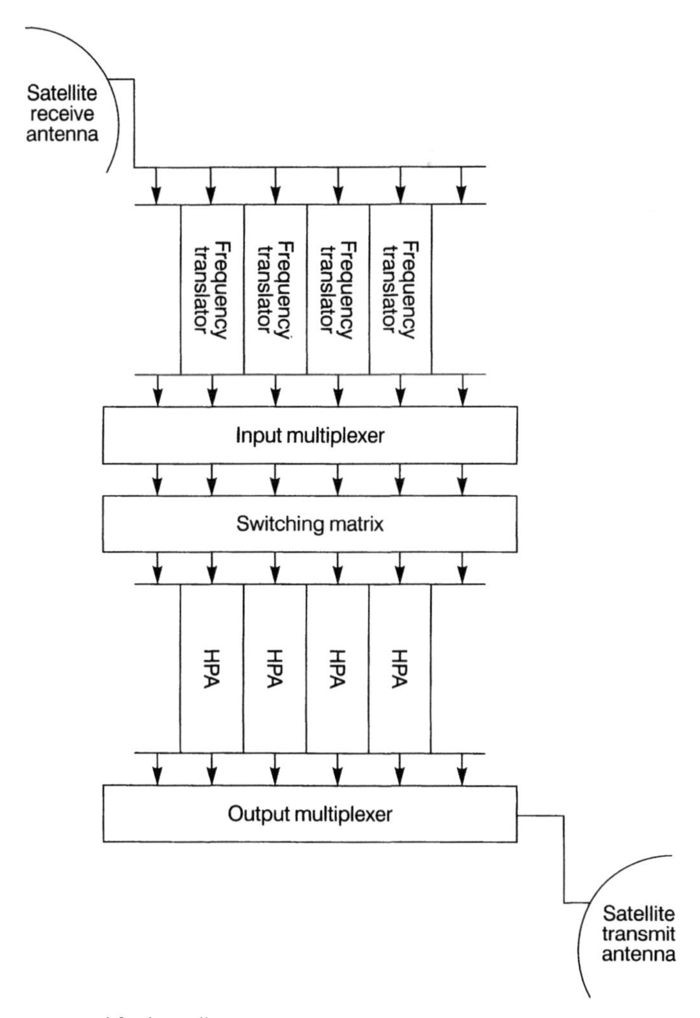

Figure 2.47 Simplified satellite overview

Figure 2.48 EUTELSAT series 2 satellite deployed. *(Photo courtesy of EUTELSAT/Aerospatiale)*

The satellite has a defined area of coverage, both for the uplink and the downlink, and the satellite operator defines these in planning a service from a satellite before it has even been launched. The coverage is published by the satellite operator in the form of 'footprints', which show the geographical coverage of each of the uplink and downlink beams from the satellite; an example of a downlink beam pattern for INTELSAT 802 is shown in Figure 2.49.

Each transponder on the satellite has a frequency plan defined for it. In fact, a transponder can have a number of frequency plans defined, and depending on the mode or type of service the satellite operator wishes to offer, a suitable frequency plan is put into operation. In general, frequency plans are not often changed on most satellites and transponders, but the satellite operator may choose to have more flexibility on some transponders to be able to respond to differing demands. A typical transponder frequency plan is given in Figure 2.50, and it shows the centre frequencies of the channels within the transponder, the size of the channels and the beacon frequency that is used as a unique identifier of the satellite. This information is obviously important for both the uplink operator and the downlink, and both footprints and frequency plans published by satellite operators are frequently referred to.

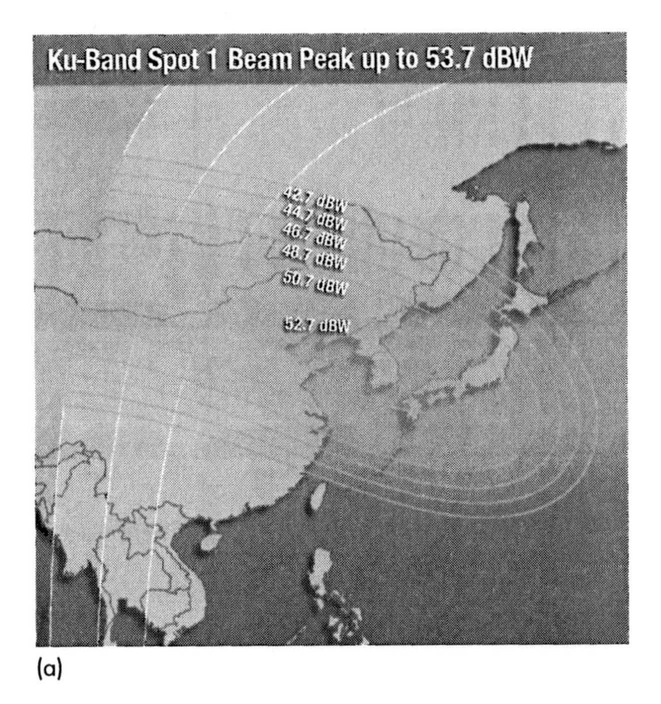

(a)

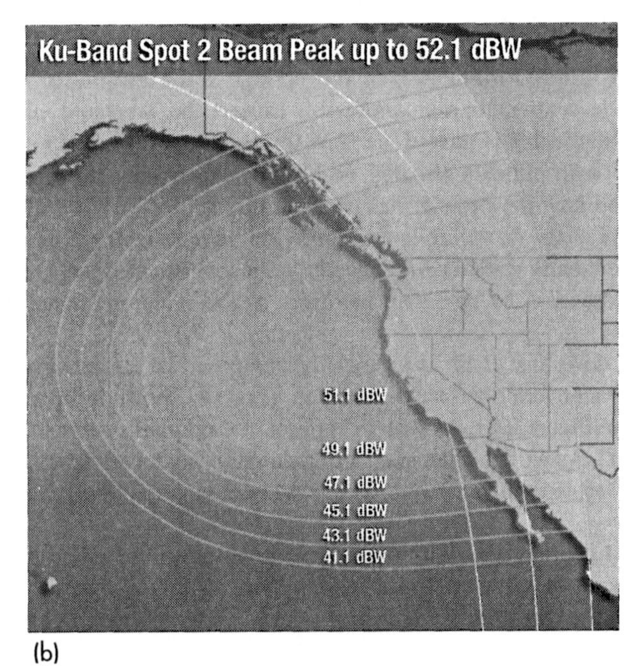

(b)

Figure 2.49 (a) INTELSAT 802 Ku-band Spotbeam 2 footprint (receive); (b) INTELSAT 802 Ku-band Spotbeam 1 footprint (receive). *(Photos courtesy of INTELSAT)*

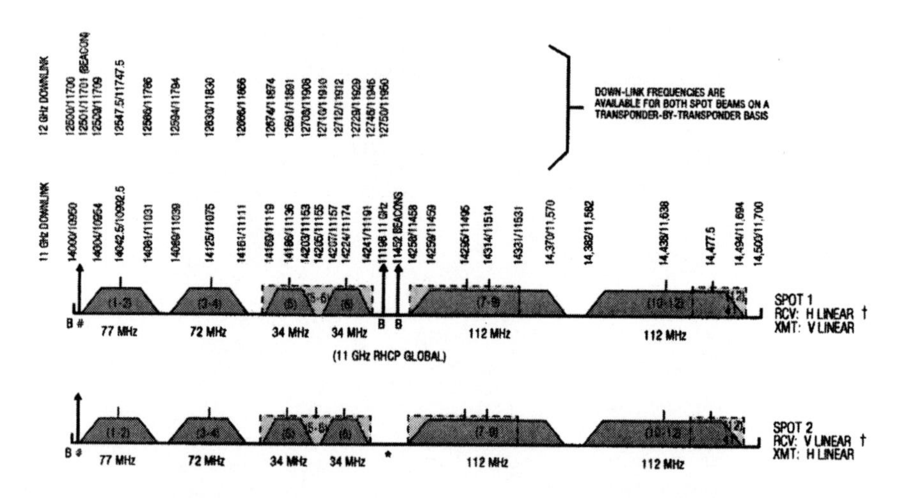

Figure 2.50 INTELSAT 802 Ku-band transponder plan. *(Photo courtesy of INTELSAT)*

2.14 The downlink

2.14.1 Background

The function of the downlink is to capture enough of the transmitted signal from the satellite to at least attain, or more preferably exceed, the 'threshold' of operation; for analogue systems, this is termed the 'FM threshold'. A digital receive system will fail abruptly with either a 'frozen' or black picture when the input signal fails (depending on how the decoder has been set up).

The system at the downlink is essentially the reverse of the uplink process, and fixed downlink earth stations are typically large sophisticated installations. We will only briefly describe the processes involved, as this is not the focus of the present book.

A typical downlink chain is essentially composed of an antenna, an LNA (or LNB), a downconverter and a demodulator. This was shown in overview earlier in the chapter, in Figure 2.21. As with the uplink, for a digital system there would also be a 'decoder'. As we shall see later, a critical factor in downlink design is the need to minimize 'system noise power', as this will potentially reduce the quality of the link.

For a fixed continuous point-to-point link application, the other critical factor is the percentage of time that the given link will be received successfully ('availability'); a typical target is in excess of 99.99%. In other words, the link will fail (termed 'outage') for no longer than an aggregated total of 53 minutes in a year. However, this is more applicable to fixed earth stations, whereas SNG uplinks by their very nature are only in one place on a temporary basis, and therefore a calculation of availability is largely academic.

2.14.2 The antenna

On a fixed earth station, the receiving antennas are usually 4–30 m in diameter, and there are typically ten to twenty antennas on a moderately sized earth station site, as shown in Figure 2.21b–d (these are sometimes referred to as 'dish farms', for obvious reasons). Antennas may either be fixed on one particular satellite or be agile and able to be freely moved from one satellite to another as required. The control mechanisms to achieve this are complex and beyond the scope of this book to describe.

As we have seen, an antenna has gain, and this is true for both directions of signal travel, both transmit and receive. The gain of an antenna in receive mode is also measured in dBi. The downlink also has to have some measure of its performance and this is termed its 'figure of merit'. The figure of merit of a downlink station depends on the antenna's receive gain (in dBi) as well as the amount of receiving system noise, which can be expressed as a temperature in degrees Kelvin (K). This figure of merit is called the 'G/T' and is expressed in units of dB/K. The larger and more positive this figure is, the more sensitive the receiving system.

2.14.3 LNA

As mentioned previously, the uplink uses an LNB as part of the satellite receiver chain. The LNB is relatively simple and inexpensive, but on a large fixed downlink earth station, a high-performance low noise amplifier (LNA) is used to receive the signal – and one antenna may have a number of these. The unit is relatively large, as to meet the performance required it needs to achieve a low 'noise temperature', and this is an important factor in link budgets. The signal from the satellite is very weak and once received it has to be amplified without adding noise.

2.14.4 The downconverter

The process of downconversion is generally the direct inverse of the upconversion at the uplink, and signals may travel 'cross-site' at an earth station at L-band, 140 MHz or 70 MHz between stages of downconversion. An example of a downconverter is shown in Figure 2.51. Again the design of the downconverter would minimize as much as possible the system noise power, and for digital operation the phase noise characteristics are also important.

2.14.5 The receiver/decoder

Fixed earth stations generally have a large number of receivers that can be switched to different antennas, and this switching forms part of the whole control

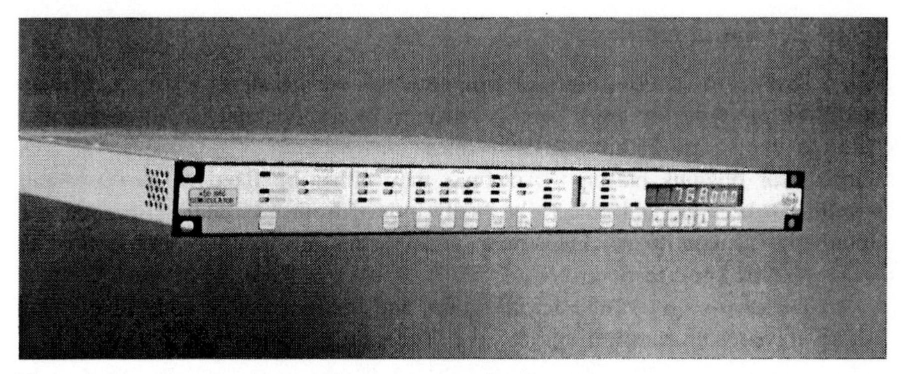

Figure 2.51 Downconverter. *(Photo courtesy of Advent Communications Ltd)*

system. The receivers are to a high specification and offer a number of configurable facilities that are not available on domestic-grade satellite receivers.

The essential process within the receiver is to filter the incoming signal to eliminate unwanted signals outside of the frequency to which the receiver is tuned, demodulate the carrier to recover the transmitted video and audio information, and reproduce a stable, 'clean' version of the transmitted picture and sound. An example of such a receiver is shown in Figure 2.52.

For analogue systems, this is essentially an FM demodulation process, whereas for digital systems the signal has to be digitally demodulated. Digital receivers may either have inbuilt digital decoders or are directly connected to separate digital decoders. The decoders take an incoming compressed digital data stream received from the satellite and produce a baseband video signal in either digital or analogue form as required. An integrated MPEG-2 digital receiver (which combines a downconverter and a demodulator, as well as a compression decoder) is commonly referred to as an integrated receiver decoder (IRD).

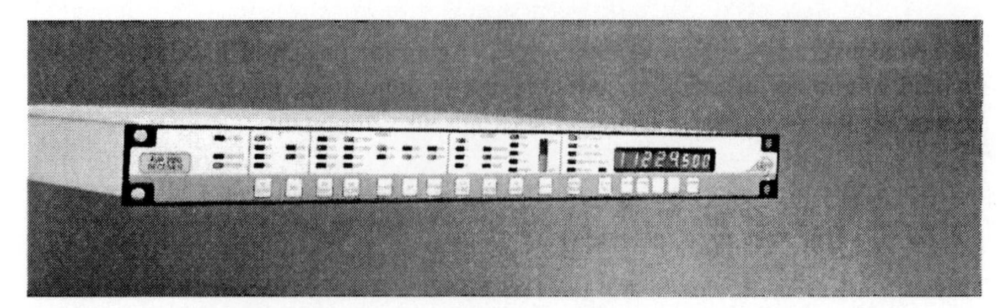

Figure 2.52 Analogue receiver. *(Photo courtesy of Advent Communications Ltd)*

2.15 Overall system performance

2.15.1 Background

The fundamental basis for assessing whether a satellite communication link will work is the calculation of the link budget. This determines whether the threshold level of the signal will be achieved – and hopefully exceeded, to give some degree of margin for factors that could not be included at the time of the calculation of the link budget. In this section, the principal elements that contribute to the link budget will be described, although the actual calculations of sample link budgets are in Appendix B for the benefit of engineers.

2.15.2 Factors to be considered

We need to consider what elements have to be included. These are:

- Uplink power
- Type of modulation
- Atmospheric losses (on both the uplink and the downlink)
- Satellite gain
- Downlink gain
- Noise and interference

We have already looked at the factors of uplink power and downlink gain (or 'sensitivity'), and will shortly consider the key satellite factors and examine the losses that affect overall system performance.

What are we trying to measure with the calculation of a link budget? Fundamentally, we are determining whether enough signal transmitted from the uplink can be received at the downlink to convey the information carried accurately. This encompasses the uplink and downlink equipment, and the uplink and the downlink paths, as well as the effects of the transition through the satellite. The link budget is a 'relative' power calculation that looks at ratios of lost or gained power, and the end result is, in fact, a set of ratios.

The link budget is similar to a financial budget, with contributions (income) and deductions (expenditure) which define the overall performance, where the currency is the decibel (dB). Because the dB is the fundamental unit of 'currency' in the link budget, we need briefly to remind ourselves why we use this unit of measurement; otherwise the contributions (and deductions) of the link budget will not carry the weight or meaning required to understand its significance. The convenience of using decibels is that multiplication and division of elements (some of which are very large numbers) can be performed by simply adding or subtracting values in dBs, as all the factors in a link budget can be expressed in dBs. As we saw earlier with the HPA, gain is expressed with a '+' and therefore losses are expressed with a '–'.

As with a financial budget calculation, depending on the level of detail, the link budget can be a relatively simple one, involving perhaps only a few pages of calculations, or it can be extremely detailed and spread over at least ten pages. The link budgets in Appendix B are simple ones.

Analogue link budgets are relatively straightforward, but digital link budgets are more complex, as the potential effects of interference from adjacent carriers, carriers on opposite polarizations, IPs etc. are potentially much more damaging to a digital signal than an analogue one. When these factors are included in the link budget, the calculations can become very complex. Link budgets are normally performed using either spreadsheets or specific application software, as the process of calculating and including each of the individual elements is very tedious if done manually.

The significant results of the link budget will be ratios that are a measure of the predicted quality of the link: the RF signal carrier to noise ratio (C/N) in the case of an analogue link, and the energy per bit (E_b) to noise density (N_0) ratio (E_b/N_0) for a digital link. In a digital link, we will also be concerned with the bit error rate (BER), i.e. the ratio of failed bits to successfully received bits.

The RF signal carrier to noise ratio can be calculated for both analogue as well as digital links. For analogue links, this is then converted into video and audio baseband signal-to-noise ratio (S/N). However, we cannot use the concept of S/N for digital video or audio signals, as these are essentially noise-free. The imperfections on the video signal are 'glitches' or 'artifacts' (see Chapter 4) which (although they should not be present in theory) result from defects either in the compression or transmission processes. Audio signals would suffer from random clicks. A better measure of digital signal quality is the energy per bit received (E_b) relative to the noise power spectral density (N_0), expressed as the E_b/N_0 ratio in dB.

This can then be converted into bit error rate (BER), which is the number of bits received in error compared to the total number of bits sent in a given time period. If the E_b/N_0 is too low (and hence the BER is too high) the decoder will suddenly stop working, giving a frozen or black picture.

So, in practice, the results will be disappointing if:

- the uplink does not produce enough power;
- the satellite does not have a sensitive receive or powerful transmit characteristics;
- the downlink antenna is not large enough (i.e. does not have enough gain);
- there is more noise and interference than expected.

In the case of an analogue SNG link, this can result in either a 'noisy' picture with a lot of interference (which can still be used if the editorial value is overriding) or no picture at all. In the case of a digital link, anything apart from total success results in just no picture at all. Digital signals suffer from what is known as the 'cliff-edge' effect, that is, either perfect results or nothing – there is an absolute divide between complete success and total failure. This is unlike analogue FM signals where the failure is gradual and it is common for links to be operated at only slightly above the FM threshold. In other words, 0 dB margin equates to no operation for digital systems, whereas 0 dB equates to degraded but possibly still usable material in analogue FM systems.

What prevents the signal being clearly received? Assuming that the uplink and downlink equipment are working correctly, then it is predominantly the factors involved in interference, noise and losses, such as the dispersal of the signal in its passage through space and the atmosphere, on both the uplink and the downlink paths.

As signals travel through space and through the Earth's atmosphere, natural factors such as noise from space (galactic noise) and the Sun (solar noise), atmospheric absorption and refraction affect them. In addition, particularly for the downlink, artificial sources of interference from terrestrial radio frequency transmissions are a potential problem. The effects vary somewhat depending on whether the link is in the C-band or Ku-band.

2.15.3 The satellite

The sensitivity of the satellite is defined by the incidental power flux density (IPFD) needed to 'saturate' the transponder on the satellite. This is, in effect, a measure of the 'illumination' required from the uplink, and is measured in dBW/m^2. An uplink must produce enough power to achieve a specified IPFD at the satellite, allowing for the location of the uplink on the satellite's receive G/T contour ('footprint') projected onto the Earth's surface. Note that, as with the downlink, the receive antenna of the satellite has a figure of merit, G/T. The satellite operator will quote figures for the IPFD of a particular transponder and this varies according to the gain setting ('gain step') of that transponder. This is not based on saturating the transponder, but on the fact that the transponder channel is 'backed-off' by a number of dBs.

As the signal is transmitted from the uplink antenna, the signal spreads out to cover a very wide area. However, as the signal spreads out, it gets weaker the greater the distance from the antenna. This is analogous to a stone being dropped into the middle of a pond and, as the ripples spread out from where the stone fell into the water, they get smaller and weaker. In engineering terms, this is termed the 'inverse-square law'. Because of the distances involved, this 'spreading loss' is a significant and constant loss in satellite transmissions, and has to be calculated to derive the amount of uplink power required to achieve the specified IPFD at the satellite. It is therefore an absolute measurement of power loss, rather than a relative power calculation that is used in the link budget. Spreading loss is not frequency dependent and is a function of the slant path between the uplink and the satellite.

On its transmission side, the satellite also produces an IPFD towards the Earth, which is a measure of its 'illumination' of the Earth's surface, and this varies according to the transmit footprint of the satellite. Again, the satellite operator publishes information relating to the IPFD, which in turn determines both the location and the size of receive antenna required to produce the correct level of input signal to the receiver.

2.16 Losses

2.16.1 Free space loss

As the signal travels through space, it suffers attenuation. The ratio of the attenuation of the signal transmitted to the amount of signal received at an antenna is termed 'free space loss' (FSL) or 'free space attenuation' (FSA). The path between the earth station and the satellite is termed the 'slant range', and as the latitudinal position of the earth station increases, the elevation angle to the satellite decreases, thus increasing the slant range. Thus, as the slant range lengthens, the loss increases. Free space loss is frequency dependent, increasing at higher frequencies, and affecting both the uplink and downlink path losses.

At elevation angles less than 5°, a significant part of the path will pass through the Earth's atmosphere and will therefore be subject to terrestrial interference and more noise picked up from the 'hot' surface of the Earth compared to the 'cold' background of space. This increases the attenuation of the signal. In addition, at such a low angle of elevation, the link may be subjected to 'scintillation'; this is a rapid fluctuation in amplitude and phase caused by the signal travelling a longer path through the atmosphere and ionosphere.

2.16.2 Atmospheric absorption

At the frequencies used for satellite transmissions, electromagnetic waves interact with gas molecules in the atmosphere causing attenuation of the signal. The effects are minimal and usually the signals suffer no more than 1 dB of attenuation, which would only be of significance if the link budget looked marginal.

2.16.3 Precipitation loss

In the Ku-band, rain is a significant factor in attenuating the signal, further dispersing the signal as it passes through water drops. Rain rather than clouds cause the problem, as even the direction of fall and shape of the water drops affect the signal. As frequency increases, the effect of rain increases – which is why at Ku-band, the effects are far greater than at C-band (and are even greater in the Ka-band). The detrimental effects of rain also vary depending on the polarization of the signal.

To aid in calculating the effect of rain the world is divided into 'precipitation zones' or 'rain climatic zones', each of which has a numerical value which is used in the calculation of a link budget. These zone values are defined by the ITU (International Telecommunications Union; see Chapter 6) and are a statistical analysis based on the frequency of rainstorms, the amount of rainfall in a year and even the type of rain. They are published in the form of tables with a letter identifying each zone, with various values depending on the percentage of time in the year when there is rainfall of a certain rate (mm/hr). It is not the amount of rain that falls in a year that is significant so much as the *rate* at which it can fall. Areas that have high rainfall values – tropical zones, for instance – have a value that effectively increases the free space loss, and vice versa for arid zones. In detailed link budget calculations, rain is accounted for on both up- and downlinks, but it is generally ignored for an SNG uplink (because of the relatively temporary nature of the transmissions). The exception is when it is located in a high precipitation zone, where there would then be a high risk of outage even during the relatively brief period it is transmitting.

2.16.4 Pointing loss

It should be clear by now that if an uplink is aiming to 'hit' a satellite with a signal, the pointing accuracy has to be very high. We have seen that the satellite is station-kept to within 0.1° of its nominal orbital slot, and therefore the uplink and downlink have to be steered to within the same accuracy. In theory, if the boresights of the earth station antennas are aligned exactly with the boresights of the antennas on the satellite, then the 'pointing loss' is zero. In practice, partly due to movement of the satellite (within the nominal 'box'), and partly allowing for some error in maintaining absolute pointing accuracy at the uplink and downlink, there is a small amount of pointing loss. This equates to some signal not reaching the satellite from the uplink, or from the satellite to the downlink. Typically, a value of 0.3 dB of pointing loss is allowed for in a link budget for a manually panned antenna at the SNG uplink, and 0.1 dB for an auto-tracking antenna at the fixed downlink.

2.16.5 Waveguide loss

At both the uplink and the downlink, there are lengths of waveguides connecting different parts of the systems together. These waveguides introduce some attenuation to the signal and, in addition, with each waveguide joint there will also be some losses. These are calculated in a link budget collectively as 'waveguide loss', and typically account for 0.5–1 dB of loss in the uplink, and varying figures at the downlink depending on the length of the 'cross-site' distances.

2.17 Noise

Having dealt with the various losses encountered that have to be included in a link budget, the concept of noise has briefly to be examined. Noise is present on the uplink signal received by the satellite and is further added on the downlink signal. As such, noise on the up- and downlinks affects overall performance. If noise added to the uplink path signal dominates the overall performance of the system, the system is described as 'uplink limited'. Conversely, if the noise on the downlink path dominates, the system is termed 'downlink limited'.

Earlier in the chapter, we looked at the concept of noise in terms of an audio signal, and this same principle applies to the satellite link. The main sources of noise are antenna noise, made up of sky and, to a lesser extent, galactic noise, and system noise from receiver thermal noise, waveguide noise and other component noise, and intermodulation noise through the whole system.

System noise is typically measured in Kelvin (K), which is a unit of measurement of temperature used in engineering calculations. At microwave frequencies, everything with a physical temperature above 0 K generates electrical noise in the receiver. As the decibel is used as the primary unit of 'currency' in link budgets, noise expressed as Kelvin can be converted to 'system noise power', measured in dBW.

The calculations involved in calculating noise powers can be lengthy and complex, and here it will suffice to say that every component in a receive downlink chain introduces some noise into the system. The aim of good downlink design is therefore to minimize these noise powers as they effectively make the downlink less sensitive. The signal being transmitted from the satellite is travelling a great distance and, as we have seen, suffers losses on its passage. The signal is therefore extremely weak when it arrives at the antenna and the addition of noise power reduces the C/N ratio, as it 'lifts' the noise 'floor'. Therefore, if the downlink design cannot minimize the noise power to a small enough degree, it can only be compensated for by increasing the size of the antenna.

2.17.1 Analogue considerations

Threshold noise

In an analogue FM link, noise appears on the picture as 'fine grain' noise. Typically, a 'unified' video S/N of 52 dB is required for a satisfactory link. However, if the system is operated at or within 2 dB of FM threshold, noise occurs in the picture as black and white dots (colloquially called 'sparklies'), where noise spikes instantaneously exceed the carrier level. This is called 'threshold noise' and the 'threshold margin' (also called the 'fade margin') is the level of downlink power we can afford to lose before the onset of threshold noise affects the picture. If the transponder being accessed is operating in the linear part of its power transfer characteristic, this also translates into the amount of uplink power that we can afford to lose.

Threshold noise can be considered objectionable – though this is a subjective judgement. If the transmission were a prestigious event with the expectation of high quality – the televising of an opera, concert or sports event, particularly on a pay-per-view basis – then this quality of link would be rejected outright. On the other hand, if the link is carrying news material, particularly of an exclusive nature, then the link will be deemed satisfactory.

Truncation noise

The C/N ratio can generally be improved by reducing the receiver IF filter bandwidth; however there is a cost here in that if the filter is too narrow, high-order 'sidebands' of the vision signal, corresponding to areas of highest deviation, are severely attenuated. This causes horizontal streaking on vertical transitions and in highly saturated colours it is visible as noise which looks like threshold noise but is confined to the area of colour. This distortion is referred to as 'truncation noise', and to avoid this the uplink must reduce deviation. This technique can also remove sparklies but leads to a gain in fine grain noise – but this is less objectionable than sparklies.

Threshold extension

The term 'threshold extension' may be heard in reference to an analogue system. In an FM signal, if the instantaneous carrier deviation is less than the allocated transmission bandwidth due to the nature of the picture information at that instant, there is in effect some 'wasted' bandwidth. This wasted bandwidth could be used if either the deviation was rapidly increased to take advantage of this, or the filter bandwidth in the receiver could be reduced to improve the C/N ratio. As improving the C/N ratio is more important than use of the bandwidth in this instance, analogue receivers have filters that track the modulated bandwidth, rapidly adjusting ('tracking') in response to the incoming FM deviation, and this process is called 'threshold extension'. Using threshold extension can typically create an improvement of up to 4 dB in the C/N ratio, which could mean the difference between being able to receive the signal on, say, a 3.7 m antenna instead of a 6 m antenna.

2.17.2 *Digital considerations*

Because of the nature of digital modulation and signals, noise is less of an issue, and as long as the noise remains below the threshold of the receiver, the output of the receiver and decoder will produce perfect pictures. As soon as the noise level rises to match or exceed the threshold level, the digital signal completely breaks up and no usable pictures are produced from the decoder. As mentioned earlier in the chapter, digital modulation and demodulation processes are particularly sensitive to 'phase noise' (most likely generated in the equipment), which has disastrous effects on the signal. The downlink chain in a digital system therefore has to minimize the amount of phase noise in particular.

2.17.3 *Interference*

Interference can come from a number of sources, and where the interference is above a certain level it needs to be included in the link budget calculations. Interference can come from signals in the same transponder (IPs) and adjacent transponders, and signals at similar frequencies but on the opposite polarization, i.e. cross-polar interference (XPI), related to XPD.

As we have already seen, the ratio of carrier to noise is expressed as a C/N figure. Similarly, the ratio of carrier to interference can be expressed as C/I. The two figures can also be added to produce an overall C/(N + I) figure. Similarly for a digital link, E_b/I_0 and $E_b/(N_0 + I_0)$ figures can be derived.

It is not proposed to go into the details of all the effects of interference here, and the impact on digital links has already been mentioned.

2.17.4 Link budget numbers

Having taken into account all the losses, the effects of noise power and interference, what is the scale of the numbers we would like to see from a link budget as a measure of the received signal? There will be some variation as to what is perceived to be 'good' depending on where in the world the signal is being received and what the expectations are, taking into account the parameters of the uplink, the satellite and the downlink.

In the case of an analogue link, the carrier to noise ratio (C/N) is the most important figure. This is the measure of received signal above the level of noise in the system. For a typical analogue Ku-band link, a satisfactory C/N is between 12 and 15 dB, i.e. the carrier signal is 12–15 dB above the noise 'floor'. This should allow some 'fade margin' to allow for some adverse weather or other unforeseen effects causing the signal to be degraded. In tropical areas, this fade margin will have to be greater to allow for extremely heavy rainstorms.

For a digital link, the energy per bit to noise density ratio (E_b/N_0) is one of two significant factors, the other being the bit error rate (BER). Generally, in a low precipitation zone, a satisfactory 8 Mbps digital SNG link (either C- or Ku-band) needs to have an E_b/N_0 of 8–15 dB. At the input to the MPEG-2 decompression process (i.e. after RS error correction), there needs to be a BER of at least 10^{-9} – this means there must be less than 1 bit error in every one thousand million bits. The BER is largely dependent on the E_b/N_0, as the nearer the carrier level is to the noise floor, the greater the likelihood of bit-errors which cannot be recovered by the error correction process. Again, margins need to be increased in high precipitation zones.

Both the C/N figure for the analogue link and the E_b/N_0 figure for the digital link must have a high enough fade margin above the threshold level of the link to ensure reliable operation. Separate figures for the uplink and downlink paths are calculated, which can then give an overall C/N or E_b/N_0.

Link budgets are not calculated before every transmission. Back in the 1980s, when satellites were very much lower in power and SNG systems were often operated at their limit to achieve a usable analogue signal, the calculation of link budgets was a frequent necessity. Nowadays, link budgets tend only to be produced either during the exploratory period of securing a suitable satellite lease, or where it is anticipated that operation of an SNG uplink on a particular satellite is likely to be marginal. This may be either due to the uplink's specification or its location on the satellite's footprint. Current generations of satellites are much higher power and because, in particular for digital transmissions, the power requirement from both the satellite as well as the uplink is lower.

Finally, a note of caution regarding link budgets. A link budget is only a theoretical calculation and it is only as good as the quality of the data used. In practice, the link can behave quite differently from the prediction. A small error in one small element of the budget can result in a significantly different answer from that expected.

A link budget can also be manipulated to produce a 'desired' figure and it should be regarded with caution if it has been prepared by someone else. On the other hand, link budgets prepared by satellite operators can often be very pessimistic, usually when satellite 'end-of-life' data are used (when the satellite will be at the bottom end of its performance), so minimizing the exposure of risk to the satellite operator. The satellite operator may wish to minimize the risk of link failure being attributed to the satellite performance, and shift the onus of failure onto the uplink and/or downlink performance.

2.18 Conclusion

In this chapter we have covered every aspect of satellite theory that affects the SNG uplink operator, and have hopefully clarified a number of issues that are not always easily understood in isolation but need to be considered in conjunction with other issues. In particular, it is hoped that the explanation of how satellites are placed in orbit gives some insight into the 'behind the scenes' aspects of satellite operation which are not necessarily obvious to users of satellites as a newsgathering tool.

We have shown that the technical parameters of digital transmission present the opportunity to access the space segment at a much lower cost, hence opening up the opportunity of using DSNG to those who previously had viewed SNG as too expensive an operation. It must be borne in mind that it is very likely in the long term that the revenue costs of operating an SNG uplink, i.e. the space segment costs, will far outstrip the initial capital cost of purchasing a system.

The deliberate minimization of the use of mathematics has also hopefully helped the non-technical reader, while Appendix B will help clarify many of the link budget calculations for the technically inclined.

References

1. Clarke, A.C. (1945) Extra-terrestrial relays – can rocket stations give world-wide radio coverage? *Wireless World,* October, 305–308.
2. INTELSAT (1998) IESS-306, Performance characteristics for television/frequency modulation (TV/FM) carriers and associated sound program (FM sub-carrier).
3. INTELSAT (1998) IESS-308, Performance characteristics for intermediate data rate (IDR) digital carriers using convolutional encoding/viterbi decoding and QPSK modulation.
4. ITU (1993) Recommendation Itu-R S.465–5, Reference earth station radiation pattern for use in co-ordination and interference assessment in the frequency range from 2 to about 30 GHz.
5. EUTELSAT (1999) EESS-400, Minimum technical and operational requirements for earth stations transmitting to leased capacity in the EUTELSAT space segment – Standard L.
6. EUTELSAT (1995) *Eutelsat Systems Operations Guide (ESOG), Television Handbook,* Volume II, Module 210.

3

Boxes or wheels: types of systems

3.1 Where to begin. . .

How do you choose an SNG system? For new users, this can be a bewildering decision with a large array of specifications and features to consider. Some of the major defining parameters are whether it should be a transportable flyaway or vehicle-based system, C- or Ku-band, digital or analogue. Less obvious is the option of choosing an INMARSAT system, which is particularly appropriate for a certain type of newsgathering. Although only briefly discussed in this chapter, INMARSAT systems are fully covered in Chapter 5.

Potential customers for any type of system have to assess its capabilities against their own operational needs. It is important to clearly identify these needs in the beginning so that an informed purchase can be made, as it will be a considerable financial investment. In this chapter, different configurations for each type of operation are examined to help the new user decide on the best system for their needs. However, as well as careful consideration of the requirements, thorough research of the market by any prospective purchaser is vital.

In addition, the signals that are to be fed to the SNG uplink have to be considered. An SNG uplink system on its own is not going to be sufficient to provide all the elements of a news field operation. There is a considerable amount of extra equipment required in addition to the basic uplink to provide all the facilities to enable 'live stand-ups' and tape feeds to be accomplished. These configurations will be examined later in this chapter.

3.2 The basics

An SNG uplink system consists of the following primary component parts:

- Antenna with mounting support
- High-power amplifier(s) (HPA)

- Upconverter
- Modulator
- Signal monitoring
- Baseband signal processing

The physical transmission components of an uplink are typically referred to collectively as the 'chain' or 'thread'. A chain typically consists of a single transmission 'path' that has one of each of the primary transmission components, i.e. modulator, upconverter and HPA. An SNG system may have some or all of its constituent components duplicated. Two or more chains can be combined to feed via a single antenna, using a 'phase combiner'. This may be to give a degree of 'redundancy' and provide immediate back-up in the event of failure; that is to say, protection against failure of either a part or the whole of the system which would make the entire system inoperable. An extra HPA may be added to increase the uplink power. Alternatively (or additionally), it may be that the system has to provide more than one transmission 'path' where there is a requirement to uplink more than one programme signal simultaneously, combining two programme signals into a single signal applied to the antenna. A single transmission chain could also achieve this by using cascaded digital encoders.

Some system variants are shown in Figure 3.1 – this figure is by no means exhaustive, as there are various permutations possible. No matter what the configuration, in news operations the factors of speed and reliability are significant issues, and the component parts have to be rugged, reliable and quick to set up and operate.

However, as might be guessed from what has already been said, a system can be configured in a variety of ways. The operational use of the system and the purchasing budget must strongly influence the choice of system. The characteristics of an SNG system are defined by:

- Type of packaging, e.g. vehicle or flyaway.
- Type of modulation, i.e. analogue or digital.
- Frequency band of operation, e.g. C- or Ku-band.
- Level of redundancy, i.e. none, partial or full.
- Number of 'paths', i.e. one or several.

3.3 Specifying a system

The type of system is crucial to successful operation. For instance, where an SNG uplink system is going to be used in city or urban areas local to the operating base, it should be considered that physical space and access are often restricted in busy streets. Flyaway systems have to be transported in a dismantled state and then re-assembled before use. A safe zone has to be created in front of the antenna, and it is desirable to provide some weatherproof protection such as a tent for the uplink operators as well as the equipment. It is therefore difficult for a flyaway to be operated quickly and easily in busy street environments, with frequent rigging and

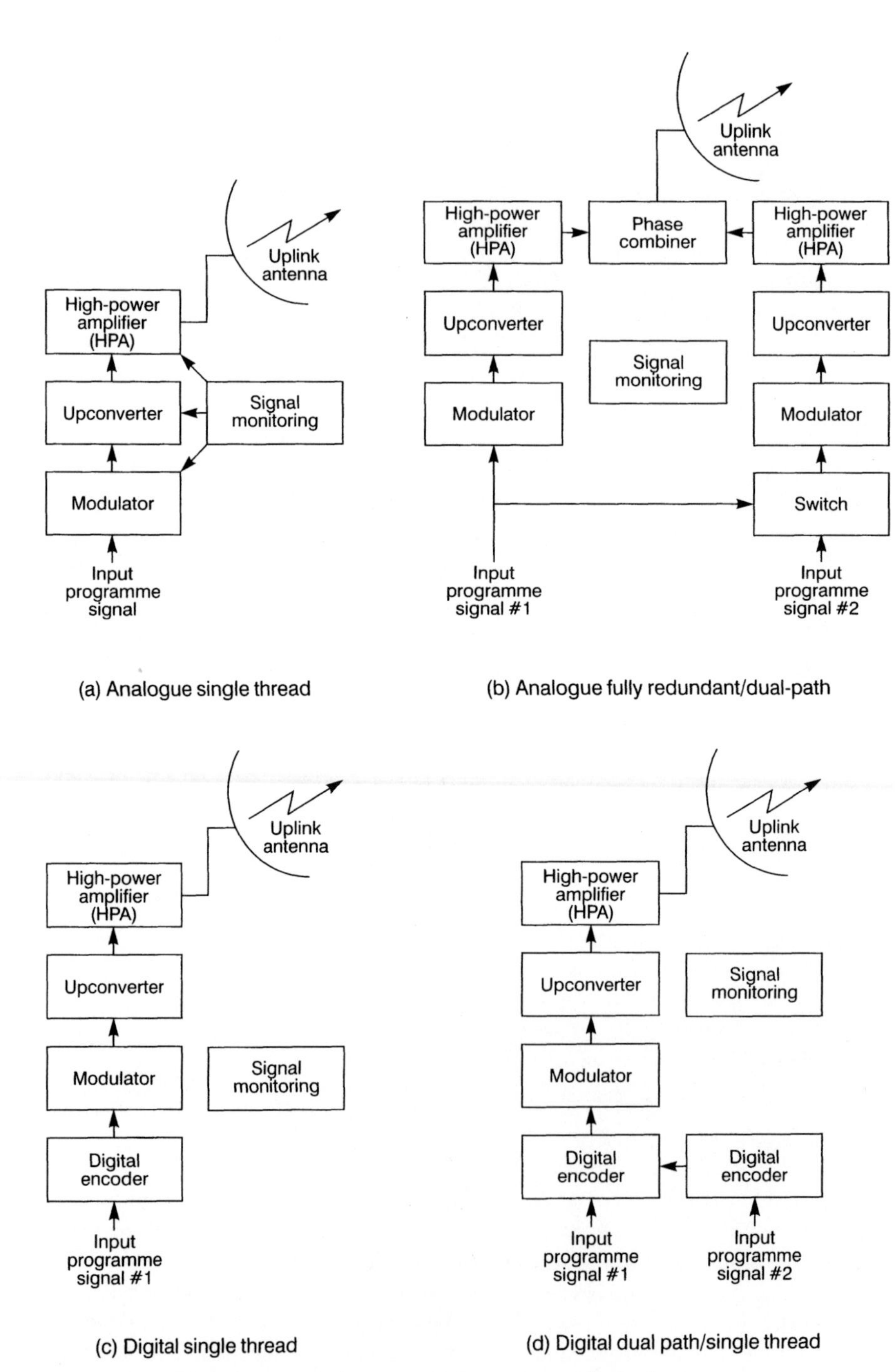

Figure 3.1 Block diagrams of various SNG system configurations

de-rigging. If transporting the system by air is unlikely to be a frequent requirement, then vehicle systems are the best choice for a restricted operating range in areas well served by road.

Buying a system of any type requires a very exact set of criteria to be examined and addressed, and these criteria need to be clarified as early as possible. It is very important to establish the operational function of the completed system and the form of the system that the manufacturer expects to deliver. The best constructors welcome a precise specification. There is a higher risk of failure when a customer simply says, 'Build me an SNG system', as the constructor may never be completely sure what the customer means. This has happened to a number of SNG constructors, and a contract entered into on this basis is more likely to be problematic on both sides. Stories are heard from both manufacturers and purchasers about misunderstandings, prefaced by 'I assumed they knew. . .', which often result in bad feeling on both sides. A customer is much more likely to be satisfied if both the technical and the operational specifications are as precise as possible. It is important that the manufacturer understands exactly how the system is intended to be used by the customer, as then the manufacturer may be able to use their experience of having built other customers' systems to satisfy the requirements. The complexity of an SNG system is such that it cannot be treated as a single article of equipment. It is unlikely that a customer would go to a manufacturer and ask for a small studio without specifying a number of key parameters. In the field, an SNG system can form a mini-studio environment, so similarly these key parameters need to be established.

How long should a specification be and how detailed? Well, the answer is as long and as detailed as possible. In fact, the maximum amount of time should be spent on deciding and writing the specification. It should state the type of operations the system is expected to cover and the key components must be identified. For instance, the specification may define particular regulatory parameters that have to be met. It may be a requirement for the manufacturer to obtain the final system registrations with satellite system operators, e.g. INTELSAT or EUTELSAT, so that the system can be used as soon as it is delivered to the customer. There may be a need for particular equipment to be integrated into the system, both bought and installed by the SNG system provider, or equipment may be 'free-issued' to the constructor by the customer to integrate as part of the project. Therefore, the basic rule is the more detail the better.

In choosing a supplier, market research is obviously very important. As well as visiting the manufacturer and inspecting the product, ask for a client list so that independent views from other customers can be obtained. The quality of build on an SNG vehicle (SNV), for instance, can be crucial to both the safety of the vehicle and its longevity. It is important to look at finished examples of the constructor's range, and checking for even the most minor details – a sharp edge here, loose or poorly fixed cabling there – can tell a lot about the standard the constructor typically builds to. It may be convenient for a constructor to promote the colour and finish of the interior trim, in the hope that the customer might not notice the fundamental flaws. Another good idea is to look at a customer's vehicle that has been in service for at least one to two years; this will give a good indication of the quality of build.

Another very pertinent way of gauging the quality of product and service from the constructor is to ask how much repeat business is placed with them: it is easy to sell something to someone once but harder to sell a second time. It is also important to measure the after-sales support. Does the constructor have 24-hour, 7 days-a-week technical support? This is particularly crucial when the system may often be operational in a different time zone to that of the constructor. Systems rarely fail Monday to Friday during office hours. Have they a system to ship replacement units to the location rapidly? Do they have replacement units 'on the shelf'? A helpful, knowledgeable voice on the telephone is important, but not much use if a replacement unit can only be supplied within 28 days – it's highly likely to be needed now!

Once the system is specified and ordered, the customer would be unwise to let the manufacturer proceed with building a system without checks and progress meetings along the way. It is highly inadvisable to change the specification during construction, and this reinforces the importance of being clear about the requirements before placing a contract with a supplier.

Many customers may not have the very critical skills of project management and the time to closely monitor the project as it progresses. Budget permitting, it is worth considering engaging the services of a professional project manager, who will closely monitor that the constructor is adhering to the specification, as well as dealing with any disputes over interpretation of the contract. The cost of using the services of a good project manager is not usually cheap and could perhaps amount to 5–10% of the total project cost. It can be regarded as a type of insurance; though, like many types of insurance, its value is only realized when needed!

3.3.1 Packaging for a purpose

By their very nature, SNG systems cannot be fixed permanently as the essential requirement is to be able to move quickly and operate in response to a breaking news story. SNG systems therefore have to be packaged in a way that allows them to be easily transported and then rapidly set up on location.

Systems are constructed in either one of the following ways.

- Flyaways: As indicated by the term, flyaway systems are designed to be broken down to fit into a number of equipment cases for easy transportation by air and then re-assembled on location.
- Vehicle-based: Here the SNG uplink is built into a vehicle (a car, van, vehicle or trailer) as a permanent installation. The system can then be driven to the news event and rapidly put into operation at the desired location.

3.4 Flyaways

Flyaway systems can potentially offer the lowest cost SNG uplink. They are therefore very appealing as an entry-level system. However, the flyaway fulfils a primary requirement for a system that can easily be transported by air and re-assembled at location typically in under an hour. Therefore, to gain the full benefit of operation of a flyaway, it should be routinely deployed in this manner. A flyaway can also be rigged and operated out of the back of a vehicle, and this is often satisfactory on a short-term basis. However, if vehicle-mounted operation is a regular requirement, the drawbacks of operating in this fashion will soon become apparent.

Their flexibility in terms of ease of transportation is of critical value to international newsgatherers, who regularly fly all over the world with such equipment in a variety of aircraft of different types and sizes. The key factors are total weight and number of cases, as news crews frequently transport these systems on scheduled commercial flights as excess baggage. Costs consequently increase greatly in direct proportion to weight and volume.

The design of a flyaway involves diametrically opposed demands. Strength has to be achieved with minimum mass while also sustaining stability. Precision and adherence to close mechanical tolerances have to be met, while all components have to be rugged. Overall size has to be minimized while also achieving maximum uplink power. The individual component parts have to be in an easily assembled form for both minimum rigging time and reliable operation – a challenging objective in view of the inherently delicate nature of some of the parts. Figure 3.2 shows a typical flyaway system.

Either the antenna is transported in a single piece or it is broken down into a number of segments or 'petals' (as can be seen if you look carefully at the antenna in Figure 3.2). It can then be mechanically re-assembled on-site onto the support or mount system (see Figure 3.3) – a reconstruction that has to be achieved to a very high tolerance. The mechanical integrity of the antenna system is crucial to its correct operation. This integrity has to be carefully monitored throughout the operational life of a flyaway antenna because of the vulnerability to damage in transportation as well as the repeated assembly and disassembly. Permanent vehicle-mounted antennas are plainly not so vulnerable.

The mount for the flyaway antenna is where wide differences can often be seen between the manufacturers – but the objective is common to all. The antenna has to be mounted so that it can easily be steered to align to the satellite correctly. At the same time, the mount has to provide maximum rigidity and stability to maintain pointing accuracy up to the satellite during operation even in poor weather. However, this physical rigidity has to be achieved while keeping the mass (and therefore weight) of both the mount and the antenna to the minimum. Nowadays the use of carbon fibre and aluminium alloys is common to achieve this design aim.

The control of the antenna is a very precise engineering requirement. The antenna has to be able to be finely adjusted in three axes, as shown in Figure 3.4 – azimuth (the rotational position), elevation (the angle of tilt) and polarization (the circular orientation of the beam, achieved either by rotating the whole antenna or the feedhorn itself), all of which have to be controlled to within fractions of a degree. There are typically both coarse and fine adjustment controls provided on the mount to achieve this. Movement on all these axes has

Figure 3.2 CML 1.5 m SNG flyaway. *(Photo courtesy of Continental Microwave Ltd)*

Figure 3.3 Assembling antenna petals. *(Photo courtesy of Harvey Watson)*

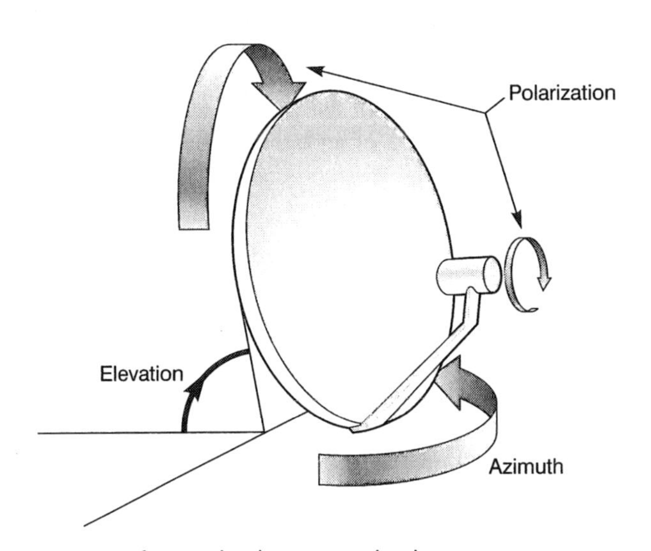

Figure 3.4 Properties of azimuth, elevation and polarization on an antenna system

to be as fluid as possible to ensure swift and accurate alignment of the antenna. The coarse controls allow rapid movement of the antenna on the mount to a point very close to the desired position, while the fine controls allow precise final alignment of the antenna towards the satellite.

When the antenna has been aligned it is critically important that the antenna remains on-station, and this is where the manufacturer's quality in the design of the mechanical locks on the azimuth and elevation controls is critically tested. Once the antenna is correctly positioned, it must be positively locked in position so that it cannot then be knocked off alignment inadvertently or by wind. These locks are often difficult to design, as they are subject to the relatively large mechanical leverage force of the antenna. Yet, even at the instant the locks are applied, the actual locking process must not move the antenna even a fraction of a degree off alignment. The locks are repeatedly used in operation and so it can be seen just how important precision engineering is, in what may seem to be a minor detail. It is as important here as in the other key aspects of the antenna and mount design.

The overall stability of the mount and antenna design is another area that has to be considered carefully. Manufacturers use different techniques, based around simple tripods, complex stabilizing frames and legs, or even interlocking the antenna and mount to equipment cases to increase the ground 'footprint' of the system. Each of these methods has its strengths and weaknesses, and it is very much up to each potential user to decide which they prefer.

The antenna must also be capable of being positioned on uneven ground and retaining pointing accuracy under quite severe weather conditions such as high winds. In aerodynamic terms, the antenna represents a sailplane, so that it is common to quote two different figures as a measurement of stability in windy conditions. One measurement is the maximum wind speed for the survival of the system, i.e. before parts may disintegrate under the wind's force. The other figure is the maximum wind speed at which the antenna will stay on-station as it faces into the wind. Side winds are less of an issue as the side profile of the antenna has low wind resistance.

3.4.1 Antennas

The antenna itself in a flyaway system is typically between 0.9 m and 2.2 m and, as previously mentioned, it is also usually segmented to allow easy transportation. Only the very smallest antennas (1 m or under) can usually be shipped as a single-piece assembly. The antenna shape also varies between manufacturers and each design has individual merits. Antennas are generally either circular or diamond-shaped (or even a flattened diamond-shape, i.e. a hexagon – see Figure 3.5), though the smaller antennas are typically circular. Diamond-shaped antennas offer improved gain through a more effective radiation area, though sometimes at the expense of sidelobe performance. In fact, circular antennas are actually very slightly elliptical in the horizontal plane to improve sidelobe performance.

The relationship between the antenna and the feedhorn is crucial as this fundamentally determines the RF performance characteristics of the antenna in defining the focus and radiation pattern. The feedhorn has to be correctly positioned so as to correctly illuminate the antenna, and typically SNG antennas are either 'prime focus fed' or 'offset fed prime focus' in type (see Figure 3.6).

Figure 3.5 Vertex 1.2 m SNG antenna. *(Photo courtesy of Vertex Communications Corp.)*

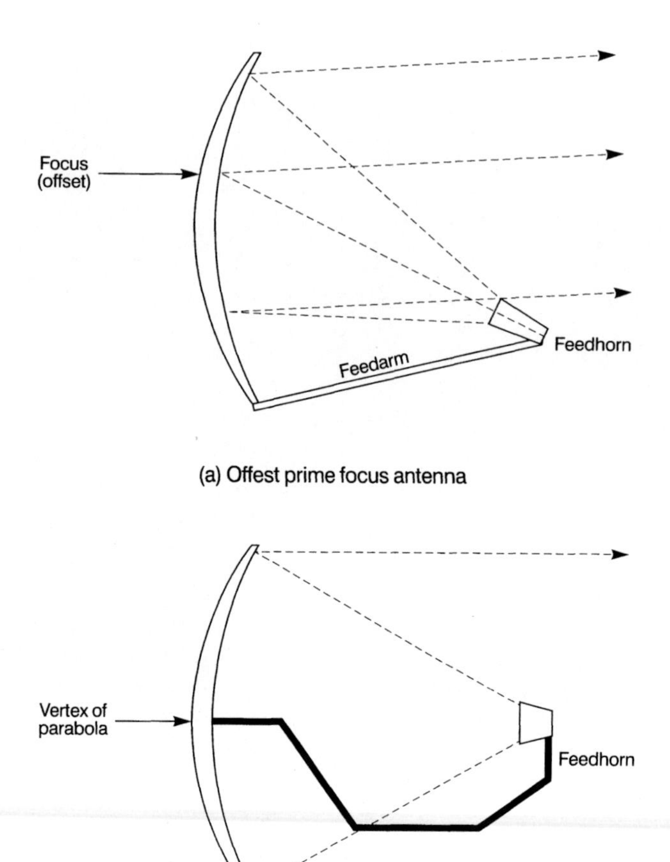

Figure 3.6 Types of SNG antenna

The use of an in-line fed prime focus is generally found on larger SNG antennas. The design of the feed arm with the antenna therefore has to ensure repeatable and mechanically precise assembly. In addition, the feed arm has to allow quick and easy connection of the RF feed from the HPA, by interconnection with a length of flexible waveguide.

3.4.2 Flight-cases

The remainder of the system consists of a number of cases of electronic equipment where the equipment is grouped in a manner set out to keep the number of cases to a minimum but also allow easy manual handling. Typically, the electronic equipment is fitted into flight-cases. This grouping also has to adhere to the functionality of the system, so that component parts that directly electrically interconnect are also physically co-located, particularly for high-power RF signals.

Historically, flight-cases were developed primarily for the military and scientific exploration industries, where expensive and delicate instrumentation has to be transported and operated in rugged environments. When SNG systems were developed, in particular by companies that were already involved in these other fields, it was natural that they should use these flight-cases for a similar requirement.

There are commonly two types of flight-case used in SNG systems. Firstly there is the type into which the antenna and mount can be dismantled and fitted into as a number of sub-systems. In this, the case acts purely as a protective shell for transportation. The other type is a case into which the electronics of the system can be fitted and also operated with the equipment in situ.

A flight-case is essentially a case that has an outer skin of (usually) metal while the contents are in some way protected against mechanical shock. This shock protection is high-density foam for the antenna and mount elements, and sprung-mounted frames for the electronics.

In addition to protecting the equipment, the other essential requirement is that the case conforms to the sizes and dimensions as outlined in International Air Transport Association (IATA) regulations. Cases are typically designed for military use and are therefore constructed to an extremely high standard. However, correspondingly the cost of the case is greatly increased over a standard transit case, and typically run into thousands of US dollars per case. This may seem very high until the cost of the equipment it is designed to protect is taken into account – which can be typically ten times the cost of the case. Figure 3.7 shows a case of typical construction.

An equipment flight-case has an internal shockproof mounted frame into which the electronics are fitted in such a way as to allow quick and easy set-up for operation at the destination. The removable front cover allows access to the front control panel, and typically has neoprene seals to protect against ingress of water and dirt during transit. The rear of the case may also be removable to allow access to rear panel connectors and controls.

The case is constructed in a manner that allows the outer skin to act as a crumple zone for the equipment inside. In addition, there is usually a pressure relief valve fitted to the body of the case to allow equalization of possible differential air pressure between the inside and outside of the case. This is typically caused by transportation in the unpressurized hold of an aircraft. If no valve was fitted, there is the potential for a partial vacuum to be created inside the case, which would make removing the end caps very difficult!

The shock mounts are typically blocks of very dense rubber, which will allow some movement under force but effectively absorb any shocks that the equipment fitted into the frame is subjected to. The frame is usually a standard 19 in

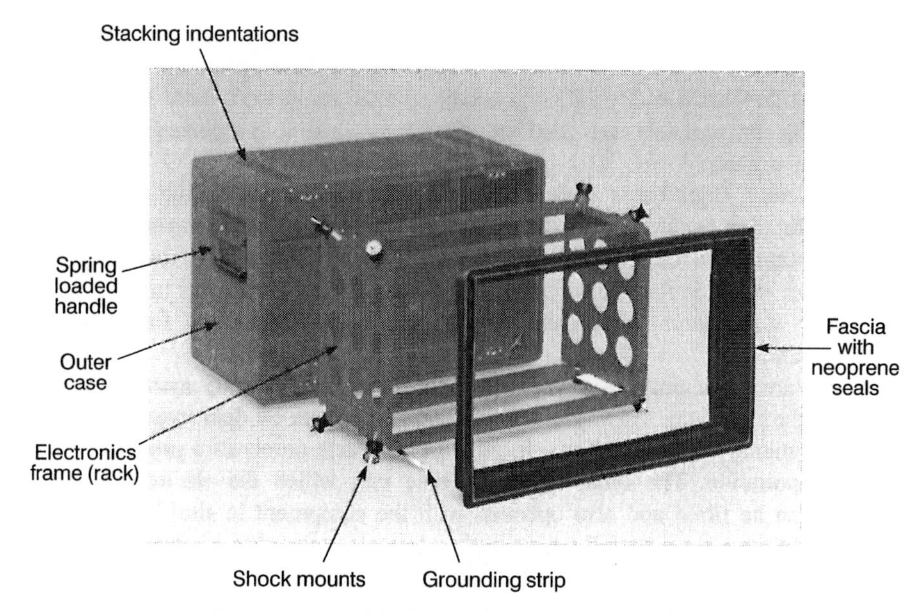

Figure 3.7 Typical construction of flight-case. *(Photo courtesy of EDAK AG)*

rack frame, to reflect the international mounting standard dimension for professional electronic equipment. Equipment modules that fit into such a frame are measured in terms of height in 'rack-height units' (RU), where 1 RU is equal to 1.75 in (44.5 mm) (see Table 3.1). There is no standard for the depth of units.

The case is usually made from aluminium alloy or sometimes polypropylene, and must be able to withstand a drop test that is usually defined at 2 m. It typically has proprietary dimples or ridges on the top and bottom of the case, which mechanically locate with corresponding recesses on the top or bottom of other cases. This is so that, when piling up a number of cases, both for operation but also in transportation, it is possible to form an interlocked stack that has a degree of stability. The corners of the case are either rounded or have some kind of protection to minimize damage if dropped on a corner, which would potentially distort the case or more importantly the internal frame.

The case also has to be able to allow the electronic equipment contained within to operate satisfactorily in extremes of temperature, high humidity, driving rain or standing in shallow puddles. Some cases are also designed to minimize the effects of any external electromagnetic interference (EMI). Taking into account all these factors, it is easy to see why the cases are so expensive, and it is a cost that has to be borne for the repeatedly reliable operation that is expected for newsgathering.

Nonetheless, the fact that the cases are so rugged does not mean that the usual standards of care in handling delicate electronic equipment can be neglected. If

Table 3.1 Standard 19 in rack-height units (RU)

Units	*Inches*	*Millimetres*
1	1.75	44.5
2	3.5	89
3	5.25	133.4
4	7.0	177.8
5	8.75	222.3
6	10.5	266.7
8	14	355.6
10	17.5	444.5

care is taken in handling as well as the use of these cases, the equipment should be fully operational when it reaches its destination. Military personnel have marvelled that SNG equipment works reliably on arrival at location, while their equipment transported in similar flight-cases does not. This is probably because there is an assumption that equipment in flight-cases can tolerate anything, but in fact care in the handling and loading of the equipment cannot be neglected.

The baseband equipment provided with an SNG uplink, either flyaway or vehicle-based, can vary widely, from simple monitoring of the incoming video to providing a small outside broadcast facility. This may include vision switcher, routing matrix, one or more equipped camera positions and comprehensive studio-remote communications.

3.4.3 Typical flyaway operation

Because there is such a wide variation for a flyway operation it is perhaps most useful to look at one particular scenario, from which other configurations can be imagined.

Basic set-up

The most common requirement is to be able to do stand-up ‘lives’ (also called ‘live-shots’) and replay tape material. This means providing a live camera position, with studio-remote communications and some audio mixing for the reporter and perhaps a guest contributor, and also a videotape recorder (VTR) to play in material live or feed back to the studio for later transmission. There needs to be a small routing switcher, able to ‘synchronously’ switch between the output of the camera and the VTR, and some vision and sound monitoring.

Electrical power

Of course, mains power may not be available or, if it is, it may not be very reliable, so a power generator is required as well. Because some of the live-shots may take place at night or in bright sun, either of which will require additional lighting for the live stand-up position, the generator will need to be capable of delivering suitable levels of power. The uplink system itself may require quite a significant amount of power, so that the generator (or generators) will need to supply typically somewhere between 5 and 20 kW, depending on the overall demand. It is also desirable to have some capacity in hand, as it is not very beneficial for the generator(s) to be running at maximum output.

The whole issue of electrical power generators and SNG uplinks is fraught with difficulties. Generators are one of the most common reasons for failures in the operation of SNG uplink systems. There is a maxim that every generator will fail at some point – a situation not helped by the less than ideal conditions they are often operated in. Manufacturers' servicing schedules should be rigorously adhered to – ignore them at your peril!

Studio-remote communications

Communications with the studio usually involve a bundle of circuits. Of vital interest are the audio circuits that provide both an audio feed of the studio output and talkback to the remote site. This is so that the reporter can hear the studio presenter's 'link' into the live piece, listen to cues from the studio control room, and also be able to conduct a two-way dialogue with the studio presenter, which requires the reporter to hear the questions from the studio. There is also often a talkback circuit from the remote site to the studio so that information can be communicated back to the studio control room – this is often referred to as a '4-wire'.

The feed of the studio output is known by a number of names, and the functions vary slightly. Examples are 'cleanfeed', 'mix-minus', 'reverse audio', 'return audio' and 'interrupted foldback' (IFB), and what is actually carried on this the circuit varies slightly according to the term used. Cleanfeed, mix-minus or reverse/return audio is usually a mix of the studio output without the audio contribution from the remote location. IFB is this same mix but with talkback superimposed on top. Switched talkback can sometimes be superimposed on the cleanfeed type of circuit as well. The exact configuration varies from broadcaster to broadcaster, with local variations.

Other communication circuits will include a number of telephone lines and a data connection with the newsroom computer system. These can all be provided via either an additional transmit/receive side-chain on the satellite uplink, an INMARSAT satphone or any available telephone landlines. The most desirable method of delivery is via the satellite uplink, as this usually provides the most direct paths to and from the studio. The circuits can be fed directly to the side-chain. They can also be fed via a statistical data multiplexer that can vary the allocation of a relatively narrow data bandwidth according to the wider

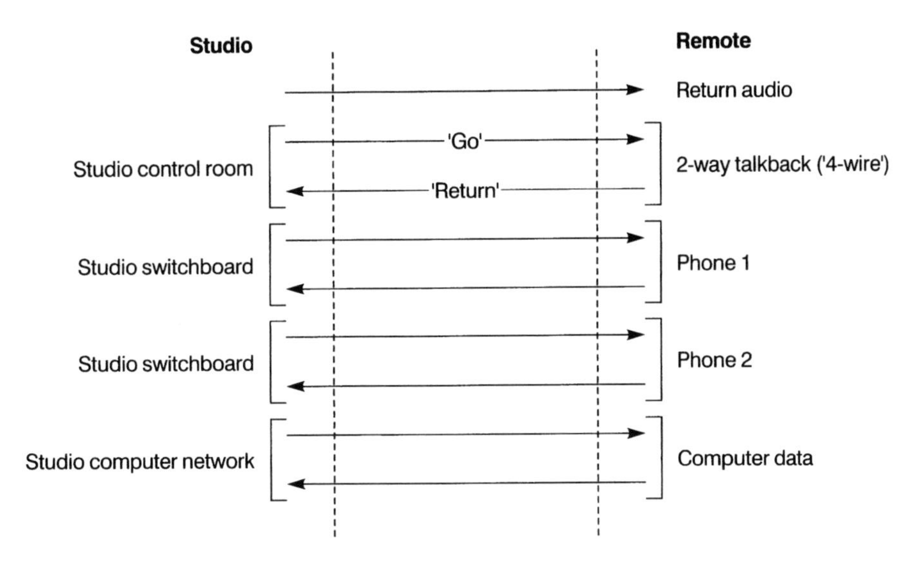

(a) Typical studio-remote communications overview

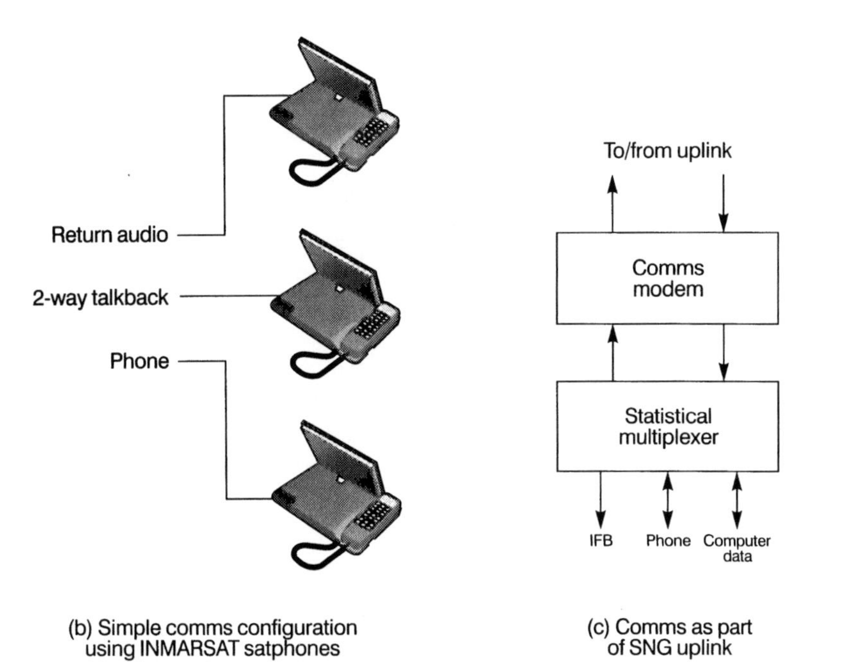

(b) Simple comms configuration using INMARSAT satphones

(c) Comms as part of SNG uplink

Figure 3.8 Comms systems configurations

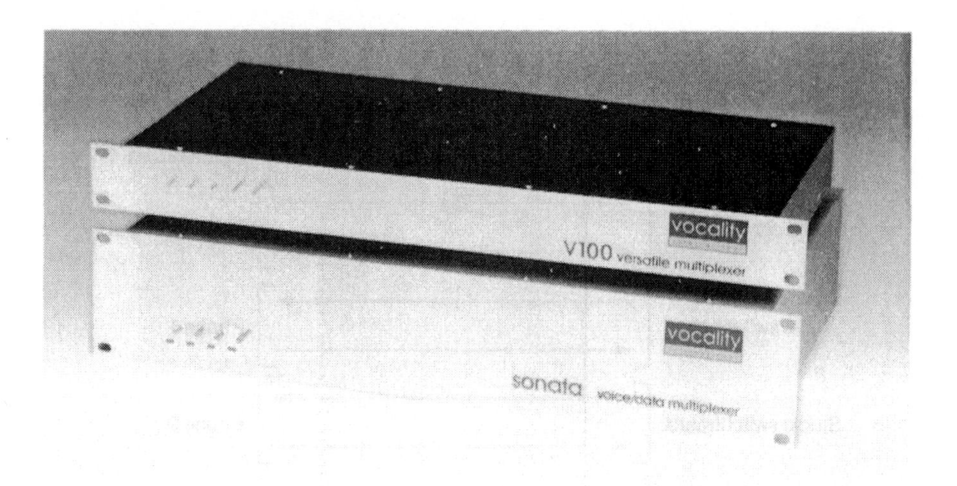

Figure 3.9 Statistical multiplexers *(Photo courtesy of Vocality International Ltd)*

simultaneous bandwidth demand of several circuits. Figure 3.8 shows a typical configuration of studio-to-remote communications, and two types of statistical multiplexers are shown in Figure 3.9.

More than ever, the connection to the newsroom computer system and providing multiple telephone circuits are proving critical. The integration of the newsroom computer system into the overall news production process means that it acts as the heart of both the newsgathering and the news output operation. Information regarding contacts, running orders, news agency 'wire' information and background information to the story are all controlled and produced by the newsroom computer system, as well as it providing rapid messaging between editorial staff and the field staff. It is also a truism that there can never be too many telephones.

No matter how simple the operation, however, this aspect of the news broadcasting chain is vital. Many operations have failed not for lack of programme pictures and sound, but because the studio-remote communications have not been working correctly.

Video editing

In addition, it might be necessary to provide a video editing facility, and if this edit suite is required to be a source to the routing switcher as well, then this will further complicate the system. The format of the video editing can obviously vary according to the organization – Sony BetaSP was the most common format, but increasingly Panasonic's DVCPro and Sony SX are now being introduced to replace BetaSP. DVCam, originally intended only as a domestic format, is also increasingly in use as a budget newsgathering format. Ultimately one format will win the day, just as Sony's BetaSP toppled Panasonic's M2 in the early 1990s, but at the time of writing this has yet to be clearly established. The appeal of the digital formats in the field is the compact laptop editors, weighing under 10 kg, which considerably reduces the average 150 kg of equipment that traditionally can make up the elements of a BetaSP edit pack. It has been the experience of many video editors that initial versions of these laptop editors are particularly deficient in audio mixing facilities, and therefore the addition of a separate audio mixer greatly improves the functionality of the laptop editor. The latest models have addressed this problem to some degree, improving the audio mixing facilities.

Whatever the format of the video editing, it is common to try to provide the ability to connect the editing as an 'online' facility to the rest of the rig. This is so that an edited piece (a 'cut') that has just been finished can be immediately played out on the satellite uplink, without the need to take the tape to a separate VTR for playout.

Interconnections and layouts

The cabling requirements are quite considerable, particularly if there is a significant physical separation between the stand-up live camera position, the edit suite, the switching/monitoring point and the satellite uplink. Typically, such an operation might take place at a hotel; this scenario is shown in Figure 3.10a. If it can be imagined that this system is rigged at a hotel, then the uplink will probably be on the roof with a power generator (see Figure 3.10b – the power generator is out-of-shot). The edit suite can be several floors below in a room, and the 'live' position may also be on the roof (Figure 3.10c) or on a room balcony, to give a visual background statement of the location of the story. The switching/monitoring point is likely to be in another hotel room which often doubles up as a production office (Figure 3.10d). All too quickly, the cabling and interconnection requirements can rise to the extent that there may be perhaps over 100 kg of cable alone to be transported to cover hundreds of metres of multiple interconnects (Figure 3.11).

It is possible to greatly reduce the weight of cable carried by using fibre-optic 'cable', which provides an impressive number of circuits down a 6 mm diameter fibre optic cable, compared to a traditional 'copper' 15 mm diameter multi-core cable. The initial high cost of a fibre-optic system, which has a small unit of active electronics to convert and multiplex/de-multiplex the electrical signals to/from light signals at each end of the fibre, can be rapidly recouped in the savings on shipping weights. A typical fibre-optic system used for SNG can carry three video channels and eight programme audio channels, as well as a number of data channels for camera control and intercom channels for the cameraman. Figure 3.12 shows a typical fibre-optic system for field use, with a drum of fibre holding around 300 m of cable and weighing approximately 20 kg – less than a tenth of the equivalent capacity of multi-core copper cable.

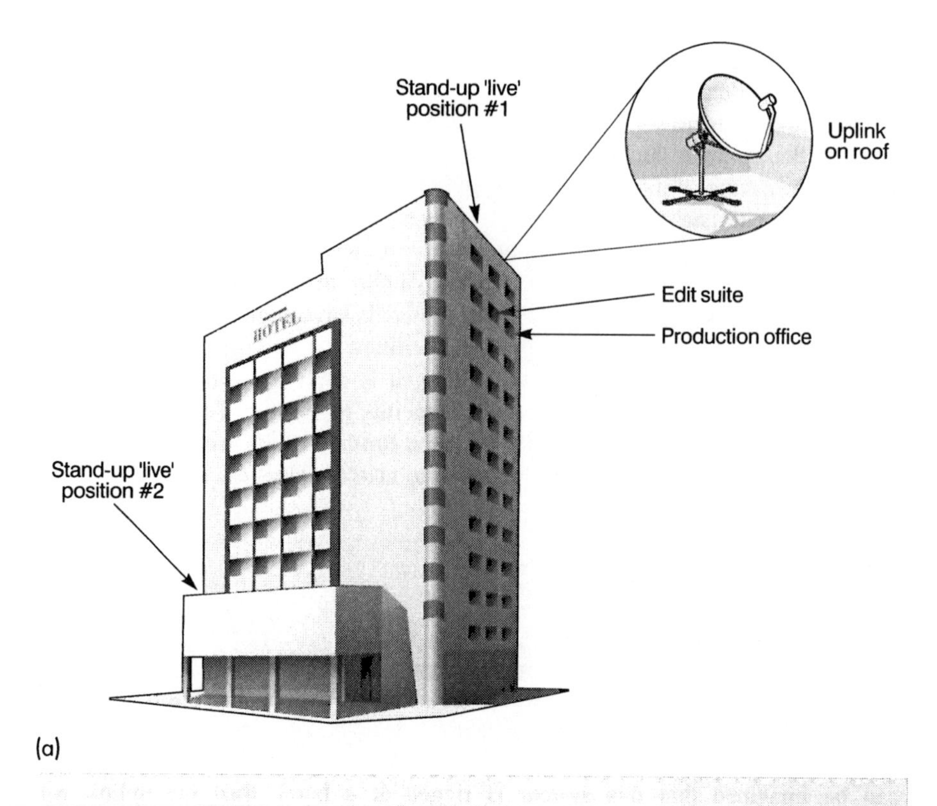

(a)

(b)

Figure 3.10 (a) Typical operation at a hotel. (b) Uplink on roof. *(Photo courtesy of Simon Atkinson)*

(c)

(d)

Figure 3.10 (c) Live position. *(Photo courtesy of Simon Atkinson)*
(d) Switching/monitoring point. *(Photo courtesy of Paul Szeless)*

Figure 3.11 Flyaway 'spaghetti'

The final picture

This whole system has to be built, tested and be operational in typically about 12 hours from arrival on location – and that doesn't include getting through the Customs formalities and travelling from the airport!

All these elements mean that the total number of cases that make up the entire system as described will probably double or triple the number of SNG uplink cases alone. This is an aspect of using SNG that is often only appreciated by those directly involved in actual deployments. However, the impression often given by SNG flyaway manufacturers is that the relatively few cases that make up their particular flyaway system is all the equipment that is required to accomplish a news assignment. If only that were true! The truth is that to mount an operation as described above, there is likely to be a total of forty to fifty cases, and the total weight of the cases between 1000 and 1500 kg (see Figure 3.13).

With the new 'lightweight' DSNG systems (see Figure 3.14), this shipped weight can be reduced, but perhaps not as dramatically as might be assumed. Although there are complete lightweight SNG flyaway systems that claim to have a shipping weight near to 250 kg, production demands can quickly increase this weight. Developments in broadcast technology are reducing the size of the equipment, but the demands on many newsgatherers now mean that the levels of service required on location are that much higher to meet the demands of continuous news provision. This in turn can mean that the system described above has to be expanded to meet

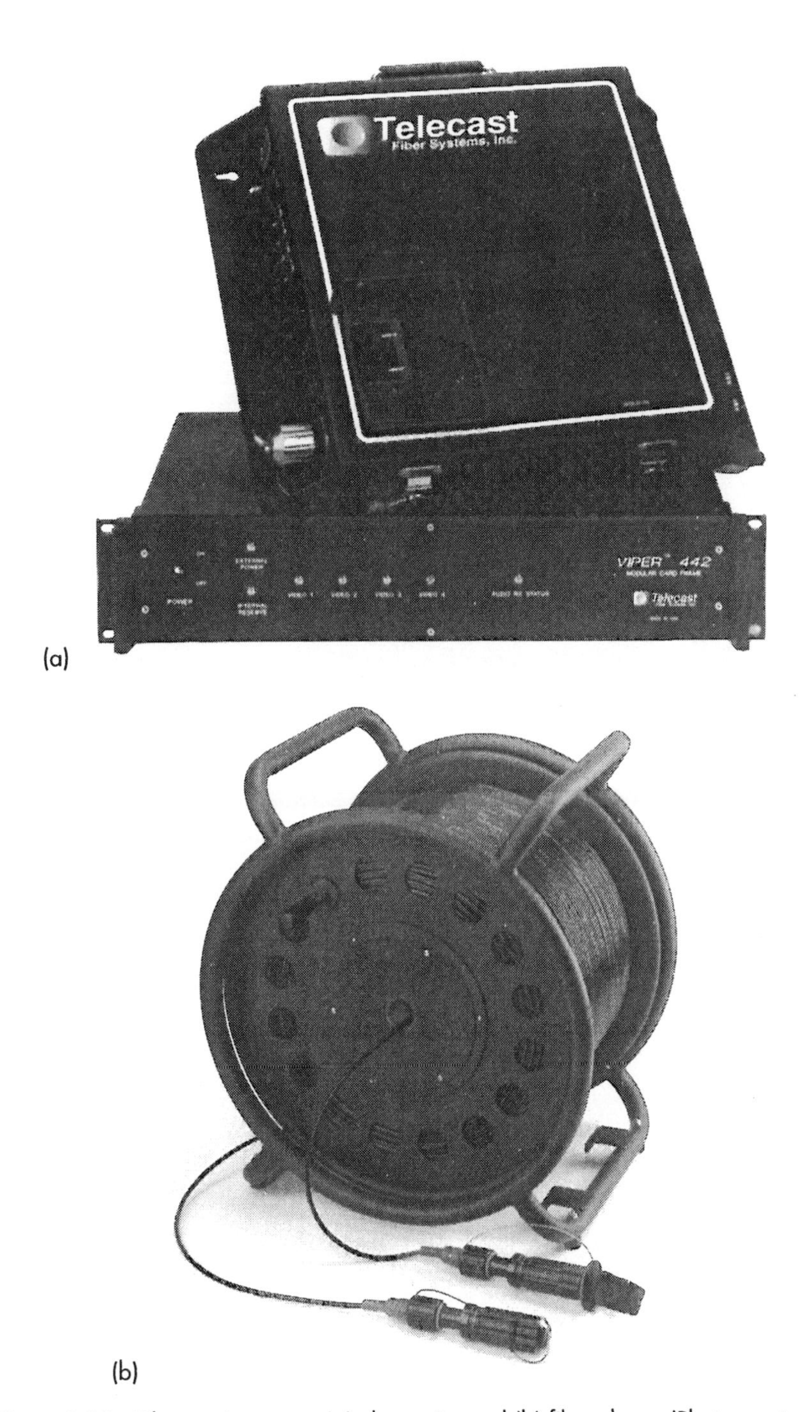

Figure 3.12 Fibre-optic system: (a) electronics and (b) fibre drum. *(Photo courtesy of Telecast Fiber Systems Inc.)*

Figure 3.13 Typical SNG flyaway system!

Figure 3.14 A lightweight flyaway. *(Photo courtesy of Harvey Watson)*

this greater demand, and the net result can be that just as much equipment has to be provided.

However, it is possible with determination – both technical and editorial – to 'travel light' and to keep the amount of equipment required to under 600 kg, and on certain stories this may be essential for logistical and safety reasons.

3.5 INMARSAT systems

INMARSAT satellite telephones can now be used relatively easily for the transmission of pictures and sound for television, as well as high-quality audio for radio. INMARSAT systems are particularly of interest for use in parts of the world where traditional means of getting material back is very difficult, or where logistics or some local political sensitivity lead to the same situation. They are frequently used in civil war zones, or where there is no national telecommunications infrastructure that will enable material to be broadcast back to the studio.

The use of land mobile INMARSAT systems for SNG has been developing since the first portable satellite telephones became available. These MESs (mobile earth stations) were the satellite equivalent of the cellular telephone but offering near-global coverage. Broadcasters could use the satellite technology to send back reports by using a compact satellite telephone – the 'satphone' as it has became commonly known. Not only that, but the technology was developed further so that a high-speed data (HSD) 64 kbps circuit could be established over analogue Inmarsat-A channels. With the right type of transportable terminal and a suitable ancillary equipment, a high-quality audio circuit can be established by a simple dial-up procedure, enabling news reports to be provided from almost anywhere in the world.

For a number of years, these compact satphones were used purely for radio SNG. A satphone could be used with a digital audio codec to produce near-studio quality audio via an HSD 64 kbps dial-up connection on Inmarsat-A and -B systems. These HSD channels have been very expensive to use, but have the benefit of being available on demand (subject to capacity being available), and the equipment has become increasingly easy to assemble and operate. This has meant that reporters could use the equipment for radio reports without the need for a technician.

Further developments led to the transmission of high-quality video over an HSD circuit. In 1993, BT and INMARSAT were able to provide Reuters Television with near-live pictures of the Whitbread Round-the-World Yacht Race. They were 'near-live', because the pictures were actually stored on an on-board computer, compressed and then sent at a slow data rate, stored at the receiving end and then uncompressed and reconstituted into real-time video. The news market viewed this development with great interest, and many felt that the further development of this 'store and forward' type of system would be the key to future newsgathering techniques.

By 1994 compact new equipment designed specifically for newsgathering applications enabled the transmission of television news pictures, albeit not in real time. The transmission of newspaper stills pictures had been possible for a long time using HSD channels. Now several companies developed digital video compression units for television newsgathering that could transmit video and audio over HSD channels in store and forward mode.

Mention is made here of INMARSAT because it does meet the need for an SNG system which is particularly suitable for use in areas where it would be logistically or politically unfeasible to use a flyaway system. INMARSAT systems are fully discussed in Chapter 5.

3.6 Vehicles

SNG vehicles can be built on a variety of types of base vehicles such as towed trailers, estate cars, MPVs, pick-up vehicles, panel vans, box-body vehicles and even combined tractor units for towing production trailers. The majority of SNG vehicles, sometimes referred to as SNVs (satellite newsgathering vehicles), are constructed on a vehicle or van between 3500 and 7500 kg (7700 and 16 500 lb) GVW (gross vehicle weight) in Europe, and up to 26 000 lb in the US. Figure 3.15 shows a range of vehicles built and operated in the US and Europe. It is common (and desirable) to have an antenna which is mounted high up on the vehicle, typically on the roof. This will place a requirement on the SNV constructor to ensure that the support structure for the antenna is appropriate for the load of the antenna and its mount.

There are two types of SNG vehicle constructor. There are those who manufacture the satellite uplink system and combine it with the rest of the vehicle system that they will build, literally as a platform for their product. The other type are those constructors who do not manufacture satellite equipment, but buy in the components of the satellite uplink system from other suppliers and integrate them along with the rest of the systems of the vehicle; these are termed 'integrators'. Either type of constructor may have their own in-house coach-building facilities, or sub-contract this aspect of the work out to a specialist coach-builder.

The primary purpose of an SNV is to allow operational deployment to be accomplished more quickly than with a flyaway, and to have a number of additional facilities built-in so that they are also rapidly available once on-site. These facilities typically include an on-board generator to provide electrical power, production facilities, signal routing and bi-directional studio-remote communications. One of the principal characteristics of achieving fast deployment is, of course, having the antenna already mounted on the vehicle. The antenna can either be manually controlled or, more usually, it is motorized so that it can be moved to the desired position from the convenience of the interior of the vehicle. When it is motorized, the system has an antenna remote controller unit (see Figure 3.16).

In deciding what can be fitted onto the vehicle, it is crucial that a weight budget is calculated and maintained in parallel with the financial budget. Numerous SNVs have suddenly been discovered to be overweight just as they are near completion, or even when they are delivered to the customer, resulting in a vehicle which is both unsafe and illegal to use on the road. The weight budget has to be calculated as a

(a)

(b)

Figure 3.15 (a) Estate car. *(Photo courtesy of Continental Microwave Ltd)* (b) Four-wheel drive. *(Photo courtesy of Continental Microwave Ltd)*

(c)

(d)

Figure 3.15 (c) European panel van. (d) Interior of European panel van.

(e)

(f)

Figure 3.15 (e) US panel van. *(Photo courtesy of Wolf Coach)* (f) Interior of US panel van. *(Photo courtesy of Wolf Coach)*

(g)

(h)

Figure 3.15 (g) Medium-size US truck. *(Photo courtesy of Wolf Coach)* (h) Interior of medium-size US truck. *(Photo courtesy of Wolf Coach)*

(i)

(j)

Figure 3.15 (i) Large US truck. *(Photo courtesy of Frontline Communications Corp.)* (j) Interior of large US truck. *(Photo courtesy of Frontline Communications Corp.)*

Figure 3.15 (k) Compact European SNV. *(Photo courtesy of Wahlberg & Selin)*

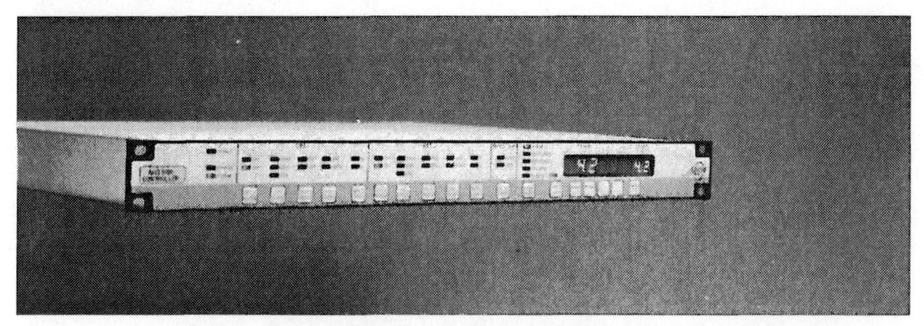

Figure 3.16 Antenna controller *(Photo courtesy of Advent Communications Ltd)*

theoretical model beforehand and then updated as each step of coachwork is carried out and as cabling and equipment are actually installed in the vehicle.

It is also vital to ensure that a more than adequate standard of documentation is supplied with the vehicle. The vehicle may well be intended to have a life exceeding ten years, and it is unlikely that the engineers who will maintain it at the beginning will be the same as those maintaining it towards the end of its life. The documentation must include maintenance information on all the systems and sub-systems, from the RF transmission path to the air-conditioning unit.

3.6.1 Choice of chassis

SNG systems can be built on a wide choice of vehicles. Trailers and estate cars are, of course, options but the majority of SNG vehicles are based on vans and customized vehicles. Within this group there is a choice between panel vans and using chassis-cabs with purpose-built bodies on the back.

The advantages and disadvantages are as follows:

- Panel vans are cheaper but the construction of the SNV is constrained by the manufacturer's model dimensions of the load space. The GVW range is limited at the top end to around 7500 kg (16 500 lb) in Europe and 10 000 lb (4500 kg) in the US, restricting the size of SNV that can be constructed. In Europe, panel vans that have a GVW of less than 3500 kg are regulated by the same speed limits as cars, which is attractive if a particularly fast operation is required. In some countries in the EU (e.g. Germany), however, this may be limited to vehicles that have a GVW of less than 2800 kg. There are usually few driver licence restrictions in this weight class.
- Chassis-cab bodies are generally more expensive to build, but the load space can be dimensioned to suit. They are the only option where a large vehicle is required, as chassis-cab vehicles range in size from a medium to very large chassis. Additionally, the life of the vehicle can be extended because the part that usually wears out is the vehicle itself. The custom box body can be constructed so that it can be lifted off and put onto a new chassis-cab, thereby extending the overall life of the system.
- Estate cars are suitable for some small systems and often look very attractive, but in reality trying to carry out repeated rapid news operations in them can be very difficult because of restrictions on space.
- Trailers can offer more flexibility by not being tied to any particular vehicle. However, it should be remembered that manoeuvring a vehicle with a trailer attached can be problematic, and the trailer will need to be fairly large if it is to include an operational area as well as the actual satellite uplink.

It should be noted that if the choice of vehicle is a panel van, as opposed to a chassis-cab, any modifications made by the SNV constructor to the base vehicle have to be made strictly in accordance with the vehicle constructor's recommendations. Most modern vans are of monocoque construction, and cutting into the vehicle structure can both invalidate the vehicle warranty and compromise the safety and integrity of the vehicle if not done in accordance with the vehicle manufacturer.

The choice of a vehicle base depends on the following:

- What is available on the market
- Budget
- Payload required

- Facilities required
- Driver licensing requirements
- Typical usage, e.g. off-road or on-road

Because there is a wide range of vehicles on which a satellite uplink system can be built, one type of vehicle will be considered for illustrative purposes. A particular type of vehicle that is currently popular in Europe, the panel van (Figure 3.17), has been selected as an example to illustrate the implementation of a system in more detail. The system shown could just as easily be found on other vehicles, and either expanded or reduced according to both the payload of the particular vehicle and the financial budget.

The SNV has a number of sub-systems fitted to it. Obviously, it has a satellite

Figure 3.17 European SNG panel van. *(Photo courtesy of Continental Microwave Ltd)*

uplink system but in addition it usually has the following installed (see Figure 3.18):

- Electrical power generator
- Electrical power distribution
- Stabilizing jacking system
- Racking and wiring for broadcast equipment
- Broadcast equipment

Figure 3.19 shows a European panel van-based SNV under construction.

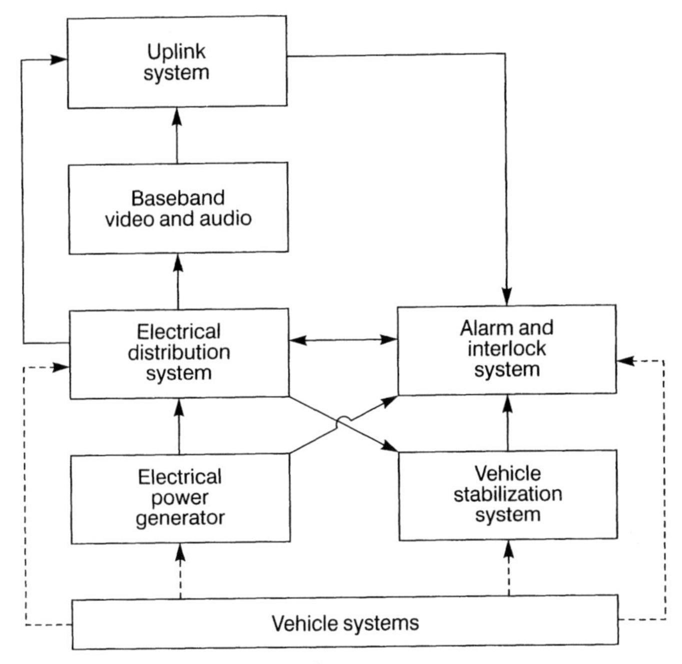

Figure 3.18 Outline systems diagram of an SNV

3.6.2 Electrical power generator

Depending on the size of the vehicle, the power generator system usually falls into one of three types:

- Self-contained generators driven by a separate engine to the vehicle road engine.
- Inverter systems, where the road engine drives a 12 V/24 V alternator which then supplies 115 V/230 V via an inverter.
- Power take-off systems (PTO) where the road engine drives a generator either via a belt or indirectly via a specially equipped vehicle gearbox.

As with powering flyaways, it should be noted that whichever type of power system is used, the whole area of electrical power on location is fraught with difficulties. The power supply to the uplink from whatever type of generating system is in use is likely to fail at some point, and some (wise) uplink operators have taken to having two sources available on a vehicle, with one acting as a standby emergency source. The reason for the likelihood of failure is that generators are notoriously unreliable, probably because of the less than ideal conditions they are often operated in.

Self-contained systems

The generator set, which consists of an electrical alternator and an engine, is installed in the vehicle in an enclosure to separate it from the rest of the vehicle for

Figure 3.19 European panel van-based SNV under construction

both acoustic and fire safety reasons. The radiator-cooling unit for the generator engine, which may be fuelled by either petrol or diesel, may be within the generator enclosure or separated off in another part of the vehicle where there is improved airflow. For example, the radiator can be mounted underneath the rear of the vehicle, or on the vehicle roof either at the front or rear. Alternatively, it may be within the enclosure and draw cooling air through the side or rear of the vehicle via vents in the bodywork.

The alternator may be either frequency or voltage regulated; generally, voltage regulation is preferable, as broadcast equipment is usually more sensitive to voltage fluctuations than frequency fluctuations. The generator system needs to be in an acoustically isolated enclosure, and particular attention needs to be paid to the mounting of the generator engine itself to minimize vibration through the rest of the vehicle. Vibration at 50/60 Hz can be particularly uncomfortable for the vehicle occupants during extended periods of the generator running.

The fuel for the generator can be drawn from either the road engine tank, a separate tank or a tank that has a connection to the road engine tank to enable prolonged running. However, if the last option is chosen, the connection must not be at too low a point, otherwise there is a risk that the generator can run the main tank dry, thereby leaving no fuel for the road engine. Where there is a separate tank, and the generator uses a different fuel to the road engine, there must be clear signage on each tank-filling inlet to avoid wrong fuel being put into a tank. In fact, it is preferable to try to avoid having a mix of fuels on a vehicle, as it is a recipe for disaster. Sooner or later, probably while under operational pressure or tired, someone will forget and put the wrong fuel in the wrong tank.

Generators need regular servicing. If they are not serviced regularly in accordance with manufacturers' recommendations, the likelihood of failure increases by a factor of ten. Ignore servicing schedules at your peril. (If you think you've read this comment before, you have, earlier in this chapter. It's very important.)

Inverter systems

This is a very popular option on the smallest SNG vehicles where the power demands are modest, say up to 4 kVA. The road engine drives a 12 V or 24 V alternator, which charges (usually) a separate lead-acid battery to the vehicle battery. This separate battery then provides a 115 V or 230 V supply via an inverter to the broadcast equipment. It is possible to use the vehicle battery to supply the inverter, but a considerably larger battery needs to be fitted. Either two separate alternators can be used – one for the vehicle, one for the inverter system – or one large output alternator can be used for both. As a powering system it is compact, relatively reliable, very cost-effective and space saving. Particular design considerations are that the required output should be easily within the maximum output of the inverter, and particular care must be taken in 'split-charging' with the vehicle battery, i.e. where one alternator is charging two batteries. There should be a properly installed split-charging device to prevent either battery from being overcharged. If a battery is overcharged there is a serious risk of fire due to expansion and overheating of the electrolyte.

Power take-off systems

Such systems are often used on vehicles that have four-wheel drive, as the vehicle gearbox sometimes has a power take-off drive as a standard feature. The PTO drive is a special selection on the vehicle gearbox while it is in neutral, where the road engine power is directed to a separate drive shaft for auxiliary devices rather than the vehicle road wheels. There is also usually an engine control system to regulate the speed of the road engine so that it runs at a speed fast enough to provide the required power. The disadvantage of this system is that, although compact and cost-effective, it is not necessarily designed to run for extended periods. Vehicle manufacturers do not usually recommend that the road engine should be operated in PTO mode beyond a certain period. There is a serious risk that both the road engine and the gearbox will overheat (even with an oil cooler) if run continuously for longer than the recommended period. It is also possible for PTO to be derived via a belt driving a supply alternator directly off the engine (as opposed to a low voltage generator as with the inverter system). Again, this is suitable where power requirements are less than 5 kVA, and there has to be close control of engine speed to prevent damage or the risk of fire.

3.6.3 Electrical power distribution

By whatever means the power has been produced, it has to supply all the different systems within the vehicle. Electrical power has to be provided with both over-current and short-circuit current protection. Safety is of paramount importance, as there are particularly lethal voltages derived in the uplink HPA. Cabling has to be of the highest standard and no compromises can be permitted. It is normal to feed the power produced from the generator to a power distribution unit (PDU), which splits the power into separate feeds to each equipment rack or sub-system in the vehicle. Figure 3.20 shows a simplified power distribution system for an SNV. Protection against over-current is provided within the PDU with the use of MCBs (miniature circuit breakers). An MCB is a resettable device that breaks the circuit under over-current conditions. It fulfils the same purpose as a fuse, but reacts much faster and is convenient because it can be reset with the push of a button. It is also highly desirable to provide protection against any part of the metal work of the vehicle becoming 'live'. This is best done by use of a residual current device (RCD) which monitors the current flow in both the live and neutral supply conductors.

In a circuit, the same amount of current should flow to the equipment from the supply as flows back. If there is a difference in the current flow between the two supply conductors, a short circuit to earth (or to a person!) is assumed, and the RCD trips. An RCD operates by sensing when the current in the phase (live) and neutral conductors within an installation are not equal and opposite. Any imbalance would imply that an additional path existed for the flow of current, invariably through the earth due to excessive leakage and/or a fault situation. Devices of this type are referred to by a variety of different names, such as RCCB (residual current circuit breaker), ELCB (earth leakage circuit breaker), GFI (ground fault interrupt) and GFCI (ground fault circuit interrupter). They do not all work in exactly the same

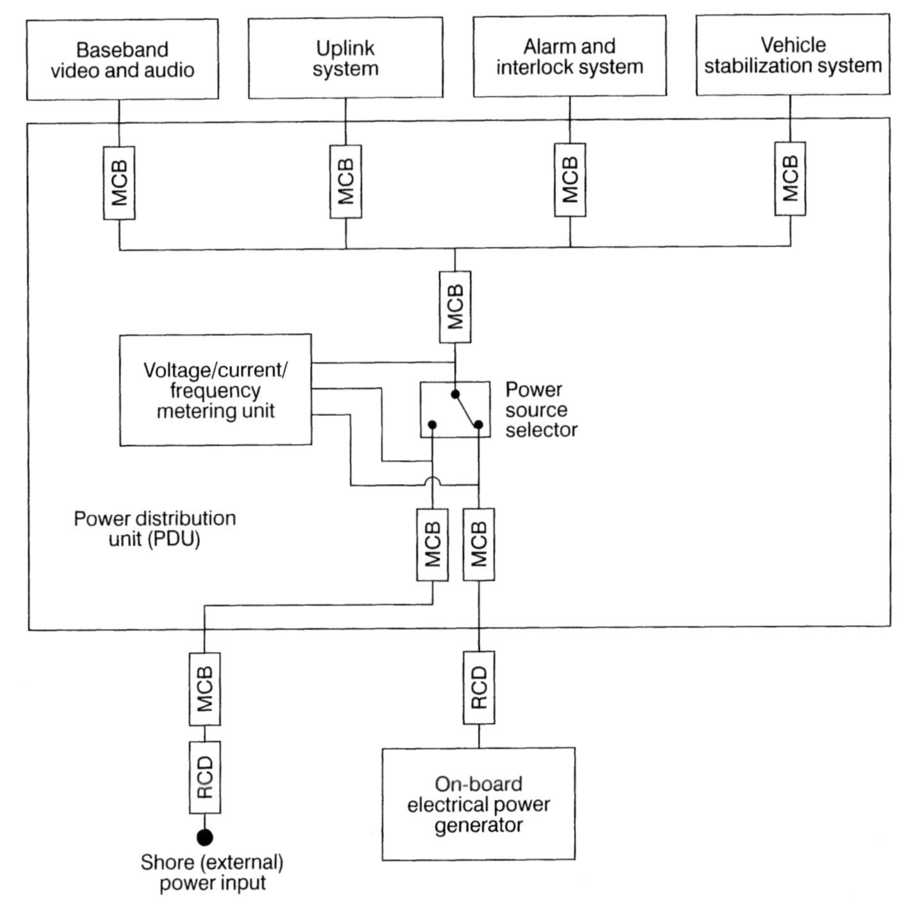

Figure 3.20 Simplified diagram of a power distribution system for an SNV

manner, as some directly measure leakage of current to the earth or ground, but the objective is the same – to protect from electrocution. Specifically an RCD-type device has a particular advantage in that it does not rely on any earth or ground wire connection being present.

There is also a potential safety issue where power is provided from an external input instead of from the on-board generator – often referred to as 'shore power'. The condition of the incoming supply cannot be necessarily known, and it is advisable to provide protection by both indication of potential phase reversal and use of an RCD. These should be at the entry point to the vehicle. The power, no matter where it is derived from, may need some 'conditioning' on the input to the uplink system, particularly the HPA, as the supply has to be very 'clean'. That is to say, there must no 'spikes' or irregularities in the supply waveform as this may, in particular, damage the HPA. Usually the HPA will have some kind of power 'conditioning' (filtering and smoothing) internally and sensing which will shut down the HPA in the event of it being supplied power which cannot be coped with by the conditioning.

3.6.4 Stabilizing jacking system

The vehicle usually requires a stabilizing jacking system to ensure that the antenna stays on-station, as any movement of the vehicle could result in movement of the antenna. Movement of the vehicle can result from either people getting in and out or wind rocking the vehicle (particularly if the wind is coming broadside onto the vehicle or from other vehicles passing by at high speed). It is common to provide either a two- or four-point stabilizing system, with two jacks at each rear corner of the vehicle, or two at the middle of the vehicle on each side, or four jacks, one at each corner. The jacks can be manually, electrically or hydraulically operated. The most sophisticated stabilizing systems have computer-controlled stabilizing of the vehicle, with an on-board sensor feeding back the status of vehicle stability.

Manual systems are usually fitted to smaller vehicles, and although some weight and cost is saved, the time taken to prepare the vehicle for operation is slower. It is possible on some vehicles not to have any stabilization if the antenna is small, as the wider beamwidth compared to larger antennas means that a slight movement of the antenna will not move it off-station. The choice of whether to use hydraulic or electrical drive for the jacks is a matter of preference. Hydraulic jacking systems generally use a single central pump driving the hydraulic rams on each jack in turn, via hydraulic piping and valves, and can often be heavy. Electrical screw jacking systems are lighter but can increase the demand for electrical power. Either type of set-up can use the sophisticated control system described, or rely on simple manual controls to achieve stability.

The aim of the jacking system is to take the movement out of the vehicle suspension system, rather than achieve absolute levelling. It is possible, particularly with hydraulic jacking systems due to their significant power, to lift the vehicle wheels off the ground, and this is undesirable for two reasons. Firstly, if the vehicle is lifted asymmetrically, i.e. one wheel is lifted clear of the ground while the other wheel stays on the ground, it is likely that the vehicle chassis will twist and be permanently damaged. This could seriously affect the road-handling of the vehicle. Secondly, if the vehicle is lifted either wholly or partly clear of the ground while parked on an incline, it is possible for the vehicle either to slide down the incline or topple to one side. This is a particular risk if the surface on which it is parked is icy or slippery.

3.6.5 Racking and wiring for broadcast plant

Another important sub-system of the SNV is the equipment racking in which the broadcast equipment is to be installed. There must obviously be enough rack space to take all the equipment required and so, at the start of the project, there has to be a calculation of the amount of rack space for the equipment needed. This underlines the need for clarity in the specification from the start, and it is one of the principal reasons why the specification should not change during the construction of the SNV.

The decisions with respect to the broadcast equipment that needs to be installed are very much a matter of preference, and there is certainly no 'standard' type of configuration. Requirements vary from user to user, and country to country, so it is beyond the scope of this book to deal with this area in detail. Suffice to say that, in pursuing the requirement to provide the satellite uplink, the broadcast function should not be overlooked. As mentioned earlier in the chapter with reference to flyaways, the SNV should also contain all the equipment that is required to accomplish a news assignment.

There also has to be careful design to fit all the required equipment into the vehicle. Constructors can be very adept at making use of every last inch of space inside a vehicle, and the racking often has to be constructed to custom specification to fit into a particular model of vehicle. The equipment racks should be as light as possible, as standard racks designed for studio installations are not usually built with weight-saving in mind. Some firms construct the racking and wire the racks outside the vehicle, as space inside the vehicle is often very cramped. The completed racks and wiring looms are then transferred into the vehicle in the final stages of build. This also allows parallel construction operations to be carried out. For instance, the uplink can be installed into the vehicle at the same time as the equipment racks are being wired outside the vehicle.

3.6.6 Weight

At this point mention must be made again of the weight budget, as it must not only allow for all the fixed equipment but also the 'loose' load. The loose load is the ancillary equipment, cable and other sundry items that can be found in the back of every SNG vehicle, which slowly get added to for the 'just in case' situations. The weight budget must account for all the installed and loose load equipment. This then has to be balanced against both the gross vehicle payload and the individual gross axle loadings.

In addition, the centre of gravity of the vehicle must be calculated and monitored throughout the construction. It is inevitable with installing broadcast equipment, the uplink chain and the antenna assembly that weight is going to be added into the upper part of the vehicle. This will raise the centre of gravity of the vehicle, which may result in the vehicle becoming unstable and adversely affecting the vehicle while it is being driven. It is a fine balance to maintain the functionality of the vehicle while not adversely affecting the handling of the vehicle.

It is an almost unavoidable fact that SNG vehicles are generally at or very close to their maximum GVW, and will be running constantly near to fully-laden over their life. This can be very wearing on the engine, transmission, steering, suspension and brakes. Running this close to the maximum GVW can shorten the life of the vehicle and vehicles need to be regularly checked for wear on all these components, as they are fundamental to safety.

3.6.7 Method of construction

There are broadly four approaches to building SNVs:

- Turnkey
- Self-build
- Part-build by a manufacturer, part-build by user
- Sub-contracting

Turnkey

For customers who have the budgetary resources to contract a fully-functioning vehicle from a constructor, the turnkey route is plainly the most obvious and attractive. However, it may still involve some allocation of the customer's own engineering resources to achieve the desired end product. It is however by far the best option for the first-time customer in the SNV market. Even if the customer is an experienced user of flyaway systems, the problems and challenges of building an SNV are different, as we have discussed.

Self-build

This approach is particularly attractive as it can potentially deliver the lowest cost vehicle. However, it should only be attempted where the user has significant engineering resources. The elements involved in the construction of a vehicle are diverse and encompass not only coachbuilding and the consequent safety issues, but also the successful specification, purchase, installation and integration of a range of equipment. The challenge faced in doing this must not be underestimated, and it is sometimes questionable as to whether the end result is as cost-effective as it might seem on first inspection. However, local circumstances may dictate this approach, or it may suit the user who is mature in the use of SNG systems and who has particular ideas on how a system should be constructed in a vehicle. It should be borne in mind though that there are few examples of vehicles built in this way that compare with the best examples from the premier constructors. In addition, the issues of warranties and regulatory approvals must be carefully considered, as should the inherent risk of a safety incident that calls into question the method of construction, and the litigation that could follow.

Part-build by a manufacturer, part-build by user

The appeal of this approach is that a specialist manufacturer handles the satellite system, coachbuilding and vehicle infrastructure, and this includes the electrical generator and the vehicle stabilization system. The user (assumed here to be expert in broadcast engineering) can then install the broadcasting equipment and communication systems, completing the project. This can work very successfully. It can be seen as a middle way between self-build and turnkey that can deliver significant cost benefits. However, the issue of project management raises its head here, as a considerable amount of hands-on management of the project is required. The successful integration of the diverse elements of the system that come from different sources can be potentially problematic. This can be virtually a full-time job, and needs careful consideration of how it is to be achieved before the project is started. If the user is able to commit to this level of project involvement, then this can be a cost-effective way of building an SNV. Alternatively, the services of a professional project manager could be used.

Sub-contracting

A variation on part-build by a constructor is that the different elements of the whole project are sub-contracted out to separate companies, each with a specialist contribution to the vehicle. For instance, one company can undertake the coachbuilding. Another company can carry out the electrical generator installation. The specialist company may not necessarily be expert on the construction of SNVs. The above observations on project management are even more pertinent here.

3.7 Type of modulation: analogue or digital

As we enter the new millennium, the decision as to whether to choose analogue or digital modulation for SNG is largely pre-determined. From 1995 onwards, the transition from predominantly analogue SNG to digital SNG in many parts of the world has been accelerated by the widespread adoption of MPEG-2 digital compression. This is because it has become a standard in almost all parts of the television broadcast process. From MPEG-2 coding in cameras to MPEG-2 manipulation, transmission and distribution through the broadcasting chain, it was natural that the SNG uplink would mirror these developments.

The availability of ad hoc satellite capacity for analogue traffic has dramatically reduced in many markets, as space segment operators in response to both market and economic forces look to maximize digital ad hoc capacity. The spectral efficiency of digital modulation means prices reduce, increasing demand and profitability for both newsgathering operations and space segment operators. However, it is just worth comparing the differences between the types of modulation.

3.7.1 Analogue

Some users may still have the analogue route open to them and it still offers certain advantages over digital modulation. These are listed below.

- The configuration of the uplink system is simpler, as there is no coding involved. There is less equipment in the chain and therefore the systems are cheaper to purchase and operate. Maintenance in the field is more straightforward.
- If the signal is reduced in quality due to atmospheric conditions or because the uplink is near the edge of the satellite footprint, there is a gradual degradation of the signal. There is some warning that the signal is being degraded on the path from the uplink to the downlink.

- Analogue transmission is a universal standard and therefore can be downlinked at a large number of locations. No special decoder is required.
- There is no picture degradation introduced as part of a coding or decoding process.
- No coding delay results in more instantaneous, natural-looking, 'live' two-ways.

Some of the disadvantages are as follows.

- Satellite capacity, either leased or ad hoc, is more expensive than digital capacity.
- Analogue satellite capacity is becoming harder to procure.
- Significantly higher uplink powers are required, achieved using either large diameter uplink antennas and/or larger HPAs.

3.7.2 Digital

The advantages here are as follows.

- Satellite capacity, whether leased or ad hoc, is cheaper and in general easier to procure. More channels can be fitted into a transponder, so spectral efficiency results in savings to the customer.
- The uplink gain required is reduced, so uplink antenna sizes and/or HPAs are smaller. Equipment size and weight can hence be reduced.
- By digital multiplexing, several programme paths can be provided more cost-effectively than multiple RF 'chains'.

As with analogue systems, there are also disadvantages.

- System costs are higher due to the increased amount of equipment (i.e. coder). This can also lead to greater risk of failure as more equipment in the chain increases potential unreliability. Upconverters in particular have to maintain a very high phase stability, particularly as the modulation schemes used in DSNG demand a degree of phase stability. In future this will become more important as DSNG modulation moves to the more spectrally efficient schemes (e.g. 8-PSK and 16-QAM; see Chapter 4).
- Hard to troubleshoot in the event of a failure. Digital equipment is more sophisticated and difficult to fault-find in the field.
- Picture quality can be degraded because of the coding process, particularly at data rates below 8 Mbps. This can particularly be a problem on pictures with a high content of fast-moving action, e.g. sports.
- Coding delay (latency) combined with satellite delay can result in live two-ways looking awkward and hesitant. This can be reduced by certain masking techniques but these require a degree of compensation in the production process, e.g. 'pre-cueing' (see Chapter 4).

3.8 Frequency band of operation

In specifying a system, be it for a flyaway or a vehicle, a key parameter is the frequency band of operation. As described in Chapter 2, the transmit frequency of C-band for SNG is around 6 GHz, and the receive frequency is around 4 GHz. This compares with 14 GHz and 12 GHz respectively for Ku-band.

Why are there two different frequency bands for SNG use? C-band was the original frequency band for satellite transmissions because of the limitations of the technology. The ITU conference WARC-77 extended the frequency bands for satellite operation to the much higher frequencies of Ku-band. WARC-77 also defined the regions of the world where these frequency bands can be used (see Chapter 6). At that time the technology for working in the Ku-band was still in its infancy, but it was recognized that with the existing satellite services in C-band there would be ever-increasing congestion. Frequency congestion was widely predicted both on the ground and in space, and experimental transmissions in the Ku-band were undertaken in both the US and in Europe during the late 1970s and into the early 1980s. It was seen that Ku-band offered a number of advantages over operating in the C-band, principally smaller antennas and HPAs. Ku-band transmissions became common for SNG, overtaking analogue C-band transmissions in the US through the 1980s, and in Europe SNG began in the Ku-band.

However, C-band has the benefit of being a more 'rugged' frequency band in terms of RF propagation, with less critical performance required of equipment and therefore a lower cost. It is little affected by weather, whereas rain is particularly detrimental in the Ku-band. With the advent of new digital coding techniques and the now rapidly increasing congestion in the Ku-band, the opportunity for using C-band for digital transmissions in a spectrally-efficient manner has emerged. As identified below, digital C-band transmissions are of special interest in certain parts of the world.

3.8.1 C-band

The use of C-band for SNG transmissions is not permitted in Europe and many other areas of the world (see Chapter 6). However, it can be used in most of Africa and Asia. The problem in some parts of the world is that this particular frequency band is widely used for terrestrial fixed microwave links. The potential for interference from what is essentially a temporary transmission from an SNG uplink is very high. Fixed networks are fundamental to the telecommunications infrastructure of a country, and SNG transmissions will not be permitted to interfere with this. Typically, a C-band uplink for a fixed application has to be 'cleared' for interference for a radius of up to 1200 km, and this can obviously take some time as international co-ordination may be required for its use.

On the other hand, the attraction of C-band for SNG transmissions is that in many parts of the world there is no ad hoc satellite transponder capacity in the Ku-band available for SNG usage. In these same areas there is not the sophisticated telecommunications infrastructure to limit C-band SNG transmissions. C-band can be used for analogue transmissions, but requires large uplink powers resulting in the requirement for large antennas and HPAs. Hence, it is preferable to use C-band for lower power digital transmissions, with antennas and HPAs similar in size to those used in Ku-band systems. Indeed, in the latter part of the 1990s, INTELSAT in particular has been keen to promote the use of C-band digital uplinks. This is because the spectral efficiency of this type of transmission matched the increasing amount of C-band capacity that INTELSAT had available in general, resulting in many more channels available for digital SNG.

3.8.2 Ku-band

The Ku-band is still by far the most common frequency band used for SNG around the world and this is due to a number of factors. In the areas of the world where there is a developed telecommunications infrastructure, such as the US and Europe, Ku-band SNG is the norm. The antennas are small and the uplink powers required, particularly on the latest generation satellites, are relatively modest. Hence very compact SNG systems, whether analogue or digital, can be constructed and easily transported between locations.

3.8.3 Dual band

By 1995 a number of manufacturers were offering 'dual-band' systems. They are termed dual band as the RF section of the system spans both the C- and Ku-bands. (Strictly speaking, they are triple band as they also cover the X-band – see Appendix A – which is used for military telecomms.) The operational advantage is obvious for flyaways, as with a single system the user has the option of travelling to a location and the choice of operating in either frequency band. The principal global newsgatherers find these systems particularly flexible. It can be decided before departure which band the system is going to operate in, or it can be left until arrival at the destination if the choice of available space segment cannot be made at the time of deployment. The additional components to enable the system to operate in both bands are relatively minor. The HPA is typically a wideband type, capable of operating in either band, and it is simply a matter of making a normally front panel selection on the HPA to change bands. This normally switches in a different set of filters for monitoring purposes. The only other components that are different are the feed arm and LNB that fit onto the antenna. The same antenna is used, though the specification of the system will change depending on the frequency band of operation.

The only disadvantage of a dual-band system is the increased cost, as a separate feed arm and LNB for each frequency band have to be purchased, and a wideband HPA is generally more expensive.

3.9 Level of redundancy

Redundancy gives protection against failure of part of the system, and is an important factor to consider if the system is being used to cover an important event or story. An SNG system may have part or all of its electronic equipment duplicated to give a degree of redundancy. Normally it is not necessary to duplicate any of the mechanical components, save for perhaps having a spare length of flexible waveguide.

The level of redundancy can be varied according to the requirement. On an analogue system, it might perhaps be sufficient to add a second HPA and upconverter – this would give partial redundancy. To achieve full redundancy another modulator should be added. The additional HPA can either be combined with the primary HPA via a phase combiner, or made available as a 'hot spare' via a waveguide switch. A hot spare is a part of the system that is switched on and in a fully operational state ready for rapid changeover in the event of a failure; conversely, a 'cold spare' is not switched on and ready!

The elements most likely to fail are the power supply within the HPA because of the high voltages generated and the amount of power produced, and the upconverter because of the frequency stability required. The upconverter is a particularly vulnerable unit in a digital system, as the both frequency and phase have to be tightly controlled to maintain system stability and operation. Additionally, for a digital system, it might be decided that the digital modulator and the encoder should be replicated to achieve full redundancy.

The decision as to what level of redundancy is required or desired is not an easy one. In such a complex array of equipment as in an SNG system, failures are inevitable at some point during the system's life, and even full redundancy can never remove that risk totally – it simply minimizes it. Companies that provide their SNG systems to third parties on a commercial basis regard higher levels of redundancy as a priority so that the exposure of risk of failure to their clients is minimized. Other organizations, particularly the global broadcasters who are primarily servicing their own outlets, prefer to reduce the capital cost, size and weight of the system and work single-threaded. They are willing to take the risk of an occasional failure in return for a lower capital cost and lower revenue costs in shipping smaller flyaway systems around the world.

3.10 Multiple programme paths

It is an increasingly common requirement to be able to provide more than one programme path from a single SNG uplink. This may be to provide differing sources of material to the destination simultaneously, either off-tape or 'live', or to service different destinations simultaneously.

For whatever reason, there are two methods of providing multiple paths. One is to provide two separate uplink chains that are combined just before the input to the antenna – this is the 'RF solution'. It has the added bonus of offering a degree of redundancy if the uplink is in a remote and not easily accessible location. The second method can only be achieved with a digital uplink – the 'digital solution' – and this is to 'multiplex' two programme paths (or 'streams') in the digital domain by cascading two digital compression encoders together to provide a single digital data stream to the uplink chain. This is termed 'multiple channels per carrier' (MCPC). In a DSNG system, there are numerous permutations possible by using a combination of both of these methods, such that you could have two RF chains, each fed by a number of digital compression coders. In the US, there are some SNG vehicles which have up to six paths available, via two RF chains each with three digital compression coders. This enables a vehicle to service both mainstream 'network' output as well as 'affiliate' station requirements on a big story.

3.11 Automatic systems

Automation has entered the broadcast process in all areas to varying degrees, and SNG is no exception. This may sound a little odd, considering the unpredictable way in which SNG systems are set up and used – after all, it is not like automating a process in a broadcast centre. But in these days of constant downward pressure of costs, combined with the element of multi-skilling that has increased in many areas, there is a natural attraction towards the use of automated processes if it saves money.

So some automation is possible, and there are two principal types of process which can be automated: acquisition of the satellite and remote operation of the system.

3.11.1 Automated acquisition of the satellite

The most difficult task in the operation of any SNG system, which by its very nature is never in one place for very long, is 'acquiring the bird'. The satellite can be an elusive object to identify at times for even the most experienced operators, but the process can be relatively easily automated.

A satellite occupies a position in space that is defined by two parameters in relation to a particular location: its azimuth and elevation. By the co-ordinated usage of three devices, an automated system can be devised to find a satellite. It has to be said that this type of automated system is only suited to a vehicle-mounted rather than flyaway system.

The three devices are a Global Positioning System (GPS) unit, a 'flux-gate' compass and a beacon receiver. They are used in conjunction with a microprocessor control system within which satellite orbital positions have been programmed. This control system can then drive the antenna to the correct azimuth and elevation for the required satellite.

GPS

The Global Positioning System (GPS) was developed in the 1970s by the US Department of Defense to provide highly accurate location data, using the US military NavStar fleet of twenty-four satellites. The first GPS satellite was launched in 1978 and the constellation of twenty-four was completed in 1993. These satellites operate in a medium Earth orbit (MEO) of 17 500 km in six orbital planes (each satellite has a 12-hour orbital period), providing global coverage 24 hours per day.

A constant stream of timing information generated from a highly accurate atomic clock on each satellite is broadcast, and GPS receivers fitted with clocks read this information. By comparing the signal from each satellite with the time in its own clock, the GPS receiver calculates the distance to each satellite and adjusts its own clock, then uses triangulation techniques to calculate its location, providing positional information accurate to 1 m for military use. This is the technology behind the ability to provide the military with the exact positional information required for modern warfare, including the precise information required by 'smart' bombs and missiles.

The service is available to be used by commercial manufacturers of GPS equipment, albeit at a 'degraded' quality that still allows positioning to be accurate to within 10 m. The satellites send out signals in the L-band, which a receiver can lock onto and determine its position. The receiver needs to receive signals from at least three different satellites in the system to be able accurately to determine its position, and a signal from a fourth to determine its altitude. These receivers are available in many countries at very economical prices.

GPS technology is widely used for all forms of navigation. It is also used in automated satellite acquisition systems to provide the longitude and latitude information to determine the 'look angle' of the required satellite for antenna positioning.

Flux-gate compass

The flux-gate compass provides an accurate bearing ('heading') and this is referenced to a nominal 'set' direction – usually the direction the vehicle is pointing in . The flux-gate compass works on the following principle.

When a magnetic field is generated in a material using an electric current, eventually the material becomes 'full' and can accommodate no more magnetic flux (magnetic flux is an expression of the magnetic field strength within a given area). If the electromagnetic coil wound around the material is formed into a ring – termed a toroid – this then forms a toroidal magnetic core. On one side of the ring, the external magnetic field – the Earth's magnetic field in the form of the magnetic force from the North Pole – 'lines up' with the internal field, increasing the overall electromagnetic field. On the other side of the toroid it lines up in reverse with respect to the internal field, decreasing the overall electromagnetic field. Thus, one side of the ring becomes 'full' of flux before the other as the internal field is increased and this difference can be sensed. It is then possible to determine which part of the ring the effect takes place in, and so the direction of the external field (North Pole) as well as its strength can be measured.

Hence, the flux-gate compass is a very accurate electromagnetic compass which has an electrical signal output giving orientation information that can be fed to a processor to be calculated with the absolute positional information from the GPS unit. Alternatively, it is possible to use two GPS receivers – particularly on a vehicle where a fixed known spacing can be set – to determine the magnetic heading.

Beacon receiver

Beacon signals were discussed in Chapter 2, and a beacon receiver is a satellite receiver that is optimized to receive the beacon signals continuously transmitted by the satellite. Beacon signals, amongst other things, uniquely identify a particular satellite. The output of the beacon receiver is fed to the control system.

3.11.2 Operation

On arriving at site, the vehicle can be readied for transmission. The required satellite is selected from a 'menu' list on the control panel, and the GPS unit and the flux-gate compass feed the positional information of the uplink to the control system. From this information, the control system can calculate the required antenna azimuth and elevation for the selected satellite.

Under the control of the automated acquisition system, which is activated by a single button press, the antenna is raised from its lowered (park) position, to a pre-determined nominal elevation. It then rotates to the satellite's azimuth and elevates to its elevation angle. The beacon receiver now starts tuning, scanning up and down a narrow range 'looking' for the satellite's beacon. On finding the beacon signal, the control system fine tunes the antenna in azimuth and elevation to optimize the signal level from the beacon receiver, ensuring that the antenna is accurately aligned.

Rather than using a beacon signal, some systems use a known strong signal from a satellite, and determine the correcting pointing angle for the required satellite by calculating the offset in azimuth and elevation from this known signal. This has the advantage that a less well-specified receiver can be used, but has the disadvantage that if the known signal is removed for some reason, the system has to be re-programmed to a different 'known' signal.

3.11.3 Advantages and disadvantages of automation

Such a system has the undeniable advantage that technology can replace the highly tuned skills of a good uplink operator in finding the satellite. Staff with lesser technical skills are just as able to press the button to set the system in motion. This, combined with other technical functions that could be automated or pre-programmed in the operation of an uplink, could very well allow the use of journalists or any other spare pair of hands to operate the uplink – though the style of operation may be subject to a satellite operator restriction, as we shall see in Chapter 7.

However, one must consider that the typical SNG system is an already sophisticated set of equipment, and the addition of an automated system for 'finding the bird' is 'just another thing to go wrong'. If the system fails to find the satellite, it could be any one of a number of processes that have failed. If such a system is used by an experienced uplink operator to save time then, if it fails, the satellite can still be found by manual methods and the antenna correctly aligned. But in the hands of an unskilled operator, the transmission will be lost.

Automation has been specified on a number of systems, but has fallen into disuse because the skilled uplink operator who is working onto a frequently used satellite can often find and align the satellite more quickly than the automated system.

Its application can therefore be limited, and it is not a cheap option to add to a vehicle. Many have decided that the money is better spent on another part of the vehicle's operation, giving more value to the operation overall. Its use is perhaps best seen on a small, journalist-operated vehicle, which has relatively simple function and usage.

3.11.4 Fully remote operation

This is an extension of the automated acquisition system to a level where the entire uplink is remotely controlled. It has been implemented both on vehicles and on some flyaway systems, where remote control can be used once the antenna has been aligned manually with some additional 'pointing aids' for the non-technical operator.

The remote control package allows control of all the key operational parameters by a separate communications channel. This channel can be provided by a plain old telephone service (POTS), an INMARSAT satphone or even a separate satellite channel that is activated via the antenna once it is correctly aligned.

The level of control can be relatively sophisticated, but as a basic minimum it will provide the following.

- Encoder control (on digital system): bit-rate, horizontal and vertical resolution, MPEG mode (4:2:2/4:2:0).
- Modem (digital): bit-rate, modulation type, FEC, carrier on/off.
- Modulator (analogue): input audio and video parameters, carrier on/off.
- Upconverter: frequency.
- HPA: mode standby/transmit.

However, as with the automatic acquisition systems, a failure of the system can be due to any number of reasons, and therefore there is a compromise that has to be accepted if opting for this mode of operation. Not many news organizations would want to deploy such a system in the hands of a non-technical operator on a 'big story', and if the use of such a system is going to be constrained by the type of story it can be used on, it can be argued that it is a false economy.

It may seem here that we are debating whether someone whose primary job is not a technical one – for example a journalist – is capable of operating an SNG uplink with the different automated controls described. That is not the issue, which is more a question of whether, if (for example) a journalist is trying to do other jobs as well as their own, something will suffer. Newsgathering is not a serial process, i.e. the jobs and tasks that have to be done do not concatenate neatly together. There are task overlaps on the time-line of reporting a story, and some even run directly in parallel.

At times, it is not a matter of 'many hands make light work', but that many hands make it happen – on-air and on time. On most stories, the journalist is busy simply keeping up with the story. On rapidly breaking events the requirement repeatedly to 'go live' for a string of different outlets can mean that the journalist is no longer effectively reporting the story as it carries on developing and unfolding, as they are still stuck in front of the camera reporting what has become 'old' news.

3.12 Conclusion

From the discussion in this chapter, it should be clear that the fundamental choice of type of system should be straightforward once the requirement has been tightly defined. It is the implementation that often causes the problems, either due to financial constraints (which are always present) or physical parameters. Clarity on what the system is expected to do and how it is to be actually used in the field is crucial before talking to any manufacturers and integrators. Is it a simple uplink or an integrated production facility that happens to have an uplink attached to it? Where will it be operated and who is expected to operate it? By being clear on these fundamental questions at the beginning of the project, there is a chance that you will end up with a system that does not disappoint.

4 Squeezing through the pipe: digital compression

4.1 Introduction

In many parts of the world the use of digital compression is now an essential characteristic of the satellite newsgathering process, so in this chapter we will look at the differing types of compression used in both video and audio. Compression is a highly complex subject and far beyond the scope of this chapter (or this book) to explain fully in detail, and as in other chapters the use of mathematics will be kept to a minimum. For those who require a more theoretical explanation of the processes, there are a large number of books and papers on the subject, a few of which are listed in the bibliography at the end of the book.

Why do we want to compress a signal? The answer is that, in an ideal world, we don't! If we had unlimited bandwidth available to us, then there would be no need for compression, for digital compression is essentially about squeezing large bandwidth signals into the typically narrower available bandwidth (frequency spectrum). On wideband fibre-optic routes there is the available bandwidth to send uncompressed video and audio signals, where cost is not the primary issue. But as we have seen in Chapter 2, satellite communication revolves around the issues of power and bandwidth, and by digitizing and compressing the signal, we can reduce demand for both power and bandwidth – and hence reduce cost.

If digital signals were used in direct replacement for the same analogue information, the resultant bit-rate at which the information would have to be transmitted would be very high, using more bandwidth than an equivalent analogue transmission, and therefore not very efficient. Hence, there is a need to strip out the redundant and insignificant information in the signal, resulting in compression.

A common analogy that illustrates the underlying concept of compression is concentrated orange juice. Consider the carton of concentrated orange juice that you can buy in a supermarket. The process began with squeezing oranges to produce orange juice, of which the major constituent is water. The water is removed to produce a concentrate that can be put into a small carton and transported to the shop for sale. The customer buys the orange juice concentrate, takes it home, and reconstitutes the orange juice by adding water back in. Effectively, the orange juice is 'compressed' (just like the signal), so that a comparatively large volume can be transported in a small carton (or bandwidth), and then the juice is reformed by adding the equivalent volume of water back in that was removed in the first place (decompressed). What is the advantage of this to the consumer? The cost of the orange juice concentrate is much less than the freshly squeezed juice and is therefore much more affordable. 'Ah, but it doesn't taste the same as freshly squeezed juice' – and that inevitably is the result of the compression processes we will be looking at. The signal at the end of the signal compression and decompression process is never quite the same as the original signal; something is lost along the way, just as with the orange juice. However, it costs less in the end and is a reasonable facsimile of the original for most people. But there should never be any doubt that a signal that has been compressed, transported and decompressed (reconstituted) will never be as good as the original.

So, compression is a necessity if we wish to reduce the amount of power and bandwidth used on the satellite, and hence reduce the cost of the space segment. The use of digital signals is a 'given', for although analogue compression has been used historically in television to a degree, digital compression is now far more efficient and cost-effective as it uses technology developed in the computer industry with all the associated efficiencies of scale.

4.2 The language of compression

The word 'algorithm' is often heard in discussion about compression, and this refers to the computational code used to compress or decompress the signal. Any technical discussion of the difference between types of compression (both video and audio) often centres around the algorithms being used. Methods of compression are often referred to as 'compression schemes'.

In digital video compression, each frame of the television picture can be considered as being made up of a number of 'pixels', where a pixel is the smallest element that is discernible (quantified) in a picture. Each pixel is 'sampled', with a 'sample' being an instantaneous measurement of the picture at a particular point in time. Each frame of a television picture is made up of 525 or 625 lines (depending on the standard used) and a picture can be divided up horizontally and vertically into a number of pixels; in a television picture, the vertical pixels correspond with the lines of the television picture. The number of pixels defines the 'resolution' or sharpness of the picture. This is the same as the way a photograph in a newspaper is made up of individual dots of ink – the more dots per millimetre, the greater the resolution or detail of the picture. If a video picture is 'progressively' scanned, each line is scanned in sequence in the frame. If the picture has an 'interlaced' structure, where each pair of television fields is divided into odd- and even-numbered lines for sending in alternate sequence (a historical method of saving bandwidth), then each row of pixels will be offset by half a pixel width. Figure 4.1 shows the principle of scanning and how the pixel structure in a video picture differs between progressive

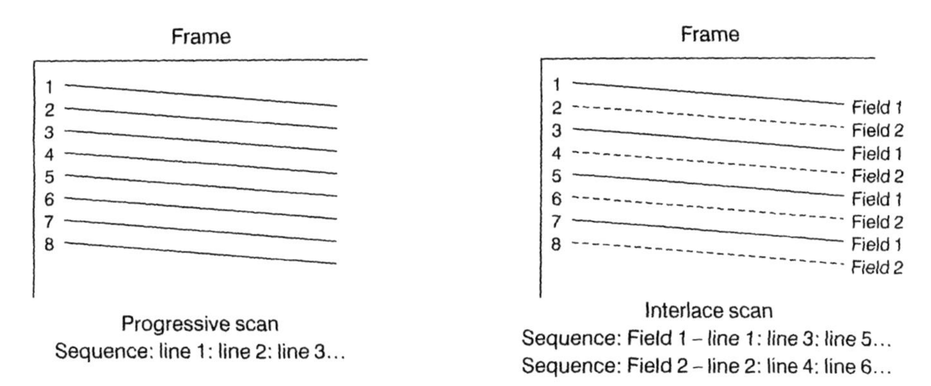

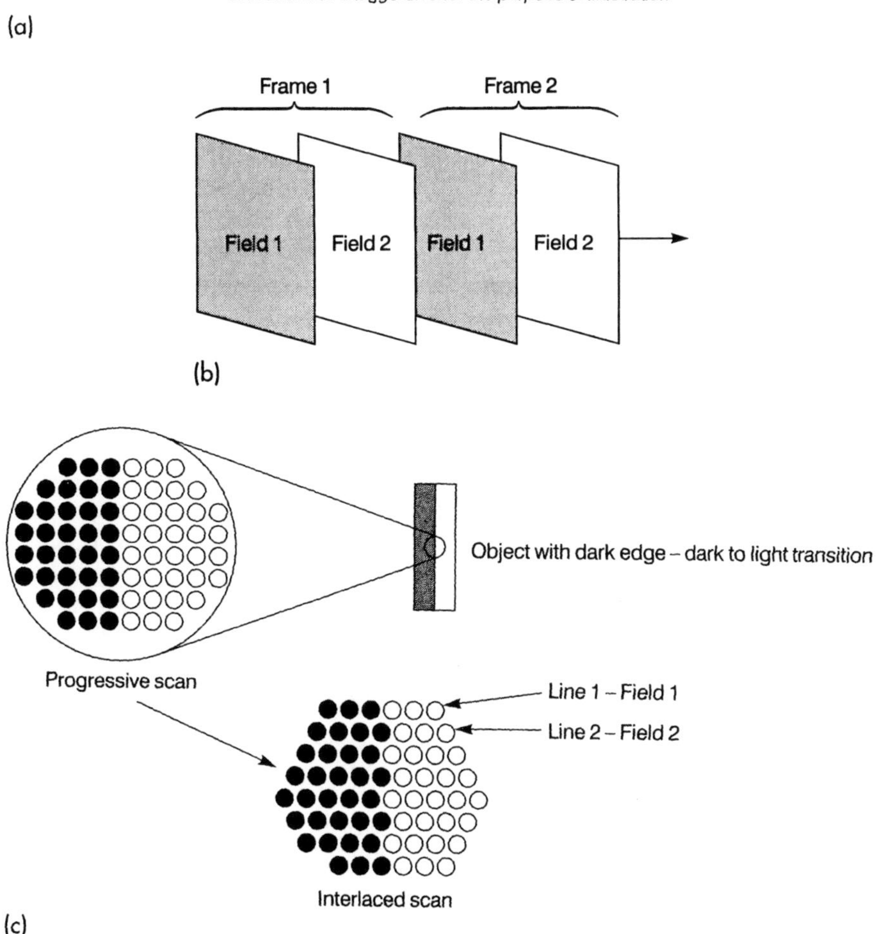

Figure 4.1 (a) Progressive and interlace scan principle. (b) Interlaced field sequence. (c) Pixel structure

and interlaced scanning. Hence, if the picture is broken up into, say, 400 × 300 pixels, the picture would have a 'blocky' look to it. If the picture was to be divided up into double the number of pixels, i.e. 800 × 600 pixels, it would look a lot smoother and sharper. (In fact, 800 × 600 pixels is a common standard for a PC monitor picture.)

Compression processes are also described as being 'symmetrical' or 'asymmetrical'. A symmetrical compression process is where the decompression process is a direct inverse of the compression process in terms of the type and number of stages. There is therefore equal complexity at each end and equal time taken to process the signal. An asymmetrical process is where one part of the process (for example, the encoding compression process in MPEG-2) is more complex than the decoding decompression process, and hence takes longer.

An inevitable result of compression, whether video or audio, is delay. There are a number of computations carried out in both the compression and decompression processes, and the higher order compression algorithms (such as MPEG-2) take longer to be completed than the more simple coding algorithms. Delay through a compression process is termed 'latency', and latency is of particular concern in DSNG operations. We will look at this in more detail later in the chapter.

4.3 Principles of compression

4.3.1 Digital sampling

A digital signal conveys information as a series of 'on' and 'off' states (analogous to a switch turning a light on and off) which are represented as the numbers '1' and '0', based on the binary number system (Figure 4.2). In a stream of 1s and 0s, each 1 and each 0 is termed a 'bit', and a digital signal is defined by the parameter of 'bits'. For instance, as shown in Chapter 2, the signal is transmitted at a certain speed or 'bit-rate', measured in bits per second (bps). The analogue signal is converted to a digital signal by a process called 'sampling'. Sampling is where a signal is instantaneously measured and the value of the signal at that instant is typically converted to a binary number.

The process of sampling is crucial to the quality of the signal that can be conveyed, as the more accurately the binary number obtained from each sample reflects the value of the signal at that instant, the more accurately the original signal can be reproduced. Therefore, if a signal is sampled at 1 bit per sample, it can only

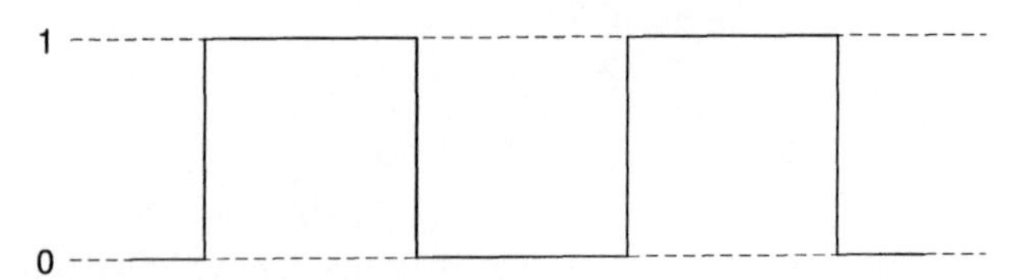

Figure 4.2 Binary digital signal

Table 4.1 Relationship between bits and values that can be measured

Bits	*Number of steps that can be measured (quantization levels)*
1	2
2	4
3	8
4	16
8	256
10	1 024
12	4 096
16	65 536

be represented as either nothing or maximum signal, because the sample is either 0 (no signal) or 1 (maximum signal), which cannot accurately represent the signal.

As there is no gradation of the signal between zero and maximum, and as an analogue signal can have a very wide range of values between minimum and maximum, more steps are required. Hence, using more bits per sample gives greater accuracy or 'resolution' of the value at the instant of sampling. Table 4.1 shows the relationship between bits and values that can be measured. In Figure 4.3 the process of sampling is shown with 1 bit, 4 bits and 8 bits, and it can be seen that, as the number of bits per sample is increased, the more precise the instantaneous value of the signal becomes.

If 4 bits are used, then a number between 0 and 15 can represent any point on the signal. If 8 bits per sample are used, then it can be represented by a number of steps between 0 and 255. If 10 bits per sample are used, the range is extended to between 0 and 1023 – remembering that the value of '0' counts as one step. A step is also referred to as a 'quantization level'.

The other key factor in sampling is how often the sample is taken – termed the 'sampling frequency'. If the signal is sampled at too low a frequency, the samples might be accurate at a particular instant but are not taken closely enough together to be able accurately to reconstruct the signal at the other end of the process. This can lead to the 'blocky' look of a picture, for instance. On the other hand, if the signal is sampled too frequently, the data rate will rise to demand a large amount of bandwidth. There is therefore a compromise to be made between the minimum number of samples required to convey the information to the brain and the data bandwidth available.

Once a signal has been digitized, the signal is a bit-stream at a given overall data rate. Within that data stream there will inevitably be sets of data that are repeated, and the repeated sections represent 'redundancy' in the signal. By extension of this argument, therefore, the absolute rate of information is lower than the overall bit-

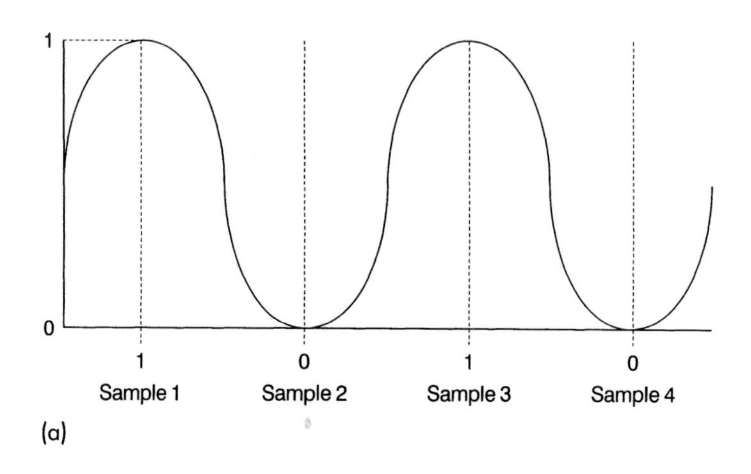

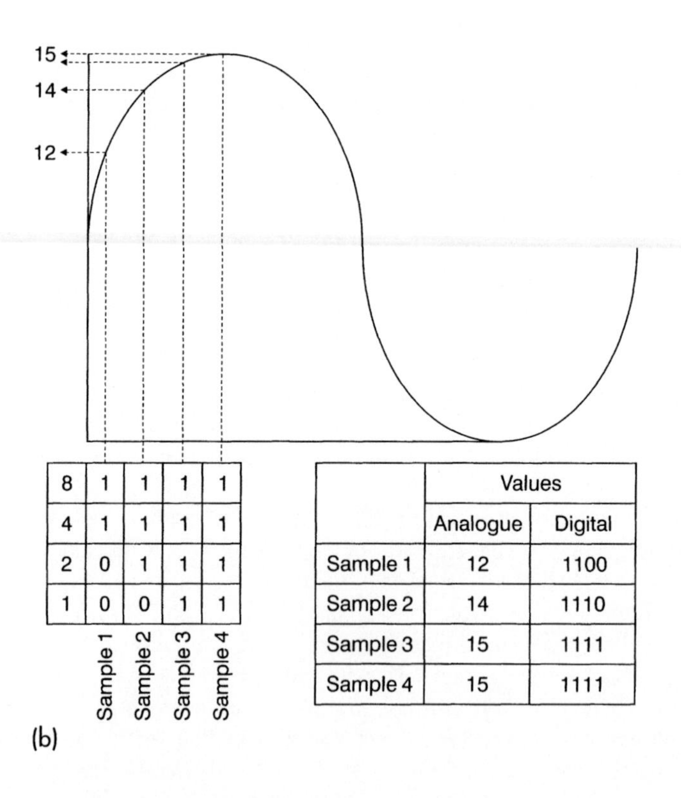

	Values	
	Analogue	Digital
Sample 1	12	1100
Sample 2	14	1110
Sample 3	15	1111
Sample 4	15	1111

Figure 4.3 (a) 1 bit sampling; (b) 4 bit sampling

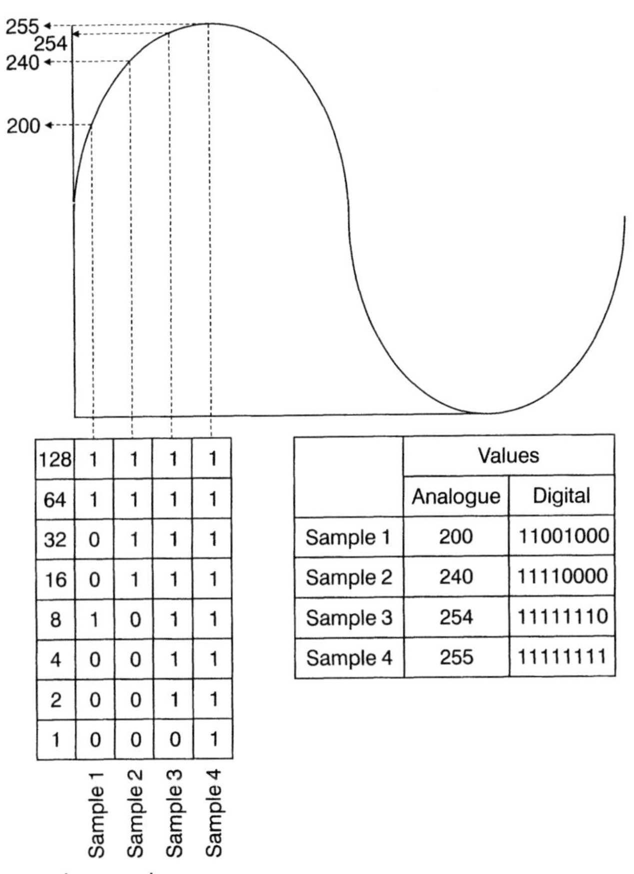

128	1	1	1	1
64	1	1	1	1
32	0	1	1	1
16	0	1	1	1
8	1	0	1	1
4	0	0	1	1
2	0	0	1	1
1	0	0	0	1
	Sample 1	Sample 2	Sample 3	Sample 4

	Values	
	Analogue	Digital
Sample 1	200	11001000
Sample 2	240	11110000
Sample 3	254	11111110
Sample 4	255	11111111

Figure 4.3 (c) 8 bit sampling

rate, and the aim of any compression system is to remove the redundancy in the information signal to minimize the bandwidth or data rate required. There are a number of mathematical analyses of a signal that can be made to determine the redundancy, depending on the compression process.

4.3.2 Types of compression

Lossless compression

Fundamentally there are two types of compression, namely 'lossless' and 'lossy'. Lossless compression creates a perfect copy of the original signal when it is decompressed at the end of the signal chain. This type of compression is commonly used in computers to achieve data compression, particularly where even the loss of a single bit would irretrievably corrupt the whole set of data. Using lossless compression, the maximum data compression ratio that can be achieved for a digital video signal is 2:1, which is far from the 33:1 or more required. Therefore we will concentrate on lossy compression as this is the method used in DSNG.

Lossy compression

Lossy compression is used in the most prominent video and audio compression schemes, and relies on the fact that the human eye and ear can tolerate some loss of information yet still perceive that a picture or sound is of good or adequate quality. For example, PAL, NTSC and SECAM analogue television systems are different variations of lossy analogue compression, as compromises are made in the amount of colour (chrominance) signal transmitted. In addition, the historical practice of 'interlacing' each pair of television fields into a single picture frame is also a method of reducing bandwidth. Each field carries half the information of each frame, so only half the bandwidth of the total picture is required when using interlacing. The reconstruction of each frame of video by combining the two fields back together again in the television receiver relies on a human psycho-visual effect, the persistence of vision. This is further enhanced by the type of display tube used where, once scanned with a field of video, the light output from the 'phosphors' which make up the screen takes a while to decay.

Lossy compression depends on human psycho-visual and psycho-acoustic models that describe human sensory perception of sight and sound. Hence, lossy compression is also referred to as 'perceptive' compression, as the result is a picture or a sound which 'tricks' the brain into thinking that it sounds and looks like the original. Referring back to the orange juice analogy, we can see that the reconstituted juice is 'perceptively' close to the original.

With the orange juice, we removed the redundant component (water) to achieve the compression. That is the essence of video and audio compression – to remove the redundancy in the signal. So how do we identify which are the redundant parts of the signal? In any frame of a television picture there are parts of the image which have the same values of brightness and colour in particular areas. So, instead of sending a repeating string of numbers that represent individual values of brightness and colour in these areas, one string of numbers can be sent that represents the brightness and colour of part of an image that is the same.

In 1948, the mathematician Claude Shannon, an expert in information theory, proposed that signals will always have some degree of predictability, and data compression commonly uses the principle of statistical analysis of a signal to predict changes. The information content of a signal is described as 'entropy', and in a lossless compression scheme all the entropy is maintained within the compressed signal. In a lossy compression scheme, some of the entropy is lost to gain further reductions in data rate and bandwidth, and the system relies on the fallibility of the human senses not to notice the loss.

We will look at the fundamentals of video compression and audio compression separately, as although some of the techniques used are common, it is easier to understand the different processes if we treat them separately. The standards used in DSNG signal compression integrate both the audio and video into a single multiplexed signal.

4.4 Video compression

The reason why video compression works so well is that there is a high level of redundancy with respect to the video data, due to the way the human psycho-visual system operates. There are three types of redundancy in video: spectral, spatial and temporal.

4.4.1 Spectral redundancy

The analogue television signal that is transmitted to the viewer is termed 'composite', as it combines the luminance, colour and timing signals into one combined signal. However, if the luminance and colour information in the signal is processed separately, before being combined into a composite signal, there is greater opportunity to exploit the redundancy of information in the colour elements of the signal.

Earlier we mentioned the ability to fool the human brain into thinking it is seeing a true representation of the original picture. This extends to how the brain is able to distinguish the brightness (luminance) and colour (chrominance) detail of a picture. The human eye, as the visual sensory receptor of the brain, is much better able to distinguish brightness differences than colour differences. This can be used to advantage in conveying colour information, as there is less precision required – a higher level of precision would be 'wasted' as the brain is unable to make use of the additional information and distinguish the difference. When digitizing a video signal, fewer samples are required to convey the colour information, and fewer samples means less bandwidth. Typically, for every two luminance samples there is only one chrominance sample required.

Let's just briefly look at the way the luminance and chrominance information is conveyed to the brain. The eye registers the amount of light coming from a scene, and visible light is made up of differing amounts of the three primary colours of red, green and blue. In terms of television, white is made up of 30% red signal, 59% green and 11% blue.

As with many aspects of television technology, the method of transmitting a colour signal is a compromise. Back in the 1950s and 1960s when colour television was being introduced, both the American (NTSC) and European (PAL) colour systems had to be able to produce a satisfactory monochrome signal on the existing monochrome television sets. Therefore, both colour systems derive a composite luminance signal from 'matrixing' the red, blue and green signals. The matrix also produces colour 'difference' signals, which when transmitted are disregarded by the monochrome receivers, but form the colour components in colour television receivers by re-matrixing with the luminance signal.

Why produce colour difference signals? Why not simply transmit the red, green, and blue signals individually? Again, this is a demonstration of the earlier use of compression in the analogue domain. Individual colour signals would each require full bandwidth and as a video signal requires 5 MHz of bandwidth, the three colour signals would require 15 MHz in total.

On the other hand, by relying on the relatively poor colour perception of the human brain, the colour signals could be reduced in bandwidth so that the entire luminance and chrominance signal would fit into 5 MHz of bandwidth – a reduction of 3:1. This was achieved by the use of colour difference signals – known as the 'red difference' signal (C_r) and the 'blue difference' signal (C_b) – which were essentially signals representing the difference in brightness between the luminance and the respective colour signal. The eye is more sensitive to changes in luminance than in chrominance, i.e. it can more easily see an error when a pixel is too bright or too dark, rather than if it has too much red, blue, or green.

Bright pixels tend to be bright in all three colours, red, green and blue, and therefore there is said to be 'spectral' redundancy (relating to the similarity between colour values at any one point in the picture). Therefore, the difference in levels of the red and blue signals will be relatively small, so sending only the difference between the colour signals and the luminance signal saves some data. The green signal can be reconstructed by deduction from the values of luminance and the colour difference signals.

Hence video presented as three separate signals – a luminance and two colour difference signals – is termed 'component' video, as opposed to composite video which has all the luminance and chrominance information combined into a single signal (along with the synchronization information).

So for historical reasons we have ended up with this convention of conveying chrominance separately from the luminance information, and this was compression in the analogue domain. This efficiency can also be used in the digital domain. Having maximized the compression by determining the redundancy in the chrominance signal (and processing the chrominance components to remove spectral redundancy is often referred to as 'sub-sampling'), the signal has to be checked for further redundant information that can be removed to reduce the data rate drastically.

Spectral redundancy is already a feature of analogue television, so in the digital domain the removal of further redundant information is carried out in two dimensions within each frame of picture, described as 'spatial' redundancy; and between successive frames, described as 'temporal' redundancy.

4.4.2 Spatial redundancy

Spatial redundancy is the relationship each pixel has with the neighbouring pixels: more often than not, two neighbouring pixels will have very nearly the same luminance and chrominance values.

Consider one frame of a picture signal. Suppose that within this frame there is a notable area of the picture that has either the same colour or brightness content, or both. The samples of luminance and/or chrominance would sequentially be the same across this area, and therefore instead of sending the same numbers for each sample, one number could be identified as applying to a number of sample points in an area where the information content remains the same. Hence we have found some spatial redundancy, and this reduces the amount of data that has to be sent. Spatial redundancy can therefore be exploited by using 'intra' ('within') coding, where compression is achieved by reducing within a single frame the amount of information that needs to be sent. Because intra coding happens within each single frame, with no reference to what has come before or what follows, it is used as the basis for compression of still pictures.

4.4.3 Temporal redundancy

Temporal redundancy is the relationship between video frames. If consecutive video frames were not similar, there would be no sense of motion. Having said that, objects that move between one frame and the next often do not significantly change shape, for example, so there is some redundancy in the data from one frame to the next that can be removed.

Consider two frames of a picture signal. In the first frame, there is an object that is also present in the second frame; for example, let us presume that the first frame has a building in it. Unless there is a shot change or a rapid camera pan, the building will be present in the next frame (though not necessarily in the same position if the camera has moved slightly). The data block that contained the information on the building in the first frame could therefore be repeated in the second frame. This is termed 'temporal redundancy'. In fact, most pictures shot in a single sequence are remarkably similar from one frame to the next, even in a moderately quickly moving sequence – only 1/25 (PAL) or 1/30 (NTSC) second has elapsed. Therefore there is a significant scope for data reduction by simply re-using sets of data sent for the previous frame, albeit that the blocks of data may have to be mapped to a different position in the frame. Coding in this manner is described as 'inter' ('between') coding, and is dependent on what has occurred before to make use of the redundancy.

4.4.4 Sampling rates

Video is typically sampled at 8 or 10 bit accuracy, meaning that each sample can have a value of up to 255 or 1023 respectively, depending on the amplitude of the analogue signal at that instant. As each sample represents a pixel, there are separate streams of samples for the luminance and for each of the chrominance difference signals.

The original standard for digital video was published as the European standard CCIR Recommendation 601 standard. The now defunct CCIR standard was subsequently adopted by the ITU as Recommendation ITU-R BT.601 and a further standard dealing with digital video interfacing, Recommendation ITU-R BT.656, was published. These were also published in the US as SMPTE 125M[3] and SMPTE 259M[4] respectively. In general, digital video at broadcast quality is loosely referred to as '601' video. In ITU-601, an uncompressed digital video signal has a data rate of 216 Mbps (sampled at 8 bits per sample) or 270 Mbps (sampled at 10 bits per sample). The choice between using 8 or 10 bits depends on the application; the original recommendation for ITU-601 was only for 8 bits, but some production systems required higher resolution for picture manipulation, and ITU-601 was subsequently revised. The ITU-601 signal at 270 Mbps is over thirty-three times the data rate of the typical 8 Mbps DSNG signal. A data rate of 270 Mbps, modulated in the same way as the typical 8 Mbps DSNG signal, would require over 250 MHz of bandwidth. It is not feasible to allocate this amount of bandwidth on commercial satellite systems (although this capacity is available on wideband fibre-optic routes – at a price).

The data rates are derived as shown in Appendix C. The specification of the number of pixels and hence the sampling rates is based on the use of multiples of a fundamental frequency of 3.375 MHz (which represents '1' in the sampling rates), which is based on a commonality of certain fundamental frequencies between 525- and 625-line systems. Notice that there is one luminance signal and two chrominance signals in the result. The chrominance signals are sampled at half the rate of the luminance signal; so for every four luminance samples, there are two red difference signal samples and two blue difference samples. This is termed a '4:2:2' signal and is shown in Figure 4.4.

In 4:2:2 the chrominance signals have been sampled at half the luminance rate in the horizontal (line) dimension, but vertically they have been sampled at the same rate. Because of what we said earlier about the brain's poor ability to discern detailed changes in chrominance, in a compression environment additional saving in bandwidth can be achieved by sampling the chrominance at only half the rate in the vertical dimension as well as the horizontal. Because a picture is made up of two fields, the chrominance samples can occur at half the vertical rate on each field. As each field is offset by a line (a feature of interlace), the chrominance samples 'line-up' when viewed over the period of two fields, i.e. a whole frame. This creates a picture that has chrominance sampled at half the rate relative to luminance in both the horizontal and vertical dimensions. When the signal is sampled in this way, it is termed a '4:2:0' signal (Figure 4.4).

We have shown how the data rate for video is 270 Mbps. In fact, because we do not need to digitize the timing and synchronization information, the actual active video data rate is lower. The active video is the part of the video signal that describes the image, in other words it is the video signal excluding the 'vertical blanking interval' and the 'horizontal blanking interval', which contain the timing and synchronization information for the television display. Consequently, as shown in Appendix C, the actual active video data rate is lower – 166 or 168 Mbps according to the standard. However, the reduction in the data rate to 166/168 Mbps by only encoding the active picture is still not enough (over 20:1), and the need for compression obviously still exists. (The same is true of 4:2:0 coding at 124/126 Mbps.)

4.4.5 Discrete cosine transform

In general, adjacent pixels within an image tend to be similar and this can be exploited to eliminate redundancy. There are a number of mathematical analyses of a signal that can be made to determine the redundancy, but the one most commonly

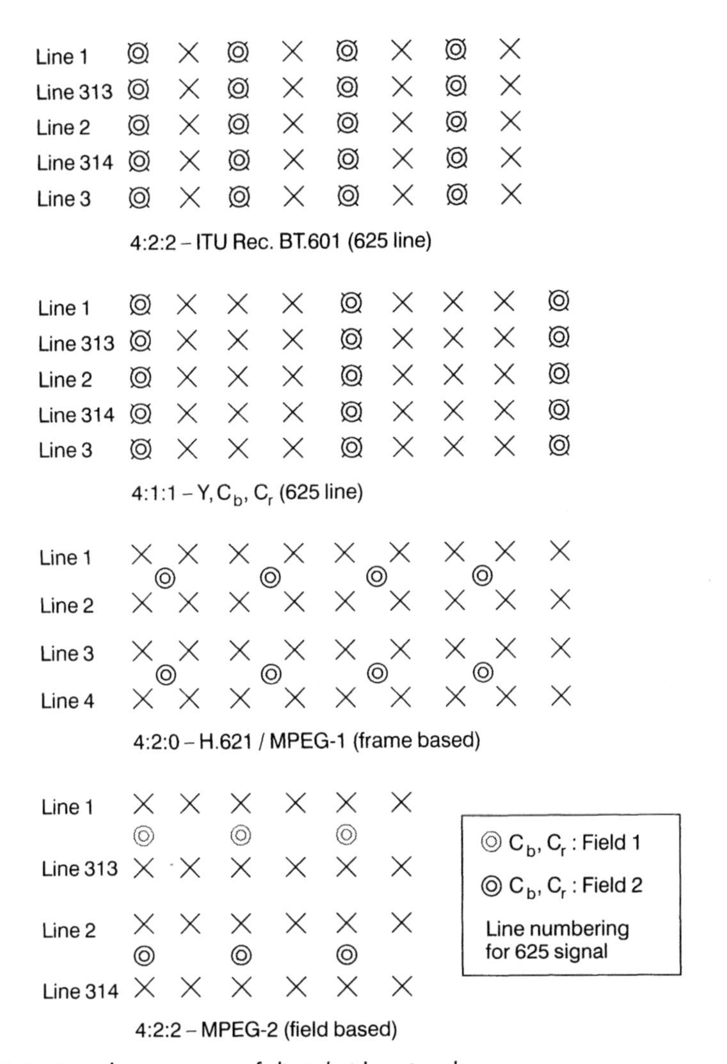

Figure 4.4 Sampling structure of digital video signals

used in video compression is the discrete cosine transform (DCT). DCT is a specific implementation of a mathematical process called a Fourier transform, involving trigonometry. The intra-frame coding technique that takes advantage of spatial redundancy is based on DCT, which represents a sequence of numbers as a waveform.

Mathematically speaking, the image is transformed from its spatial representation to its frequential (frequency) equivalent. In other words, instead of looking at the image as a freeze-frame of a scene within a space, the image is analysed by looking at how individual objects appear in terms of their detail, or frequency. Each element in the picture is represented by certain frequency 'coefficients'. A coefficient is a mathematical term for a number with a particular algebraic significance. As a sequence becomes closer to an actual cosine wave, then a single coefficient becomes very large and the remainder of the coefficients become smaller.

Most of the energy in an image is contained in the low-frequency coefficients – these are objects with large surfaces such as sky, walls or expanses of ground – and these are the parts of the image that the human eye focuses primarily on, because of the energy concentrated in this area. Detail in an image is expressed by the high-frequency coefficients – these are objects with fine detail, such as hair, grass, leaves and tree branches. Interestingly, the low-frequency coefficients contain more information (entropy) than the high-frequency ones.

In the DCT process, blocks of 8 pixels horizontally by 8 pixels vertically are converted to a set of frequency coefficients that describe that block. It has been determined that picture information can be described by a set of 64 discrete frequency steps in both the horizontal and vertical planes, so each pixel has a value describing the 'amount' of horizontal and vertical frequencies corresponding to one

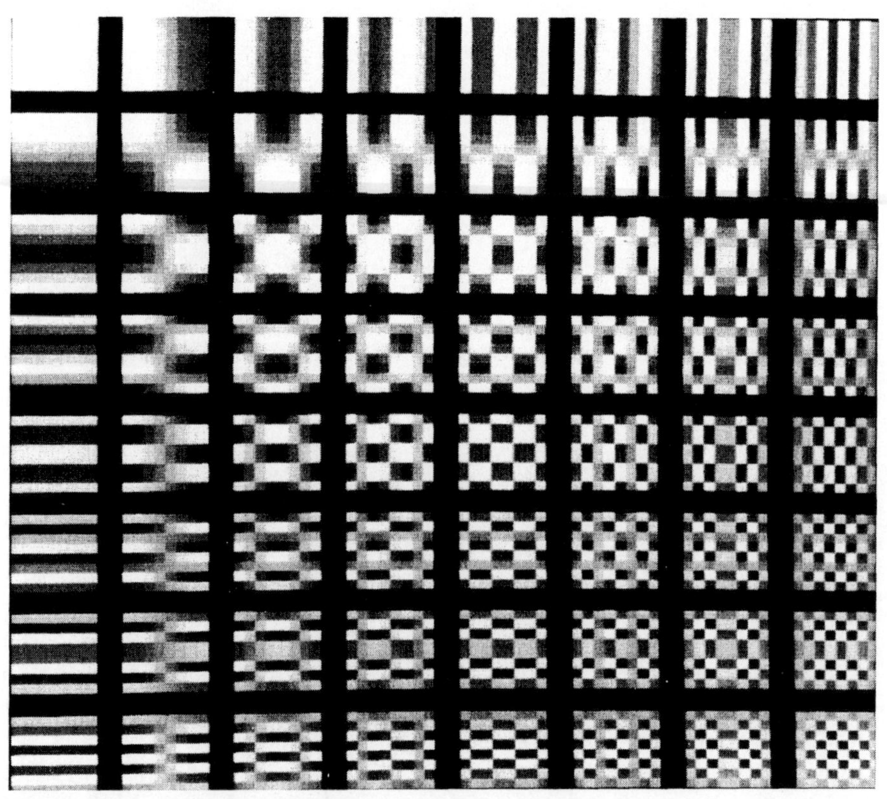

Figure 4.5 DCT frequency coefficients. *(Reproduced with permission from Watkinson, J. (1999) MPEG-2. Focal Press)*

of these 64 steps. There are therefore 64 frequency coefficients in each 8 × 8 pixel data block (as shown in Figure 4.5). The very first value in the top left describes the average brightness of the whole 8 × 8 pixel block (termed the 'DC' coefficient). The coefficients to the right of the DC coefficient (datum point) are the increasingly higher horizontal spatial frequencies. The coefficients that are below the DC datum point are increasingly higher vertical spatial frequencies, and therefore looking below and to the right see increasingly higher coefficients in each of the horizontal and vertical planes. Each pixel's DCT coefficient is calculated from all other pixel values in the block, so using an 8 × 8 block saves time, and the smaller the difference between one pixel and its adjacent pixels, the smaller its DCT value. The larger coefficient values describe the lower frequency elements of the 8 × 8 pixel block, because of the value to the brain of low-frequency information. The high-frequency coefficients tend to have values towards zero, as there is usually less information at these frequencies in most frames, and this information is least useful to the brain.

It is important to note that the DCT process (and the following quantization process) is carried out separately on 8 × 8 pixel blocks in the luminance (Y), red difference (C_r) and the blue difference signals (C_b), producing blocks of samples for each.

At the decoder, an inverse DCT is applied and the actual DCT process is essentially lossless. However, there is information loss because of the quantization stage that occurs next. The complexity of the DCT process is such that we cannot further describe it here without delving deeper into mathematics, but hopefully you now have an outline understanding of the basic principle.

4.4.6 Quantization

It is in the quantization stage that the significant part of the compression process occurs. As we have seen, the signal has been digitized into a series of samples, and the DCT process has converted this stream of samples into a set of frequency coefficients. Quantization now assigns a specific number of bits (resolution) to each frequency coefficient.

Each 8 × 8 pixel data block is compressed by assigning fewer bits to the high-frequency coefficients (remembering that we have said that higher frequency information is less obvious to the viewer). This process does discernibly affect the overall quality of the picture and it is a lossy process, i.e. the information that is lost cannot be reconstructed at a later stage. There may also be a decision-making process within the encoder that reduces the high-frequency (near-zero) coefficients to zero – this is one of the characteristics that can differentiate between one manufacturer's encoder and another.

Some blocks of pixels need to be coded more accurately than others. For example, blocks with smooth intensity gradients need accurate coding to avoid visible block boundaries. To deal with this inequality between blocks, the algorithm may allow the amount of quantization to be modified for certain blocks of pixels.

4.4.7 Variable length coding

Following quantization, there is a final process of manipulation of the bit-stream called 'variable length coding' (VLC), where further efficiencies are used to reduce the amount of data transmitted while maintaining the level of information. The quantized DCT frequency coefficients are re-ordered into a form that is easier to manipulate. The values are read out of the 8 × 8 pixel data block in a 'zig-zag' fashion, as shown in Figure 4.6, and the aim is to maximize the likelihood of a run of zeros or a run of low value coefficients. The very first value read out is the DC coefficient (the average brightness of the whole block). The rest of the coefficients describe the amount of horizontal and vertical, and coarse and fine detail in the block. The resulting block has groupings of coefficient values interspersed with zero values, many of which often tend to have a predictable pattern. Starting with the first coefficient, which has the largest value (as it represents the average brightness of the

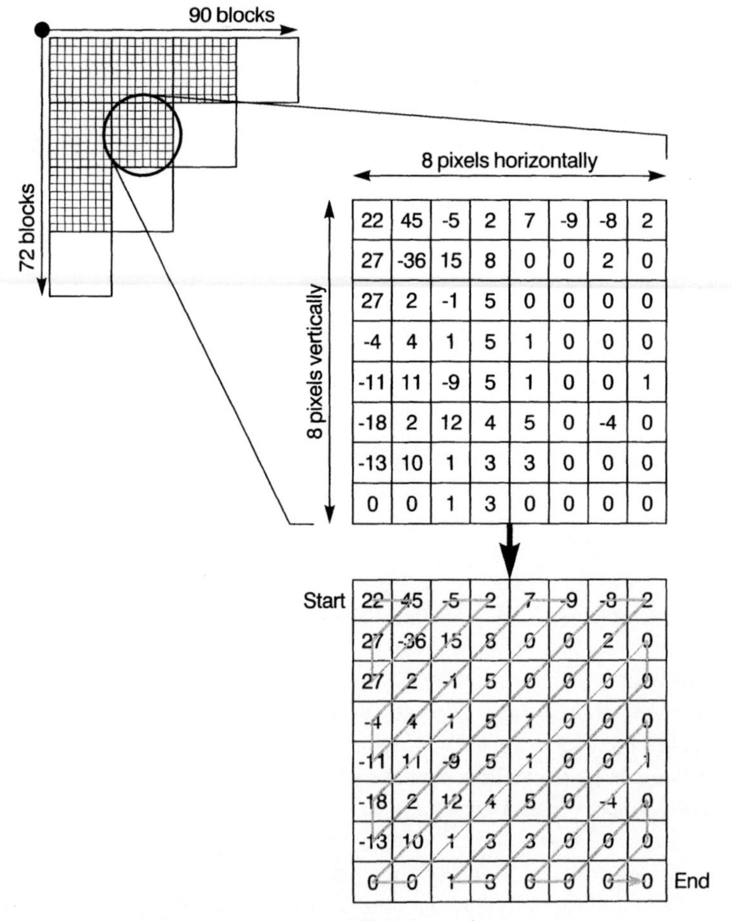

Figure 4.6 Quantized DCT frequency coefficients and processing

whole frame), the coefficients are 'read out' in a zig-zag sequence, and when only zeros are left to be sent, an 'end of block' code is sent instead. A single number gives the run-length of the number of zeros in the sequence, giving a count of the zeros instead of listing them.

With variable length coding, the aim is to send the minimum number of bits, and therefore to use a fixed length code in which small numbers are prefixed with many zeros is not desirable. Variable length codes are thus devised such that the source values that need to be sent most often use fewer bits in transmission, and values that are sent on only a few occasions use codes which have many bits. A famous type of variable length coding is Morse code, where the letter 'E', which is a common letter in English text, is sent as a short code, i.e. as a single 'dot' (1 'bit'), whereas the letter 'Z', which occurs infrequently in English, is sent as 'dash dash dot dot' (4 'bits'). The longest codes are for punctuation, which have six 'bits', but then these occur very infrequently in the average message.

Variable length coding therefore distinguishes between short runs of zeros, which are more likely than long ones, and small coefficient values, which are more likely than large values. Code words are allocated which have different lengths depending upon the probability with which they are expected to occur. The values are looked up in a fixed table of 'variable length codes', where the most probable (frequent) occurrence of coefficient values is given a relatively short 'code word', and the least probable (infrequent) occurrence is given a relatively long code word. The process is also known as 'entropy coding'.

There are therefore two types of data compression occurring in variable length coding. The sequence of zeros has been shortened to a number describing how many zero values there are, rather than sending a string of consecutive zero values. Secondly, the translation of patterns of the significant value coefficients (i.e. those that describe low-frequency information and which have some predictability) into short codes achieves a further saving in the amount of data sent.

As with the DCT process, this has been much simplified to enable you to gain some insight into the process, and any further explanation would take us deeper into the realms of mathematics. As we shall see later, a typical frame is sampled in 4:2:2 broadcast quality video as 720 × 576 pixel resolution. In either 1/25 or 1/30 of a second, this results in 6480 of these 8 × 8 pixel blocks being processed for luminance, and 3240 pixel blocks for each of the two chrominance signals (as these are at half the resolution and therefore sampled at half-rate). There is a total of around 830 000 samples per frame with 4:2:2 sampling, to which the DCT, quantization and variable length coding processes are applied between 25 and 30 times per second. This plainly takes a phenomenal amount of processing speed and power.

4.4.8 Hierarchy of compression

Compression schemes are devised for much wider application than simply that of data transmission through communications channels. Therefore, although a certain type of compression scheme could be used, for example, for SNG, it might not be suitable for picture manipulation. Due to the economies of scale in commercial manufacturing, the most flexible type of compression standard, which can be used for the greatest number of applications, is the one that will gain the widest acceptance.

The simplest form of compression would consist of simply looking at the difference between two frames and then sending only the information that describes the difference. This is called 'differential' coding. The disadvantage of this would be that if the first frame is lost, or the sequence is interrupted in any way, the signal is irrecoverable, as the very first frame is the reference frame and without that, no sense could be made of the following frames. Another disadvantage of differential compression is that it cannot be easily edited, which is a common requirement in manipulating compressed video signals for broadcast. Therefore, there has to be a more sophisticated way of compressing the video information.

The next form of compression would be simply looking for redundancy within each single frame and sending such frames in a continuous sequence. Two kinds of redundancy, spectral and spatial, can be handled inside each frame without reference to any other frame. Compression techniques based on these types of redundancy are therefore called 'intra-frame coding'. Intra-frame coding techniques can be applied to a single frame of a moving picture (or to a single still image).

The higher forms of compression use the most complex algorithms, which also attempt closely to predict data for any objects that have moved between one frame and the next. This is called 'motion prediction' and is the most taxing aspect of video compression, as it demands the greatest computational power.

4.4.9 Composition of the bit-stream

The composition of the bit-stream has a hierarchy, which occurs in the following order (Figure 4.7).

- Group of pictures: a particular sequence of pictures.
- Picture: a frame of information of a particular type (I, P or B).
- Slice: a string of macroblocks across the picture.
- Macroblock: made up of a group of blocks.
- Block: made up of a group of pixels.

In digital video compression, using three types of information frames within a sequence reduces redundancy. The sequence forms the 'group of pictures' (GOP).

I-frames (Intra-coded frames) must be used at the beginning of any video sequence. An I-frame completely describes the picture and only has compression applied within (intra) the frame, and has no dependence on any frames before or after. I-frames are essential at the beginning of a GOP, and must occur regularly, otherwise errors occurring in a predicted frame will progressively worsen. For the purpose of production processes, the I-frame also allows random access, but in a transmission chain, such as an SNG path, this is not of any significance. The I-frame

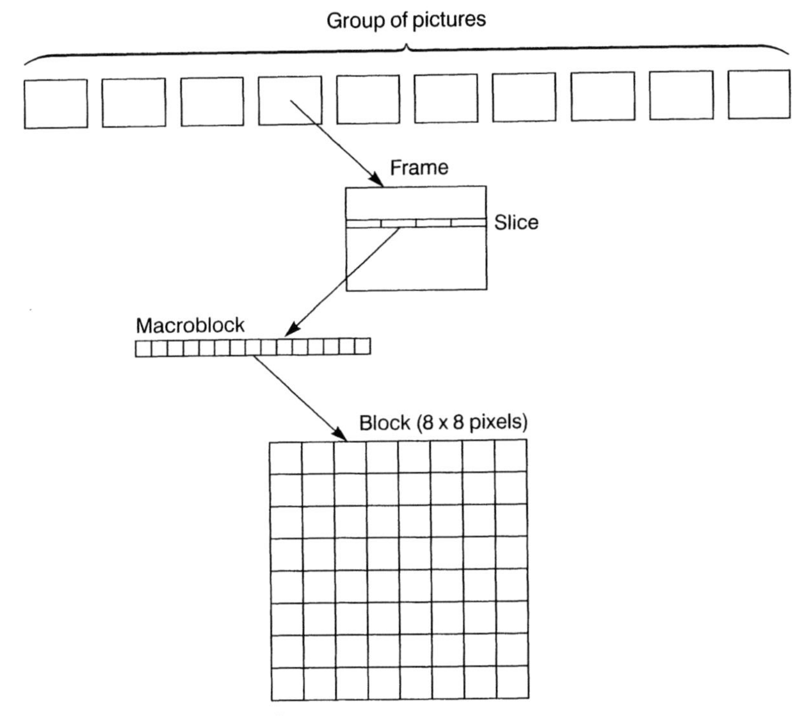

Figure 4.7 Bit-stream hierarchy

can also constitute a reference frame, necessary for acting as an 'anchor' in the sequence used to predict B-frames.

P-frame (Predicted frame) is encoded using forward motion prediction. The P-frame is calculated by applying motion prediction to a previous frame and deriving a difference signal (inter-coding). A P-frame is forward predicted from the last I-frame or P-frame, i.e. it is impossible to reconstruct it without the data of another I- or P-frame. The P-frame is also an anchor frame, necessary for acting as a reference frame to predict B-frames.

B-frame (Bi-directional [predicted] frame) is encoded as a combination of forward and backward motion prediction information. Therefore, the use of B-frames implies out-of-order encoding, as the encoder can only encode a B-frame after encoding the requisite previous and future frames. We will consider this shortly when we examine the coding and transmission ordering of frames.

A B-frame can be considered as the average of two P-frames – one which has already been encoded and sent (i.e. in the past) and one that has been encoded but not yet sent (i.e. in the future) – used as predictors for the current input frame to be processed. B-frames are useful, for instance, when objects in a picture alter during a sequence. A B-frame yields much better compression than a P-frame which, in turn, yields much better compression than an I-frame. This is because B-frames take advantage of maximum temporal redundancy, while I-frames cannot take advantage of any temporal redundancy. Generally speaking, a B-frame can be about one-third the size of a P-frame in terms of data.

Quality can also be improved in the case of moving objects that reveal hidden areas within a video sequence. Backward prediction in this case allows the encoder to make decisions that are more 'intelligent' on how to encode the video within these areas. Also, since B-frames are not used to predict future frames, errors generated will not be propagated further within the sequence. B-frames, while enabling the highest level of compression, have the disadvantage that a future frame must be decoded before the present frame, a process that introduces further delay.

4.4.10 GOP sequences

A GOP contains a combination of I-, B- and P-frames. Some compression algorithms (e.g. MPEG) enable the encoder to choose the frequency and location of I-frames. This choice is based on the application's need for random accessibility and the location of scene cuts in the video sequence. For instance, in post-production where random access is important, I-pictures are typically every frame. The number of B-pictures between any pair of reference (I- or P-) frames can also vary. For example, many scenes have two B-frames separating successive reference frames (I,B,B,P,B,B,P).

An example of a typical GOP is shown in Figure 4.8, where there is a repeating sequence of twelve frames, beginning with the I-frame. The sequence runs I,B,B,P,B,B,P,B,B,P,B,B and then repeats with the next I-frame. We have already seen that a B-frame is encoded as a combination of forward and backward motion prediction information, but it is difficult to see from the above sequence how use can be made of a B-frame. The secret is that, although the sequence is encoded as above, it is actually transmitted in a different order. There is a 'coding order' and a 'transmission order', and in Figure 4.9 the coding and transmission ordering difference is shown.

It can be seen that the first P-frame in the sequence is sent before the first pair of B-frames, as the B-frame information cannot be used before the P-frame to which it is linked has been decoded. Backward prediction requires that the future frames that are to be used for backward prediction be encoded and transmitted first, out of

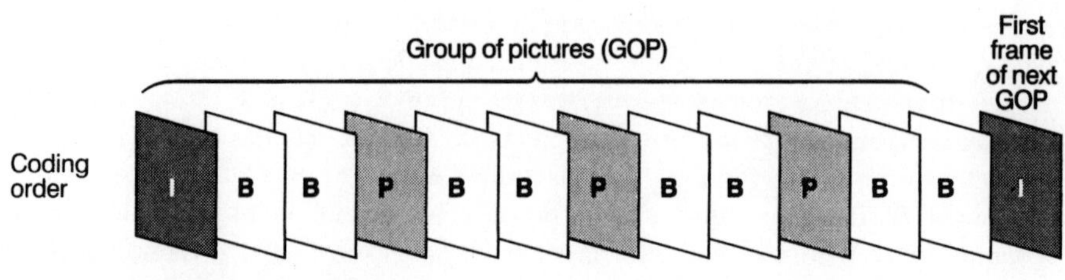

Figure 4.8 Typical GOP sequence

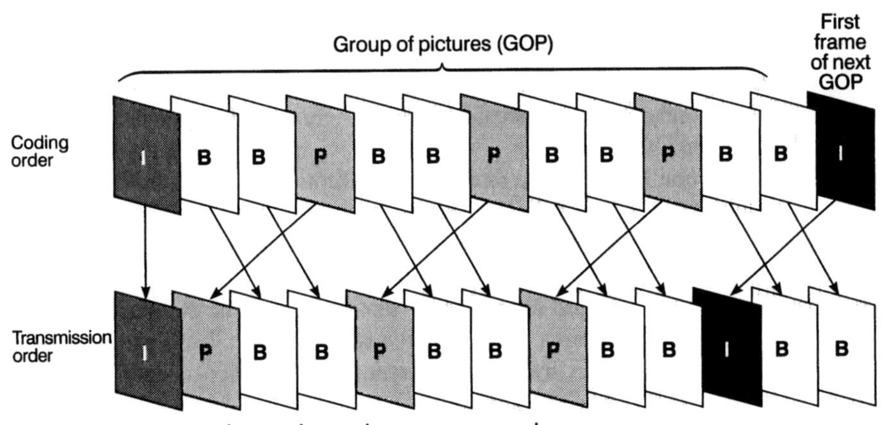

Figure 4.9 GOP coding order and transmission order

order. This picture 'shuffling' in the transmission adds delay to the overall codec process and therefore increases the latency, which as we have already discussed is particularly undesirable for live 'two-way' interviews in news broadcasts.

It also results in both the encoder and the decoder requiring increased amounts of memory (or 'buffers'). In the coder this is to store these frames after coding so that they are available for transmission as and when required, and in the decoder so that they can be stored until they are required to reconstruct the coding sequence. This increases the complexity and cost of the equipment.

4.4.11 Motion prediction

The movement of objects over a period of time is represented in a frame structure by a part of the picture being in a different place from the previous picture, and in a different place again in the next picture. Generally, the actual object does not change shape radically, if at all, from one frame to the next. It is simply the same object in a different position. In the context of the overall frame rate, the actual shift in position from one frame to the next is not very great, even for very fast-moving objects. Therefore, the block of data that describes that object could be 're-used' from one frame to the next by simply placing it in the new position. In the encoder a copy of each previous frame is 'held' (stored) while analysing the current frame for the position of objects 'seen' in the previous frame. Thus, by carrying out some intricate maths, it is possible to calculate where the position of that object will be in the following frame to quite close accuracy, and to send only the co-ordinates of the new position of the object. This is called 'motion prediction' and this process achieves further compression.

In motion prediction, the offset between the two blocks of data is known as a 'motion vector', which indicates in which direction horizontally and vertically the macroblock must be brought from the next frame so that a match is made. Motion prediction occurs in both B- and P-frames, and the macroblock structure is used as part of the prediction process for movement. A macroblock is made up of a number of luminance blocks, and although it can vary depending on the video compression coding standard, the block is typically 8 × 8 pixels in dimension. These are again coded in a particular manner according to the compression standard and are typically based on DCT.

Motion estimation is not applied directly to the chrominance data, as it is assumed that the colour motion prediction can use the same motion information as the luminance, i.e. the chrominance elements will move in the same way as the luminance. The error between the current macroblock and the similar macroblock in the previous frame is encoded along with a motion vector for the macroblock. A sequence of macroblocks form a 'slice', and the coding of the slice is used to further reduce redundancy in frames by using differential compression coding within the frame.

The degree and complexity of the motion prediction can vary widely between compression standards and even between manufacturers using the same compression process. Although a full exhaustive search over a wide area of the frame yields the best matching results in most cases, this performance comes at an extreme computational cost to the encoder. As motion estimation is usually the most computationally intensive process of the encoder, some lower cost encoders might choose to limit the search range or use other techniques, usually at some cost to the video quality.

4.5 Compression standards

Video compression uses inter-coding, intra-coding, or a combination of both, depending on the standard. There are five principal compression standards in use in SNG, although one (MPEG-2) is now the pre-eminent standard. The standards are:

- H.261/3 and H.320 (video-conferencing)
- Motion JPEG (a derivative of JPEG and not strictly a defined standard)
- ETSI (ETS 300 174; ITU-T J.81)
- MPEG-1
- MPEG-2

Most of these standards are interrelated, as historically MPEG-1 was derived from H.261 and JPEG. H.263 then followed, based on H.261 and MPEG-1, and adding some enhancements of its own. On the other hand, MPEG has some enhancements not found in H.261 or H.263. MPEG-2 is a further derivative of MPEG-1.

MPEG supports a range of frame sizes and frame rates: in MPEG-1 the most common are 352 × 240 pixels at 30 fps and 352 × 288 pixels at 25 fps; and in MPEG-2 the most common standard definition is 720 × 486 pixels at 30 fps (720 × 576 pixels at 25 fps).

Video-conferencing using H.261 supports Quarter Common Intermediate Format (QCIF) at 176 × 144 pixels and Common Intermediate Format (CIF) at 352 × 240 pixels, at frame rates of 7.5–30 fps. H.263 adds a smaller size (128 × 96 pixels) and larger sizes (704 × 576 and 1408 × 1152 pixels).

In case you wondered what happened to MPEG-3, it was originally intended to cover high-definition (HD) video standards, but MPEG-2 was found to be perfectly adequate after some further development. The MPEG-3 project was subsequently abandoned. There is also another standard currently being developed – MPEG-4 – but this is not aimed at meeting the needs of DSNG. Although we will look at all of the standards listed above, we will concentrate on MPEG-2, as this has now become the *de facto* standard for the vast majority of DSNG transmissions.

4.5.1 H.261/3

The coding standards for video-conferencing – H.261/3 and H.320 – used for some forms of DSNG are covered further in Chapter 5. The H.261 standard[5] is very similar to the MPEG standard, but only I- and P-frames are used, resulting in less encoding and decoding latency delay. DCT is still used to reduce spatial redundancy, and forward motion prediction is used for motion compensation.

One feature that H.261 and H.263 have (unlike MPEG) is the ability to vary the frame rate within a video sequence. This is important in video-conferencing because an encoder must be able to both lower the frame rate to maintain reasonable visual quality and dynamically adjust to sudden changes in video content in real time.

4.5.2 Motion JPEG

The acronym 'JPEG' stands for Joint Photographic Experts Group, which is a body established under the auspices of the International Standards Organisation (ISO) and the International Electrotechnical Commission (IEC) to develop digital compression standards for still images. It is beyond the scope of this chapter to explore JPEG in detail, as we are really only interested in Motion JPEG, but suffice to say that it has developed a standard for anlaysing and compressing still images for electronic transmission. No doubt you have come across computer image files with the suffix '.jpg', and this is a common computer transfer standard for compressed image files (though there are many others).

The JPEG compression standard[6] is a lossy compression standard, typically achieving a data compression ratio of 20:1, and is at its most effective in terms of compression when working on the type of information in photographic images as opposed to the detail found in line drawings. It is a symmetrical compression process, with equal complexity at each end of the processing chain. Unlike the other video compression schemes described, there is no audio compression facility in this system because of its target application.

A derivative of JPEG is Motion JPEG. This is not a standard, but is a means of using JPEG coding techniques to compress and transmit television pictures. In Motion JPEG, the still-image JPEG compression process is applied to each video field, one after the other, in real time, and a Motion JPEG stream therefore consists of I-frames only. It uses DCT-based analysis, and images can be coded at a range of resolutions. Because it is not a standard, the way in which this is done is often proprietary and therefore there is usually a requirement to have the same manufacturer's equipment at both the transmission and reception points. As we have just mentioned, JPEG works well on photographic (i.e. natural) images as opposed to computer-generated imagery, and hence its application for use with 'natural' television pictures.

4.5.3 ETSI

The beginning of the development of the ETSI digital compression pre-dated that of MPEG and is of historical interest because the emergence of DSNG in Europe in 1994 depended on the video compression system which is generically known as the '8 Mbps ETSI' standard. ETSI compression started to be used in a large percentage of satellite contribution links in Europe through the 1990s, and was adapted for use on DSNG systems as it only imposed a short processing delay on the signals. While the ETSI standard defines a transmission data rate of 34 Mbps, Thomson (France) developed a further proprietary version of the standard for low bit-rate digital video transmission that became available in 1994 – Thomson were one of the leaders in the European high-definition television project (FLASH-TV).

In 1985, the EU initiated the RACE programme (Research and development in Advanced Communications technologies in Europe) to stimulate the development and the introduction of integrated broadband communications (IBC) by 1995, when it was terminated. It was a broad umbrella R&D programme of projects that was orientated according to equipment at customer premises, and not exclusively satellite-related. As the integration of services could only be achieved through digitization, RACE restricted its scope to digital imaging, and the research dealt with digital TV/HDTV systems, video telephony and digital videotape recording. The objectives of RACE included advancing transmission techniques and applications, improving ground segment technology, and achieving compatibility of satellite and terrestrial transmission systems.

Within RACE there was a project called FLASH-TV (FLexible Advanced Satellite system for High-quality TV), aimed at digital HDTV contribution links rather than direct broadcast. The main objective of FLASH-TV was to provide broadcasters, news agencies and telecom operators with a flexible, cost-effective and advanced means of setting up and operating high-quality digital video transmissions. This was planned to use video compression with transmission rates of between 34 and 70 Mbps, and reception with antennas of less than 4 m diameter at Ku-band in the range 34–70 Mbps. Applications included HDTV outside broadcasts and distribution of movies to cinema theatres for projection. The transmission data rate was to be adapted to the available margin by a low-rate feedback link from the receiver (in essence, bit error rate monitoring at the receiver).

One of the leading partners in the FLASH-TV project was the CCETT (Centre Commun d'Études de Télédiffusion et Télécommunication), the research laboratories of France Telecom and TDF (Télédiffusion de France). The CCETT focused on the transmission aspects of FLASH-TV as well as being involved in another RACE project, HIVITS (HIgh quality VIdeotelephone Television System). HIVITS was a contributor to the preparation of the H.261 videoconferencing standard, and some of the concepts resulting from this work directly influenced the development of both the ETSI and MPEG standards in digital compression.

From the RACE projects in 1992 came the standard defined by ETSI in ETS 300 174[7] for 34 Mbps and 45 Mbps digital television coding for inter-studio contribution links, designed for HDTV (although still usable for SDTV). The ETS 300 174 standard is based on 4:2:2 sampling and a DCT-based algorithm (the audio element of the programme signal is not compressed). Unlike MPEG-2, it is almost symmetric with virtually the same processing power at the encoder and decoder, and designed as we have said for applications where there were few reception points, such as contribution links and feeds to cable head-ends (which provide distribution on cable systems).

Thomson (France), a partner in the FLASH-TV and HIVITS projects, and leading the design of the HDTV codec, were convinced by the early 1990s that a DCT-based algorithm with motion compensation was the technology to master to develop a transmission product range. The key reasons were the video performance at less than 10 Mbps and the implementation techniques developed within the RACE projects with the integrated circuit technology available at the time, which allowed the development of professional quality products. From a compression algorithm point of view, transmission of an HDTV signal at 34 Mbps is equivalent to the transmission of an SDTV signal at 8 Mbps (as it has roughly the same compression ratio). Thomson developed in parallel the 34 Mbps HDTV codec for FLASH-TV, and the SDTV 8 Mbps product that pioneered DSNG in Europe (see Figure 4.10). This used an improved algorithm based on motion-compensated inter-field processing, and made it possible to offer 8 Mbps, 2 × 8 Mbps and 17 Mbps proprietary modes. The principal feature of the 8 Mbps adaptation of the ETSI standard was the introduction of this motion-compensated field prediction (the ETSI 34 Mbps standard had only a field prediction without motion compensation), with a GOP sequence of I,P,P,P,P with an associated 'rate control' algorithm. This was not part of the ETSI standard because rate control was only part of the encoder decision process and thus made the system proprietary.

During this period in the US there was little interest in DSNG. The early development of DSNG was pioneered in Europe, with a number of broadcasters and service providers adopting the Thomson 8 Mbps ETSI units for use with the new small Ku-band (and subsequently C-band) DSNG systems. It was not until after 1997, when MPEG-2 had been developed to the point that it became usable in DSNG applications, that ETSI 8 Mbps usage began to decrease in favour of MPEG-2 – though it was still in use in many parts of Europe in 1999.

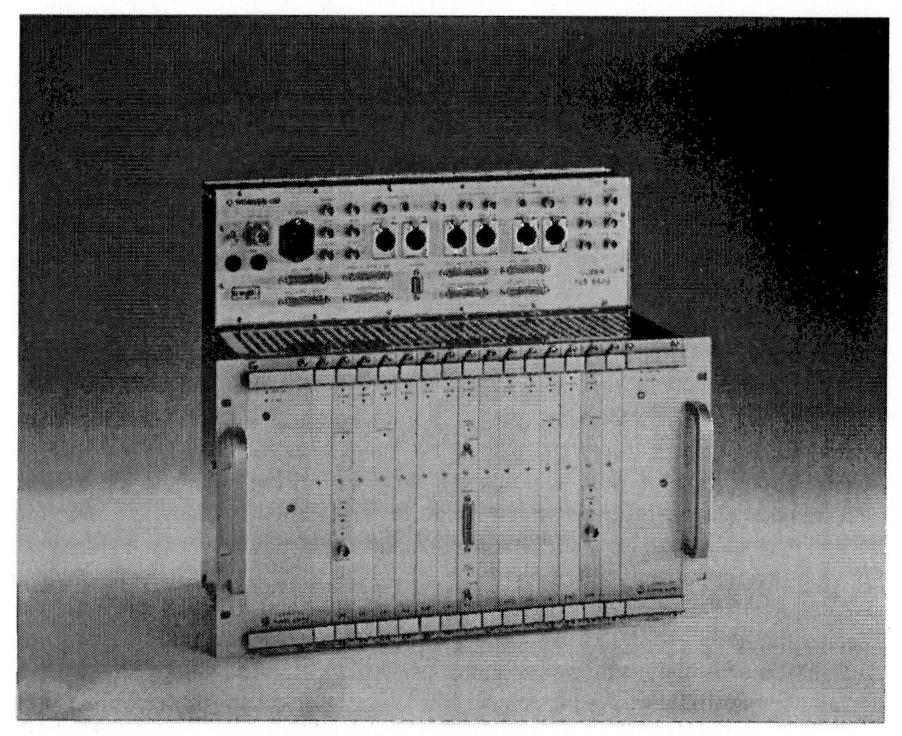

Figure 4.10 Thomson ETSI 8/17 Mbps encoder *(Photo courtesy of Thomson Broadcast Systems)*

4.5.4 MPEG

Similarly to JPEG, the acronym MPEG stands for Motion Picture Experts Group, a sister organization to JPEG established in 1988, again under the auspices of ISO and the IEC, to develop digital compression standards for video and audio material. Its work is not exclusively within the field of broadcasting – its remit is much wider than that, and it is currently working on standards that can be used with the Internet.

MPEG standards, and particularly MPEG-2, are truly global standards, used in a very wide variety of applications from computers to HDTV, with a range of bit-rates from 2 Mbps to 80 Mbps. MPEG-2 has an architecture to support this range, as we shall see shortly.

The MPEG standards define the way the audio and video are compressed and then interleaved for playback – and this is the key point. MPEG standards do not define how the signal should be encoded, but how the signal must be decoded. This is a subtle shift in the normal process of standardization, where the defining standard normally sets the originating process, which then dictates how a signal is subsequently dealt with.

MPEG is backward-compatible, i.e. a valid MPEG-1 bit-stream is also a valid MPEG-2 bit-stream. Particular characteristics of all MPEG bit-streams are that they are variable bit-rate and randomly accessible (this second factor is particularly important in post-production, as an MPEG bit-stream can begin playing at any point). All MPEG specifications are split into four parts: systems, video, audio and conformance testing. We will only focus only on video, which is Part 2 of all MPEG standards.

4.5.5 MPEG-1

The first project for MPEG was the development of a compression standard that could be used with new computer storage media such as CD-ROM and interactive CD (CD-i) formats, and was published in 1993 as the MPEG-1 standard[8]. This allowed video and audio compression coding up to around 1.5 Mbps, with a video data rate of approximately 1.3 Mbps and audio typically at less than 0.3 Mbps; and in terms of overall quality it approximates to pictures on the VHS tape format.

MPEG-1 was originally optimized to work at video resolutions of 352 × 240 pixels at 30 fps (NTSC) and 352 × 288 pixels at 25 fps (PAL). This is referred to as 'source input format' (SIF) and is comparable to the CIF standard used in video-conferencing (see Chapter 5). It has no direct provision for interlaced video applications, such as those used in broadcast television, and its use with interlaced television signals is contingent on a certain amount of pre-processing.

The MPEG-1 standard defines what constitutes a valid bit-stream, and in addition defines requirements for a 'constrained parameters bit-stream' (CPB). The CPB is a subset of MPEG-1 valid bit-streams that conform to other requirements, e.g. 1.5 Mbps bit-rate. Hence, reference to the use of MPEG-1 usually implies CPB. MPEG-1 divides every frame up into individually coded 8 × 8 pixel blocks, and then 16 × 16 pixel macroblocks. Hence, each macroblock consists of four blocks, and spectral redundancy is reduced by using 4:2:0 sampling.

As a computer standard, MPEG-1 suffered slow acceptance because the personal computers that were available at the time it was introduced were not fast enough to decode the 25 or 30 frames per second display rate. Now, processors for PCs are fast enough to cope with this demand, and MPEG-1 is used for a multitude of applications, from Internet video streaming to use in satellite newsgathering via INMARSAT (see Chapter 5).

4.5.6 MPEG-2

The MPEG-2 standard was published in 1995 by ISO/IEC[9], intended for digital television broadcasting, and is the dominant compression standard for DSNG. The ITU subsequently adopted the standard[10], and MPEG-2 equipment began to be developed and manufactured in 1995. MPEG-2 deals with issues not covered by MPEG-1 – in particular the efficient coding of interlaced video and the principle of 'scalability'. As already mentioned, MPEG-2 has an architecture to support a wide range of picture quality and data rates.

Table 4.2 MPEG-2 levels: picture size, frame rate and bit-rate

Level	*Max. frame width (pixels)*	*Max. frame height (lines)*	*Max. frame rate (Hz)*	*Max. bit-rate (Mbps)*
Low	352	288	30	4
Main	720	576	30	15
High-1440	1440	1152	60	60
High	1920	1152	60	80

Unlike MPEG-1, which only has non-interlaced frames, MPEG-2 can deal with interlaced video frames. As we discussed earlier in the chapter, interlacing the frames has the effect of splitting each frame into two fields, such that half the picture is displayed at a time. Thus, 25/30 frames per second is displayed as 50/60 fields per second. MPEG-2 processes fields and performs all the functions on each field rather than each frame as in MPEG-1.

4.5.7 Profiles and levels

MPEG-2 consists of many different types of service, classified under 'profiles' and 'levels', which allow 'scaling' of the video. The profile defines the data complexity and the chrominance resolution (or bit-stream scalability), while the level defines the image resolution and the maximum bit-rate per profile: this is illustrated in Table 4.2, which shows the profiles and levels for both HDTV and standard definition (SD) TV.

Simple profile (SP)

The simple profile uses no B-frames, so there is only forward prediction, and consequently there is no re-ordering required for transmission. This profile is suitable for low-delay applications such as video-conferencing, giving a coding delay of less than 150 ms, and where coding is only performed on a 4:2:0 video signal. It is not commonly used in DSNG applications.

Main profile (MP)

The main profile uses B-frames, and is the most widely used profile in both production operations as well as DSNG. Although by using B-frames the picture quality is increased, coding delay is added to allow for the frame re-ordering. Main profile decoders will also decode MPEG-1 video, and most MPEG-2 decoders support MP@ML.

MP@ML refers to 720 × 486 pixel resolution video at 30 fps (NTSC) and 720 × 576 pixel resolution video at 25 fps (PAL), both at up to 15 Mbps. These pixel sample specifications are different from the ITU-601 standard shown in Appendix C because the pixel spacing in ITU-601 (a digital television standard only) is different between the vertical and horizontal planes, resulting in 'rectangular'

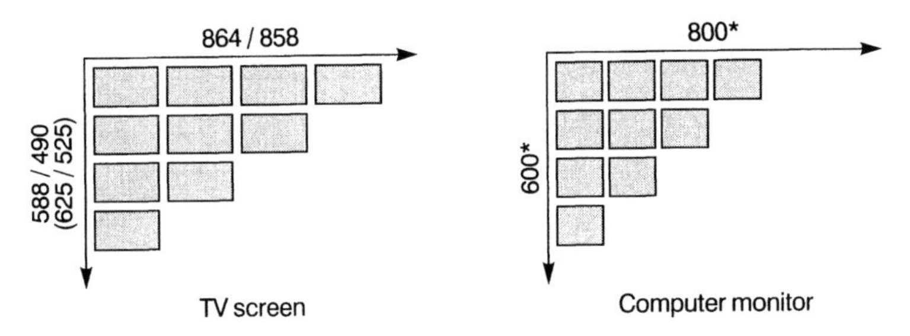

Figure 4.11 Pixel layout difference

pixels. In computer graphics, the practice has always been to create a symmetrical relationship and therefore equal resolution in both horizontal and vertical planes, giving 'square pixels'. Hence the sample rates in MPEG (which was originally aimed at the computer technologies) have been adjusted from the ITU-601 specification to give the same resolution in both planes. The difference between TV and computer display pixel structures is shown in Figure 4.11.

The coding can either be 4:2:0 MP@ML or 4:2:2 (the 4:2:2 profile is sometimes referred to as 422P@ML, or 4:2:2P), and 4:4:4 for high-quality applications (not DSNG). The MPEG-2 macroblock structure used in 4:2:0, 4:2:2 and 4:4:4 is shown in Figure 4.12.

MPEG-2 coding can reduce the 168/166 Mbps uncompressed digital video data rates by a factor of up to around 25:1, using the techniques described, down to about 3–15 Mbps. Some of the current generation of DSNG encoders offer both 4:2:0 and 4:2:2 sampling. Originally it had been assumed that 4:2:2 sampling always gave inferior results to 4:2:0 for broadcast applications at bit-rates below 10–15 Mbps, particularly after evaluation by the EBU and CBC (Canadian Broadcasting Corporation)[11,12]. However, recent studies[13] have indicated that 4:2:2 may always produce better pictures, certainly down to 4 Mbps for 625-line systems and 2 Mbps for 525-line systems. At much lower bit-rates, the impairments introduced by the MPEG-2 coding and decoding process become increasingly objectionable, and it is generally accepted that MPEG-2 is unusable below 2 Mbps.

4.5.8 DVB and MPEG-2

In Chapter 3, we discussed the option of 'multiplexing' two programme paths (or 'streams') in the digital domain by cascading two digital compression encoders together to provide a single digital data stream to the uplink chain. The MPEG standard defines the method of multiplexing and programme-specific information (PSI) tables which enable the individual 'packetized elementary streams' (PES) to be transmitted and decoded correctly. The DVB (digital video broadcasting) standard has enabled this to be implemented in a transmission process for the

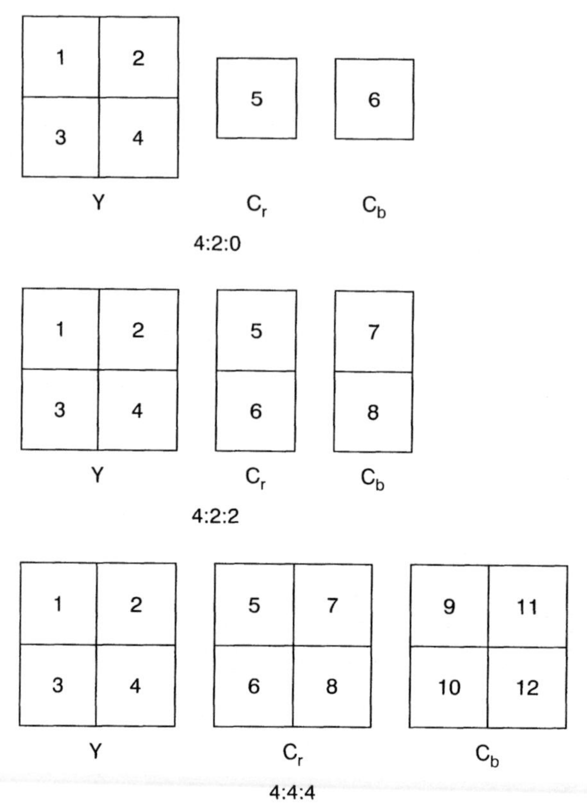

Figure 4.12 Colour sampling: macroblock structure

consumer that can be used by manufacturers to produce equipment that will inter-operate between manufacturers.

The DVB Project was officially inaugurated in 1993, born out of the European Launching Group for Digital Video Broadcasting initiative. The Project consists of a voluntary group of more than 200 organizations that have joined forces to develop standards for DVB. The MPEG-2 standard has been adopted as the DVB compression standard for broadcast distribution and transmission, and part of that standard covers the combining of programme streams.

The European DVB Project covers a range of transmission methods, including terrestrial cable and satellite. It has derived a number of standards that have been endorsed by both ETSI and ITU. The first specification was for the satellite delivery of DVB signals, entitled DVB-S[14]. This defined, for the first time, different tools for channel coding (see Chapter 2) which later on became important for all other transmission methods as well. Channel coding is used to describe the processes used for adaptation of the source signal to the transmission medium, and principally covers the compression, error correction and modulation processes.

The DVB-S specification enabled the start of DTH services via satellite in many parts of the world. Under this specification, a data rate of typically 38 Mbps can be accommodated within a satellite transponder channel bandwidth of 33 MHz. A typical satellite uses eighteen such transponders and thus may deliver 684 Mbps to a small (60 cm diameter or less) satellite antenna for DTH services. As a DTH standard, a key feature of DVB is programme security, so that only those who have paid for a service may receive it. As a result the specification has provision for conditional access (CA) using a 'common scrambling algorithm' (CSA). This is a powerful tool to make secure scrambling of multiplexed programme channels ('transport streams', or TSs) or individual programme channels ('packetized elementary streams' or PESs) possible. Scrambling is used on some DSNG feeds where there is a fear that the material may be 'hijacked' by the unscrupulous, or where the material contained in the transmission has a high degree of exclusivity.

MPEG-2 transmissions are either transmitted as single channel per carrier (SCPC) or multiple channel per carrier (MCPC) feeds. However, at an individual programme channel level, both techniques use the same method for building a data stream containing the video, audio and timing information. In SCPC, the combination of compressed video and audio forms a single transport stream and includes timing information to allow the audio and video to be synchronized together before modulation to DVB standard is carried out (Figure 4.13).

Although not as likely a problem as in the analogue domain, it is considered desirable to add an energy dispersal signal to the transport stream (based on a pseudo-random 'noise' signal) to distribute any constant peaks of energy. In MCPC, multiple PESs are multiplexed together into a larger stream and the service information (SI) stream gets added, resulting in the final MPEG-2/DVB multiplex transport stream that is uplinked to the transponder on the satellite.

As many of the MPEG-2 encoders used for DSNG are derivatives of those used in programme distribution, the ability to multiplex streams together is either an integrated part of the encoder or, more unusually in a DSNG system, an external multiplexer can be used. The streams are combined at the uplink, and the downlink is able to de-multiplex the transport stream back into separate streams using the Packet IDentifier (PID), which is the channel identifier containing all the navigation information required to identify and reconstruct a programme stream. PID values

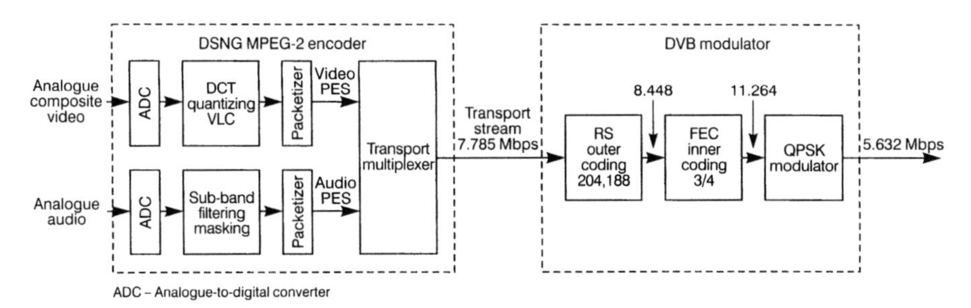

Figure 4.13 MPEG-2 encoder and DVB modulator (8.448 Mbps signal)

are contained in the PSI tables encoded as part of the stream, and these values (typically three- or four-digit numbers) are entered into the MPEG decoder. (This is the same process as using a DVB receiver/decoder at home.)

4.5.9 Interoperability

An important aspect of MPEG-2 and DVB is 'interoperability'. One of the aims of the DVB standardization process is to enable any signal processed by an encoder which is DVB-compliant to be decoded by a DVB-compliant decoder made by a manufacturer different to that of the encoder. There are so many variables in the process – symbol rate, video coding rate, video resolution and FEC rate – that this is not easy to achieve. We need to remember that although the MPEG-2 standard has certain fixed parameters, there is wide scope for manufacturers to use different algorithms to improve the quality of the video.

In DSNG, there have been several studies carried out to establish the level of interoperability between decoders manufactured for use in DSNG applications. These studies have principally been carried out by INTELSAT[15], at the behest of ISOG (Inter-Union Satellite Operations Group). Other studies have also been carried out by the EBU[16].

These interoperability tests have resulted in the ISOG 'standard mode' for DSNG operations:

Video encoding rate:	7.5 Mbps
Coding standard:	MPEG-2 MP@ML (4:2:0 sampling)
Audio data rate (stereo):	256 kbps
Audio sampling rate:	48 kHz
Transport stream (excluding RS encoding):	8.448 Mbps
Reed-Solomon code rate:	204/188
Inner FEC code rate:	3/4
Transmitted symbol rate:	6.1113 Msps

There is a slight variation to this, where the transport stream rate of 8.448 Msps is quoted including RS encoding; in which case, the resultant transmitted symbol rate is 5.632 Msps. It has been decided that in order to enable automated interoperability and the setting of the receiving parameters, this information should be signalled in the SI data stream of the transport stream. A specification for DVB signals used for DSNG has also been issued by ETSI[17].

A wide range of different manufacturers' encoders and decoders have been tested, both in the laboratory and via satellite, in several of rounds of tests, and the point has been reached where almost all the coders and decoders were fully interoperable within the test parameters. At the time of writing, the investigations by INTELSAT's labs are ongoing, with the aim of testing with 4:2:2 MP@ML sampling and possibly at other (lower) bit-rates.

4.5.10 Artifacts

When pushed to the limit, compression codec processes can produce visual defects, termed 'artifacts'. These include a 'pixellation' effect commonly referred to as 'blockiness', momentarily showing as rectangular areas of picture with distinct boundaries. The visible blocks may be 8 × 8 DCT blocks or misplaced macroblocks, perhaps due to the failure of motion prediction or because of transmission path problems. Other artifacts may be seen near the edges of objects, such as 'blurring' caused by reducing horizontal and/or vertical resolution. Another artifact is 'mosquito noise', where quantizing step changes between scene changes cause quantizing errors, seen as a fine patterning of black dots in certain parts of the picture, usually around the edges of objects in a scene. However, one of the most commonly seen artifacts in a compressed video signal is quantization error, often obvious on finely gradated areas of colour, where there are seen to be distinct transitions from one shade to another in the same hue. This is caused by the limitations of the quantization of the chrominance signal.

H.261 will be the most sensitive to any of these problems, followed by H.263 and then MPEG, although the artifacts are rarely seen in normal MPEG formats at typical bit-rates (apart from quantization error) unless there is a major disruption to the data flow in the transmission path.

4.5.11 Delay and latency

The interaction in a live 'two-way' interview requires that both the questions and answers are delivered as smoothly as possible, but the latency of the compression process added to the fixed satellite delay means that these interviews often have an awkward hesitancy about them. This can be masked to a degree by imaginative techniques used both in the studio and out in the field, but there is always the evident hesitation between interviewer and interviewee.

These compression delays have reduced as the computational advances in processing speed have increased, and combined with the viewer's growing acceptance of these delays to the point where they are hardly noticed, the problems are diminishing as time passes. Some coders also offer a facility to improve the latency at the expense of movement compensation – the so-called 'interview' mode. This is selected via the front-panel menu on the MPEG-2 coder, and effectively reduces the GOP sequence to eliminate B-frames and change the P-frame order. This then reduces the overall processing time of the signal. If the coder abandons the DVB specification, it can reduce the latency to below 200 ms but at the expense of interoperability, so that the system then becomes proprietary in this mode, requiring the same manufacturer's decoder at the downlink.

4.5.12 Choosing an MPEG-2 encoder for DSNG

The choice as to which manufacturer's encoder is best suited for DSNG is becomingly increasingly difficult, as the differences between the encoders for this type of application are diminishing. The choice will essentially centre around the following key points (and in no particular order of significance).

Cost

The price of MPEG-2 products has dropped dramatically recently, and this is particularly important when considering the purchase of an encoder. As mentioned previously, MPEG processing is asymmetric and this is reflected in the relative pricing of the encoders and decoders. At the time of writing, the list price of many encoders was dropping to well below US$50 000, and decoders cost around $\frac{1}{10}$ of the cost of an encoder.

Picture quality

This essentially comes down to a choice between whether the encoder offers 4:2:2 or 4:2:0 sampling, and which 'chip-set' the encoder uses. A number of models now offer both 4:2:2 and 4:2:0 sampling, which can be selected via the front panel, and from the results of the EBU/CBC tests, there seems little doubt that although 4:2:0 has been the standard for DSNG, there is a transition to 4:2:2 underway.

There are two principal manufacturers of MPEG-2 chip-sets that are widely used (although not exclusively): C-Cube Microsystems Inc. and IBM. Some manufacturers offer encoders with either one of these chip-sets, or their own proprietary chip-set. The choice between which is best is hotly contested, and the way the motion vector search function is implemented is often the most significant differentiating factor; hence it is for the end-user to make up their own mind.

Latency

Particularly important for news operations with their high proportion of 'live' interactive interviewing is the degree of latency introduced by the encoder. This was a significant issue in the past, where there was a large disparity between some manufacturers' products, but the gap is closing. Bearing in mind the inherent satellite delay, it is important that the encoder has an 'interview' or low-latency mode, which although not suitable for fast-moving sports action, sacrifices the amount of motion processing to achieve low delay for the typical news 'talking-head' scenario. Interview modes typically reduce the latency to under 250 ms.

Interoperability

It is essential for most DSNG system operators to have equipment that offers the maximum degree of interoperability, as the type of decoder used at the downlink is often not known. Therefore, the end-user should carefully view independent tests, such as those carried out by the EBU and INTELSAT.

Ease of use

A very important factor that distinguishes the suitability of an MPEG-2 encoder for use in the DSNG field is how easy it is to operate. All encoders have an LCD screen on the front panel, with a control menu structure that is accessed by software-defined buttons at the bottom or side of the display. MPEG-2 encoders that have not

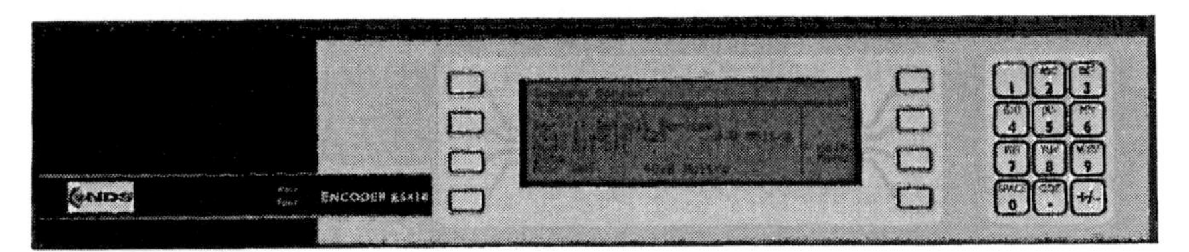

Figure 4.14 Front panel of DVB MPEG-2 professional encoder *(Photo courtesy of Tandberg Television)*

been particularly well adapted from a fixed-distribution application will not offer easy access through the front panel menu structure to those controls which the DSNG operator will need frequent and rapid access to. For instance, the following controls should be at or near the top level of the front panel menu (see Figure 4.14):

- Video bit-rate and symbol rate.
- Horizontal and vertical resolution.
- Delay mode (normal/'interview' low-latency mode).

It should be noted that the measurement and interpretation of bit-rates and symbol rates is complex, usually necessitating the use of 'look-up' tables, and the encoder display needs to be viewed with caution when displaying the transmission parameters.

Some DSNG MPEG-2 encoders are now available in compact sizes which offer the modulator installed inside to form a single unit. Again, some essential controls should be available at or near the top level of the front panel menu:

- Carrier on/off
- Level control
- FEC selection/Modulation mode.

Integrated multiplexer

An increasingly important consideration for DSNG operators is the ability to uplink more than one programme signal at a time. As has been previously discussed, one of the two methods for achieving this is in the digital domain by multiplexing MPEG-2 signals together into a single transport stream. An encoder that offers this facility integrated within the unit, so two or more MPEG-2 encoders can be 'cascaded', is plainly going to be more attractive to the end-user where size and weight of equipment is a constant issue.

These points are given as an outline guide – the end-user may very well have other essential issues that they regard as important.

Having examined the issues of video compression, we now need to look at how bandwidth is saved in the audio domain. All bar one of the standards so far described also have an audio specification (the exception is JPEG/Motion JPEG), but we will be focusing on MPEG audio standards (which is Part 3 of all MPEG standards).

4.6 Audio compression

A number of digital audio compression systems aimed at dealing with both speech and music signals are 'perceptual' audio coders, rather than so-called 'waveform coders'. In a perceptual audio coder, the codec does not attempt to retain the input signal exactly after encoding and decoding; rather its goal is to ensure that the output signal sounds the same to a human listener. The primary psycho-acoustic effect that the perceptual audio coder uses is called 'auditory masking', where parts of a signal are not audible due to the function of the human auditory system. The parts of the signal that are masked are commonly called 'irrelevant', as opposed to parts of the signal that are removed by a source coder (lossless or lossy), which are termed 'redundant'.

4.6.1 Psycho-acoustics

Psycho-acoustics plays a key role by exploiting this reduction of 'irrelevance', i.e. compressing parts of the signal that are imperceptible to the human ear. The human ear hears frequencies between 20 Hz and 20 kHz, and the human voice typically produces sounds in the range 400 Hz to 4 kHz, with the ear most sensitive in the range 2–4 kHz. The low frequencies are vowels and bass, while the high frequencies are consonants. The dynamic range (i.e. quietest to loudest sounds) of the human ear is about 96 dB (around four thousand million to one), with the threshold of pain at around 115 dBA and permanent damage at around 130 dBA. The dBA is a unit of measurement of sound pressure level, which is what the ear is sensitive to. It is therefore important to consider these frequency and dynamic ranges when determining the most suitable type of audio coding and compression to use for audio broadcast purposes.

4.6.2 Why compress?

It could be considered that audio does not need to be compressed as it uses relatively little bandwidth compared to video. CD audio, for instance, is not compressed, and has 44 100 samples per second (44.1 kHz sampling), with 16 bits per sample and two channels (stereo), which gives a data rate of 1.4 Mbps. It can also be argued (and is hotly by audiophiles!) that audio suffers when it is compressed. However, in MPEG-1 audio evaluation tests, expert listeners listened to audio in optimal listening conditions, with 16 bits per sample of stereo sampled at 48 kHz compressed by 6:1 to 256 kbps, and they could not distinguish between coded and original audio clips.

Therefore, in the context of transmission, it is both desirable and possible to compress the audio signal, with no discernible detrimental effect, to reduce the overall data rate of the programme signal. There are several techniques for the real-time digitization and compression of audio signals, some having been defined as international standards, while some remain proprietary systems (e.g. Dolby AC-3). We will only be considering the international standard systems.

4.6.3 Sampling, quantization and modulation

The processes of sampling and quantization used in audio compression are, in principle, the same as in video compression. The audio signal has to be converted from analogue to digital as the first stage of processing, and this is done in the same way as described earlier. Because of the higher 'resolution' required for audio due to the human psycho-acoustic characteristic, the audio is typically sampled with 16 bits per sample. In the psycho-acoustic model, the human ear appears to be more discerning than the human eye, which only requires a relatively limited dynamic range of dark to light. The typical sampling frequencies for audio are 32, 44.1 or 48 kHz (although there are a range of lower and higher sampling rates). After quantization the coded bit-stream has a range of rates typically from 32 to 384 kbps (although again this range is extended both lower and higher for certain applications).

There are a number of different quantization and modulation schemes, and firstly we will look at the waveform coding techniques before considering higher order perceptual coding.

PCM

Pulse code modulation (PCM) encoding is the simplest form of audio coding (digitization and compression), and is a waveform coding scheme. PCM encoding samples the amplitude of an audio signal at a given sampling frequency and quantizes these amplitudes against a discrete number of pre-determined levels. If these quantization levels are uniformly spaced then the encoding is said to be uniform PCM encoding.

However, if the quantization levels have been logarithmically spaced, thus enabling a larger range of values to be measured, then the coding is said to be μ-law or A-law, depending on the type of logarithmic transformation used. Because of the logarithmic nature of the transform, low-amplitude samples are encoded with greater accuracy than high-amplitude samples. The μ-law and A-law PCM encoding methods are formally specified in the ITU-T G.711 standard (see Chapter 5).

H.261/263

PCM encoding methods encode each audio sample independently from adjacent samples. However, usually adjacent samples are similar to each other and the value of a sample can be predicted with some accuracy using the value of adjacent samples. The adaptive differential pulse code modulation (ADPCM) waveform coding method computes the difference between each sample and its predicted value and encodes the difference (hence the term differential). Fewer bits are needed to encode the difference than the complete sample value and compression rates of typically 4:1 can be achieved.

The ITU-T G.722 standard is one of the most common ADPCM encoding methods used over both ISDN and satellite circuits. A coding delay of about 6ms makes it very suitable for live interviews and voice feeds over a satellite circuit. However, the audio bandwidth of 7.5 kHz is generally considered too narrow for music and so its use is limited to voice applications (see Chapter 5).

4.6.4 MPEG and perceptual coding

In order to achieve higher compression ratios and transmit a wider bandwidth of audio, MPEG audio compression uses the psycho-acoustic principles we outlined earlier, and in particular a technique known as 'masking'.

Imagine a musical ensemble comprising several different instruments and playing all at the same time. The human ear is not capable of hearing all of the components of the sound because some of the quieter sounds are hidden or masked by the louder sounds. If a recording was made of the music and the parts that we could not hear were removed, we would still hear the same sound but we would have recorded much less data. This is exactly the way in which MPEG audio compression works, by removing the parts of the sound that we could not hear in any case.

Imagine a strong tone with a frequency of 1000 Hz and a second tone nearby of 1100 Hz and 18 dB lower (a lot quieter) than the first. The human ear will not be able to hear the second tone because it is completely masked by the first 1000 Hz tone. If a third tone of 2000 Hz, also 18 dB less than the first, is introduced this will be heard because, the further away from a sound, the less the masking effect. In effect we can discard the second tone without affecting the sound heard by the human ear, thus reducing the amount of information that would need to be coded and recorded.

In MPEG processing, the frequency spectrum is divided into 32 sub-bands and calculates the masking effect of each sub-band on adjacent sub-bands. The MPEG coding also takes into account the sensitivity of the human ear to different frequencies. The ear is more sensitive to low frequencies than high frequencies and is most sensitive to frequencies in the range 2–4 kHz – the same range as the human voice. It would then make sense to have narrower sub-bands, thus giving more precision, in the lower frequencies and wider sub-bands in the higher frequency range.

This psycho-acoustic model analyses the input signals within consecutive time blocks and determines for each block the frequency components of the input audio signal. Then it models the masking properties of the human auditory system and estimates the just noticeable noise level, sometimes called the 'threshold of masking'.

MPEG uses a set of 'bandpass' filters to confine quantization noise within each sub-band and dynamically allocates bits to each sub-band. If sub-band energy falls below the psycho-acoustic masking threshold, then that part of the signal receives zero bits; otherwise it is allocated just enough bits to keep the quantization noise in the sub-band just below the psycho-acoustic audible threshold. Masking not only occurs at different frequencies (frequency masking) but also before and after a loud sound (temporal masking). Pre-masking is only effective 2–5 ms before an event but post-masking can have an effect up to 100 ms after a sound event.

The MPEG decoder is much less complex, because it does not require a psycho-acoustic model: its only task is to reconstruct an audio signal from the coded spectral components.

4.6.5 MPEG standards

Both in MPEG-1 and in MPEG-2, three different layers are defined, each representing a group of coding algorithms; the layers are denoted Layer I, Layer II and Layer III. The different layers have been defined to give, as with MPEG video, a degree of scalability to cope with the wide range of applications that can make use of the standards. The complexity of the encoder and decoder, the encoder/decoder delay and the coding efficiency increase when going from Layer I via Layer II to Layer III. This is summarized as follows.

Layer I

Layer I has the lowest complexity and is specifically suitable for applications where the encoder complexity also plays an important role. The Layer I psycho-acoustic model uses only frequency masking, with 32 sub-bands each of 750 Hz. The audio data are divided into frames, each containing 384 samples, with 12 samples from each of the 32 filtered sub-bands. There is no temporal masking, i.e. taking into account the pattern of loud and quiet signals. It is a simple implementation, giving the lowest compression ratio (1:4), and is used in consumer audio systems. Its main advantage lies in the low cost of implementation.

Layer II

In this layer, the psycho-acoustic model uses a more complex frequency characteristic. Layer II requires a more complex encoder and a slightly more complex decoder, and is directed towards 'one to many' applications, i.e. one encoder serves many decoders. Compared to Layer I, Layer II is able to remove more of the signal redundancy and to apply the psycho-acoustic threshold more efficiently. Again there are 32 sub-bands, each of 750 Hz; but in addition, there are three temporal frames ('before', 'current' and 'next'), giving some temporal masking. Layer II gives greater compression (1:6 to 1:8) and is found in numerous consumer and professional applications. It is common for Layer II to be used for DSNG applications.

Layer III

This is the most complex of the MPEG audio layers and here the psycho-acoustic model uses improved band filtering to model human ear sensitivities (non-equal frequencies) and temporal masking effects; it also takes into account stereo redundancy and uses variable length coding. Layer III is directed towards lower bit-rate applications due to the additional redundancy and irrelevancy extraction from enhanced frequency resolution in its filtering. It gives the greatest level of compression (1:10 to 1:12) at the expense of complicated encoding and decoding, which increases the processing delay.

MPEG-1 audio coding provides both single-channel (mono) and two-channel (stereo or 'dual-mono') coding at 32, 44.1 and 48 kHz sampling rates. The pre-defined bit-rates range from 32 to 448 kbps for Layer I, from 32 to 384 kbps for Layer II, and from 32 to 320 kbps for Layer III.

MPEG-2 audio coding provides extension to lower sampling frequencies, providing better sound quality at very low bit-rates. It extends to the lower sampling rates of 16, 22.05 and 24 kHz, for bit-rates from 32 to 256 kbps (Layer I) and from 8 to 160 kbps (Layer II and Layer III). It also provides for multi-channel sound. MPEG-2 audio supports up to five full bandwidth channels plus one low-frequency enhancement channel (such an ensemble of channels is referred to as '5.1'). This multi-channel extension is both forward- and backward-compatible with MPEG-1. Typical MPEG-2 transmissions use 48 kHz sampling, giving a frequency range of 40 Hz to 22 kHz.

4.6.6 MUSICAM and DAB

The ISO/IEC MPEG established in 1988 considered digital compression standards for audio as well as video material. Meanwhile, under the auspices of a technical body called Eureka – an international consortium of broadcasters, network operators, consumer electronic industries and research institutes – the digital audio broadcasting (DAB) standard was being developed to replace conventional broadcast radio technology (analogue AM and FM). DAB is the most fundamental advance in radio technology since the introduction of FM stereo radio, as it gives listeners interference-free reception and high-quality sound, and the potential for enhanced text and information services.

This project was called Eureka 147, and the one of the results of the project was a digital audio compression standard called MUSICAM (Masking pattern adapted Universal Sub-band Integrated Coding and Multiplexing). Meanwhile, another international group had devised a coding system called ASPEC (Adaptive Spectral Perceptual Entropy Coding). The International Standards Organisation (ISO) held a competition to select a world standard and MUSICAM, designed and manufactured by CCS (US), was judged to be the best overall system. Both the MUSICAM and ASPEC systems were submitted to ISO/MPEG, and ISO/MPEG subsequently took the best features of both systems to form the basis for MPEG audio Layer I and II. Since the publication of the MPEG-1 audio standard, the original MUSICAM algorithm is not used anymore, although MPEG Layer II is often generically (but not strictly correctly) referred to as MUSICAM.

4.7 Conclusion

The impact and significance of digital compression technology should now be clear, and as was said at the beginning of the chapter, for the reader interested in the technical detail of these systems, there are a selection of papers and books available on the subject. Video compression is becomingly increasingly common in broadcasting generally, and a very large proportion of SNG operations are now in the digital domain.

There are drawbacks to using digital compression, though these are of lesser importance in the realm of satellite newsgathering. The effect of compression on picture content varies widely depending on the content, and there is still no widely accepted scientific method of measuring picture quality – the human eye has been found to be the best guide, but it is a subjective evaluation. However, suffice to say that for most news organizations, the advantages of the cost of utilizing compression far outweigh what are considered esoteric concerns over issues of quality.

However, a secondary issue in video compression which is worth mentioning is the outcome of 'concatenation' through differing MPEG-2 systems. In a programme path in a broadcasting environment, from DSNG uplink to distribution to home, that is increasingly in the digital domain, there is concern about the effects of a programme signal being passed through a number of MPEG-2 processes. The signal can be coded, decoded and recoded a number of times, and each time a different DCT, quantization and GOP structure will be applied to the signal. This could potentially result in the finally delivered programme signal being turned into a digital 'soup', full of artifacts and lacking resolution and adequate motion processing.

The EU funded a project called ATLANTIC (Advanced Television at Low bit-rates And Networked Transmission over Integrated Communication systems), under the ACTS (Advanced Communications Technologies and Services) group. ATLANTIC was to sponsor research and development in technology which would enable the signal to be kept in compressed MPEG-2 format throughout the complete broadcast chain. The project was completed in 1998, and from the work carried out primarily by the BBC and Snell & Wilcox, a UK broadcast equipment manufacturer, there is now technology available on the market called MOLE™ , which will carry the original coding profile as an extra data signal buried invisibly in the chrominance component of the transport stream. Subsequently, MPEG-2 decoders and encoders that have the MOLE™ facility can then use the same coding profile on subsequent coding/decoding processes. Other manufacturers such as Sony and Thomson have adopted complementary approaches, and it is hoped that a SMPTE standard will be ratified soon.

However, although this is a concern, it has to be said that for the average DSNG operator it is of minimal interest – particularly as the solution will result in increased cost of equipment. There is also no guarantee so far as the DSNG operator is concerned that equipment that can make use of the additional coding information will be present further downstream in the programme chain.

Although the refinement and improvement of MPEG-2 equipment continues, and it has become the workhorse standard in the digital broadcast industry, the work of MPEG is not finished. MPEG-1 was designed for CD-ROMs and MPEG-2 was designed for digital broadcasting. MPEG-3 was originally meant for HDTV, until it was found that MPEG-2 covered that requirement. MPEG-4 is the next level of MPEG due to emerge, to be followed by MPEG-7. However, neither MPEG-4 nor MPEG-7 offer any application to newsgathering, so MPEG-2 as a standard is likely to remain the workhorse of the DSNG compression business for the foreseeable future.

Finally, it is worth briefly mentioning 'metadata'. Metadata is programme content description information that is carried as an auxiliary data channel within the MPEG-2 signal. The overall concept is that metadata can carry a whole host of information about the programme, including archive information, origin of material etc. – the level of information is almost boundless, and it is very much up to the end-user to define what information is carried with the signal. In the context of newsgathering, it has been suggested that metadata could be used to carry information relating to the material from the field which is often lost (who shot the pictures, where, when etc.). However, in the natural semi-chaos that often surrounds newsgathering in the field, it is not easy at the moment to see how this could practicably be implemented. It would be possible to transfer the time, date and location information automatically by recording a signal from a GPS (Global Positioning System) receiver, for instance – provided the GPS receiver was integrated into the camcorder. Means to achieve the insertion of metadata in the field will surely be found as the pressure to improve archival information increases. One of the great debates in the new digital broadcasting environment is how all the material that can be readily accessed in the medium of the video programme server can be indexed and found again – but that's another story!

References

1. ITU (1998) Recommendation ITU-R BT.601–5, Studio encoding parameters of digital television for standard 4:3 and wide-screen 16:9 aspect ratios.
2. ITU (1998) Recommendation ITU-R BT.656–4, Interfaces for digital component video signals in 525-line and 625-line television systems operating at the 4:2:2 level of Recommendation ITU-R BT.601 (Part A).
3. SMPTE 125M-1995, Component video signal 4:2:2 – bit-parallel digital interface.
4. SMPTE 259M-1993, 10-Bit 4:2:2 component and 4fsc NTSC component digital signals – serial digital interface.
5. ITU (1993) ITU-T Recommendation H.261 (3/93), Video codec for audiovisual services at p × 64 kbit/s.
6. ISO/IEC (1994) ISO/IEC Reference 10918, Digital compression and coding of continuous-tone still images.
7. ETSI (1992) ETS 300 174, Network aspects (NA): Digital coding of component television signals for contribution quality applications in the range 34–45 Mbps.

8. ISO/IEC (1993) ISO/IEC Reference 11172, Coding of moving pictures and associated audio for digital storage media at up to 1.5 Mbps.
9. ISO/IEC (1995) ISO/IEC Reference 13818, Generic coding of moving pictures and associated audio information.
10. ITU (1995) Recommendation ITU-T H.262 (7/95), Information technology – generic coding of moving pictures and associated audio information.
11. Caruso, A., Cheveau, L. and Flowers, B. (1998) EBU Technical Review 276, MPEG-2 4:2:2 profile – its use for contribution/collection and primary distribution.
12. Caruso, A., Cheveau, L. and Flowers, B. (1998) The use of MPEG-2 4:2:2 profile for contribution and primary distribution, IBC '98, Amsterdam.
13. Cheveau, L. and Caruso, A. (1999) EBU Technical Review 279, Comparison between 4:2:2P and 4:2:0 for 525- and 625-line pictures.
14. ETSI (1994) EN 300 421, Digital broadcasting systems for television, sound and data services – framing structure, channel coding and modulation for 11/12 GHz satellite services.
15. INTELSAT (1996/1998) INTELSAT/ISOG interoperability tests.
16. EBU (1997) Digital satellite news gathering using MPEG-2 MP@ML – EBU interoperability tests.
17. ETS (1999) EN 301 210 digital video broadcasting (DVB), Framing structure, channel coding and modulation for digital satellite news gathering (DSNG) and other contribution applications by satellite.

5

Have phone, will travel: INMARSAT services

5.1 Introduction

The INMARSAT system was originally designed for maritime communication, but its use has now expanded to other market sectors, including newsgathering. In recent years, the INMARSAT system has become a vital tool for global newsgathering, used not only for person-to-person voice communication in 'keeping in touch', but media material transfer applications as well. High-quality still pictures, studio-quality audio for radio reports, and even television pictures using 'store and forward' video units can be sent from almost anywhere on the globe. This versatility, combined with increasingly compact equipment, has enabled INMARSAT to become a significant force in newsgathering. It now plays its part in delivering news stories into people's homes around the world.

The INMARSAT system consists of four geostationary satellites that cover most of the Earth's surface (Figure 5.1) and have a range of services with different capabilities. However, all services offer the same common core feature: dial-up on-demand access, limited only by satellite capacity and coverage, using portable mobile terminals. In this chapter we will examine the application of INMARSAT systems to newsgathering, and in doing so we will cover some basic aspects of ISDN, video-conferencing and store and forward file transfer, as well as features of the different INMARSAT services. So to begin with, let's look at the story behind the establishment of the INMARSAT system.

5.2 History

In 1973, the Assembly of the Inter-Governmental Maritime Consultative Organisation (later to become the International Maritime Organisation, or IMO) voted to convene a special conference. This conference was to discuss the possibilities of a truly global satellite communications system and try to achieve international agreement for such a system. There was a growing need for a system to meet both

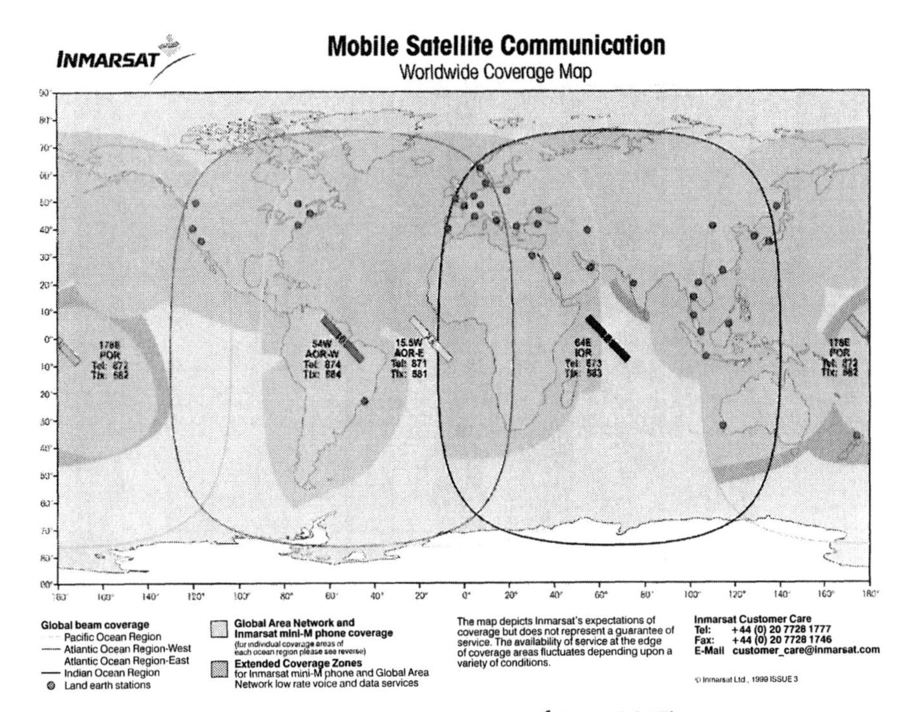

Figure 5.1 INMARSAT coverage map. *(Courtesy of INMARSAT)*

the demand for emergency distress communications and routine business communication between shipping lines and their vessels at sea. In 1975 the first of a number of conferences began the work which led to the INMARSAT Convention of 1976, and then to the establishment of International Maritime Satellite Organisation – INMARSAT – in 1979. The organization currently has 86 member countries, with its headquarters in London, and the INMARSAT system is used in more than 160 countries.

Each member country (the Party) appoints a Signatory, typically the principal public telecommunications operator (PTO) in the country, who invests in INMARSAT and provides INMARSAT services to the end-user (the customer.) In addition to the satellites, the INMARSAT system comprises land earth stations (LESs) which are operated by Signatories and which provide the link between the satellites and the terrestrial telecom networks. There are more than 150 000 mobile earth stations (MESs) in use, produced by a number of manufacturers around the world. INMARSAT sets the specifications for mobile earth stations and every manufacturer must meet INMARSAT's stringent type approval procedures before its equipment can access the satellite system.

However, the INMARSAT system did not actually commence operations until 1982. INMARSAT began providing the world's first mobile satellite communications system for the maritime market with the 'Inmarsat Standard-A system'. The INMARSAT satellite system originally consisted of two leased satellites from

Figure 5.2 INMARSAT 3 satellite. *(Photo courtesy of INMARSAT)*

Marisat, operated by Comsat General, two MARECS satellites leased from the European Space Agency (ESA), and some capacity leased on an INTELSAT V series satellite. INMARSAT now operates four INMARSAT 2 satellites and five of the INMARSAT 3 series (Figure 5.2).

Coverage is centred on the three major ocean regions by using satellites as shown in Table 5.1. Over time a number of additional satellites have been brought into service. There are currently nine satellites in use, with four in primary use (the others are in-orbit spares), providing rapidly accessible links from virtually anywhere in the world. LESs (originally known as coast earth stations, or CESs) were also established on all the continents to handle the traffic between the satellite and ground segment and connect to the international telephone network; there are now over forty around the world (see Figure 5.1).

The system grew quickly: in 1985 it was extended yet again to provide land mobile services, primarily for the truck and railway markets; by 1990 it had been further developed to provide aeronautical services. The system operates with C-band links between the satellite and the LESs, and L-band links between the mobile terminals and the satellites. The earlier satellites used global beams, each covering approximately one-third of the Earth's surface, but the current satellites use both global (INMARSAT 2 series) and 'spot' beams (INMARSAT 3). Spot beams concentrate coverage in particular areas, primarily intended for the smaller, lower

Table 5.1 INMARSAT satellite constellation

Satellite	*Launched*	*INMARSAT deployment*	*Orbital position*	*Coverage*	*Operator*	*Status**
MARISAT F1	1976	1982	15°W	AOR	Comsat	D/O**
MARISAT F2	1976	1982	72.5°E	IOR	Comsat	D/O
MARISAT F3	1976	1982	176.5°E	POR	Comsat	D/O
MARECS A	1981	1988	177.5°E	POR	ESA	D/O
MARECS B2	1984	1988	26°W	AOR	ESA	D/O
INTELSAT V MCS A	1988	1989	63°E	IOR	INTELSAT	D/O
INTELSAT V MCS B	1988	1989	18.5°W	AOR	INTELSAT	D/O
INTELSAT V MCS C	Lost on launch					
INTELSAT V MCS D	1988	1989	180°E	POR	INTELSAT	D/O
INMARSAT 2 F1	1990	1990	179°E	POR	INMARSAT	Spare
INMARSAT 2 F2	1991	1991	98°W	AOR-W	INMARSAT	Lease
INMARSAT 2 F3	1991	1991	65°E	IOR	INMARSAT	Spare
INMARSAT 2 F4	1992	1992	109°E		INMARSAT	Lease
INMARSAT 3 F1	1996	1996	64°E	IOR	INMARSAT	
INMARSAT 3 F2	1996	1996	15.5°W	AOR-E	INMARSAT	
INMARSAT 3 F3	1996	1996	178°E	POR	INMARSAT	
INMARSAT 3 F4	1996	1996	54°W	AOR-W	INMARSAT	
INMARSAT 3 F5	1996	1996	25°E	AOR-E	INMARSAT	Spare/ lease

*Most operational satellites also provide spare capacity.
**D/O = de-orbited.

power mobile terminals, and allow re-use of the scarce frequency resource allocated for INMARSAT.

As of January 1999, there were in excess of 24 000 Inmarsat-A, 8500 Inmarsat-B, 53 000 Inmarsat-C, and 39 000 Inmarsat Mini-M terminals in service.

5.2.1 Privatization of INMARSAT

In September 1998, the INMARSAT Assembly agreed by consensus that INMARSAT would be privatized, and INMARSAT became a private company in April 1999. The new structure comprises two entities: a private company that will seek an initial public offering (IPO) of shareholding by mid-2001 and an intergovernmental body to ensure that INMARSAT continues to meets its public service 'lifeline' obligations, in particular the Global Maritime Distress and Safety System (GMDSS). For the first couple of years, share trading will be largely restricted to existing shareholders, although there is provision for the introduction of strategic investors. INMARSAT are seeking up to US$500 million of investment to develop services.

INMARSAT plan that, with the transformation into a limited company, they will move forward with a streamlined decision-making process. This will provide the commercial flexibility to respond more rapidly to changing market needs in an increasingly competitive environment, and be less burdened than it was as an IGO with the bureaucratic overheads that inevitably resulted. (This mirrors what is happening with INTELSAT.) INMARSAT want the new structure to enable them to operate on an equal footing with their competitors, and to expand in the future, supported by a business that is strong and developing.

5.3 Types of INMARSAT systems

In the description of the different services below, we will concentrate on the use of the land mobile application. For convenience, we will use the term for the land mobile earth station (LMES) used by all newsgatherers: the ‘satellite telephone’ – often referred to as the ‘satphone’. Note that there are INMARSAT systems not covered here as they are not used for newsgathering.

5.3.1 Inmarsat-A

The original Inmarsat Type-A system is still widely used, despite the fact that its replacement – Inmarsat-B– has been operating for almost a decade (see Figure 5.3). Inmarsat-A uses analogue modulation, though it can pass digital signals via a modem as with any analogue public telephone circuit. It handles voice, telex, data and fax.

The data service is further available as high speed data (HSD), with data rates of either 56 kbps or 64 kbps, with add-on digital modulation/demodulation. It is available on Inmarsat-A satphones as an option either in one-way (simplex) or two-way modes (duplex). The advantages of simplex are that it is lower in cost per minute as well as the lower purchase cost of the satphone than the duplex mode models. It is advantageous where an HSD path is only necessary in one direction, usually from mobile to base. Some store and forward video units will operate on Inmarsat-A services – these units are described later.

An Inmarsat-A satphone is quite bulky, around the size of a large suitcase. It typically weighs around 35 kg (77 lb) and has an antenna that is approximately 90 cm in diameter. The antenna stows inside the case, and the handset can usually be operated remotely from the antenna typically up to 30 m (sometimes more depending on the particular model). This is achieved by either extending the antenna or the handset away from the base unit. Most manufacturers have now ceased making satphones for the Inmarsat-A system, concentrating on the Inmarsat-B system.

(a)

(b)

Figure 5.3 (a) Inmarsat-A satphone; (b) Inmarsat-A satphone in use. *(Photos courtesy of Nera Ltd)*

5.3.2 Inmarsat-B

This is the replacement service for Inmarsat-A, launched in 1992, and it uses digital modulation. It features all the services as for Inmarsat-A, though with full duplex HSD as a standard feature. Inmarsat-B also has the added advantage that a satphone can have multiple 'identities'. This means that there are separate numbers for voice, fax and data, and that there can be several for each type. Typically, a satphone can have two voice numbers, two fax numbers, a medium speed data (9.6 kbps) number and an HSD number, though only one service can be used at any one time.

The satphones are around half the size and weight of those for the type-A system, although the antenna are approximately the same size (see Figure 5.4). Prices have also fallen significantly compared to the original cost of an Inmarsat-A. The cost of satellite time per minute has also decreased.

Store and forward units generally only operate with type-B satphones, but it is also possible for two or more phones to be configured to provide multiples of 64 kbps for increased data capacity. This will give improved picture quality in 'live' mode, and faster transfer times for video files. However, this currently has to be done with extra equipment and configuration, and is only available as a custom package from specialist companies.

Satellite costs vary between US$2 and US$12 per minute, depending on time of day and whether the service being used is voice/fax, low-speed data (9.6 kbps) or HSD (56/64 kbps). Purchase prices for the satphones start from US$15 000.

5.3.3 Inmarsat-C

This is a messaging-only service introduced in 1986 for the maritime market. In the early 1990s it was expanded into the land mobile market with small briefcase terminals operating at a maximum data rate of 600 bps, and a maximum file size varying from 8 to 32 kB depending on the model and the size of antenna used. Its use for newsgathering is limited as it has no voice service – it is a 'keeping-in-touch' service, and newsgatherers are generally better served by the 'Mini-M' service. The units cost under US$2500 and satellite costs are around US$0.10 per bit, depending on time of day.

5.3.4 Inmarsat-M

This was introduced in 1993, just prior to the Inmarsat-B launch, and it was aimed at the business traveller market. An Inmarsat-M satphone is the size of an executive briefcase – and some companies actually built them into genuine briefcases (see Figure 5.5). It is therefore much smaller than an Inmarsat-B satphone. It is also a digital system, developed in parallel with the Inmarsat-B system, but has restricted data transmission capability. The maximum data rate is 4.8 kbps, and it is primarily intended to provide voice and limited data transfer, such as for e-mail or bulletin-board services. It can support fax, although still only at 2.4 kbps. It also has separate number identities for voice, fax and data; but yet again, only one of these can be used at any one time.

(a)

(b)

Figure 5.4 (a) Inmarsat-B satphone; (b) Inmarsat-B satphone in use. *(Photos courtesy of Nera Ltd)*

(a)

(b)

Figure 5.5 (a) Inmarsat-M satphone; (b) Inmarsat-M satphone in use. *(Photos courtesy of Nera Ltd)*

The cost is significantly lower than the -A or -B satphones, both in terms of cost of the phones and call charges, but these satphones have now largely been superseded by the 'Mini-M' satphones. The Inmarsat-M service is still available, but only pre-owned satphones are available for purchase and are available for under US$1500. Satellite costs typically vary between US$3 and US$5 per minute, depending on time of day.

5.3.5 Inmarsat Mini-M

This digital service was introduced in 1996 as the successor to Inmarsat-M. It is marketed by INMARSAT as 'Inmarsat-phone' (which has not been a successful brand name), and the satphones are approximately the size of a notebook computer (see Figure 5.6). The satphone has the option of using a SIM card so that service is bought and access-controlled in the same way as GSM. The available services are voice, data and fax, but with no HSD service. The cost of both the satphones and call charges have been falling – they currently cost under US$4000, and satellite costs vary between US$2.50 and US$4 per minute, depending on time of day.

From the point of view of keeping in touch in hazardous areas, the Inmarsat Mini-M satphone has a distinct advantage for newsgatherers. It is available in vehicle-mounted versions from a few manufacturers, and this allows a small and relatively discreet auto-tracking antenna to be easily fitted to the roof of a car, with the handset and control unit inside the car. This has a great advantage in that the satphone is permanently rigged, and calls can therefore be made or received on the move, rather than having to stop and get out of a vehicle to set the satphone up before use.

5.3.6 Inmarsat-M4

This new HSD service was introduced in late 1999 as a parallel service to Inmarsat-B (although in many ways it can be considered a successor). Although 'M4' (Multi-Media Mini-M) was the project title while the system was in development, it is marketed by INMARSAT as 'Global Area Network' (GAN) – but for the time being, the M4 tag appears to be sticking within the industry. The M4 satphones are the size of a laptop computer, weighing under 4 kg (10 lb), and offer voice services at 4.8 kbps (as with Inmarsat-M and Mini-M), and fax, file transfer and high-speed data at up to 56/64 kbps (as with Inmarsat-B) (see Figure 5.7). As with Inmarsat-B, there can only be one channel working at any time, so for instance it is still not possible to have a voice call and a fax call running simultaneously.

The pricing of the satphones is US$10 000 or less, and satellite costs are between US$7 and US$8 per minute. It has been tailored for use by the oil and gas industries, resource exploration and broadcast users, rather than for maritime users. M4 terminals offer the prospect of revolutionizing newsgathering by INMARSAT, as these units are very compact and promise to marry up particularly well with the new notebook store and forward units dealt with later in this chapter. The units also offer a standard ISDN (S0) interface, enabling standard ISDN applications and hardware

(a)

(b)

Figure 5.6 (a) Inmarsat Mini-M satphone; (b) Inmarsat Mini-M satphone in use. *(Photos courtesy of Nera Ltd)*

Figure 5.7 Inmarsat-M4 satphone in use. *(Photo courtesy of Nera Ltd)*

to be used with the HSD service. The satphone features a SIM card so that service is bought and access-controlled in the same way as GSM, and with the built-in lithium-ion battery, offers more than 100 hours of standby time. After the initial launch, INMARSAT plan to introduce a packet-data service. There are four manufacturers who will have M4 satphones on the market by mid-2000: Nera (Norway), Thrane & Thrane (Denmark), STN Atlas (Germany) and Ottercom (UK).

The coverage of Inmarsat-M4 is, like Mini-M, limited to continental landmasses only, but it is predicted that it may develop into oceanic coverage at some point (although, as with Mini-M, there is some coverage spillage into parts of the oceanic regions).

5.3.7 Inmarsat-D+

In 1999, INMARSAT introduced the Inmarsat-D+ service. Inmarsat-D+ is a global mobile messaging system that offers bi-directional store and forward short message data communications (it is an enhancement of Inmarsat-D, introduced in 1998, which offered only uni-directional messaging). It offers a range of options to the person originating the message, including four grades of call priority, distribution in more than one ocean region, delivery confirmation and time-of-delivery notification. The remote mobile terminal is a unit the size of a portable CD player with a compact antenna, priced at US$500–700, intended primarily for fitting onto a vehicle or boat. Its use for newsgathering is as a 'keeping-in-touch' facility, particularly useful in hostile environments and war zones.

The mobile unit can store and display up to forty messages of up to 128 characters each. Messages from the mobile remote location are generally sent in the form of macrocode (a short code which represents a longer message). This can comprise up to 64 bits of data, enough for GPS position reports, alarms, telemetry data, or application-defined macrocodes. In addition, each message is automatically identified with a terminal, and receives a date/time stamp when it passes through the LES, a few seconds after being sent.

The 'home' terminal would typically be a PC, which is connected to the service typically via the Internet, which communicates with the mobile remote unit with specifically designed application software. Messages to the mobile remote location can include simple commands, longer configuration instructions, polling of position, or status and text messages. Up to 2000 bits of information can be sent in a single to-terminal message, though it is more economic to minimize the amount of data using the macrocodes.

The system currently has limited availability, with service only being offered by BT in the UK and KPN Station 12 in the Netherlands, and is still in the process of being developed.

5.4 Impact of INMARSAT services on newsgathering

Only Inmarsat-A and -B satphones could be used for radio newsgathering contributions up until the mid-1990s (though not for television newsgathering), but their use was limited to a degree by the size of the units. The development of the briefcase-size Inmarsat-M in 1993, and then the even more compact Mini-M in 1996, meant that radio news reports could be filed from many places where the heavier and more bulky (and conspicuous) Inmarsat-A and -B terminals could not be used. In recognition of the fact that these smaller units were designed for a larger mass business-traveller market, INMARSAT actively sought wider approval for the importation and use of these systems in countries that had previously been resistant to such use. Gone was the need in many countries for special permissions for to use these satphones and pay high fees for importation licences. Partly these controls had been for political control, partly due to protectionism. The PTOs of these countries were protective of revenues from lucrative international call traffic through their own national telephone systems.

By 1994 compact new equipment designed specifically for newsgathering enabled the transmission of television news pictures, albeit not in real time. The transmission of newspaper stills pictures had been possible for many years via analogue channels. Several manufacturers developed digital video compression units for television newsgathering that could transmit video and audio over HSD channels in 'store and forward' mode. The broadcast news market felt that the further development of this store and forward type of system would be a key to future newsgathering techniques. Store and forward units are used by the major global newsgatherers, including the US, Japanese and European networks, as well as the principal international television news agencies.

The dramatic impact of these store and forward units is the ability to transmit news pictures from places where:

- there is no infrastructure for television transmissions;
- local politics or the overall logistics prevent the deployment of conventional SNG flyaways;
- the news story has not developed to the point where it is considered cost-effective to deploy an SNG flyaway.

5.5 Trans-border issues

Although technically it is possible to use INMARSAT satellite communications virtually anywhere in the world, some countries either do not permit its use, or if they do, make it prohibitively expensive to do so (see Chapter 6). INMARSAT is encouraging those countries to remove or reduce the regulatory barriers that restrict or prevent the use of its equipment within their borders.

Some countries prohibit any use of mobile satellite communications equipment, while others permit it only in particular circumstances, such as for disaster relief or emergencies, or in limited geographical areas. High licence fees, taxes and customs duties are all factors intended to put off many potential users – in some countries the annual licence fee can be thousands of US dollars. Additional type-approval of INMARSAT equipment is sometimes demanded, even though the equipment has already been type-approved and meets internationally recognized standards. Often these regulatory barriers exist because the country does not have a policy or regulatory framework covering mobile-satellite services or because they fear bypass of their terrestrial network (even in areas where there is no network to bypass). Some countries are concerned that mobile satellite earth stations may be used for criminal activity. Another reason for regulatory barriers is concern about interference to other telecommunications equipment, but spectrum sharing studies show that this is not an issue.

In 1997, the International Telecommunications Union (ITU) hosted a meeting of eighty-eight administrations, satellite system operators, manufacturers and service providers which agreed a Memorandum of Understanding (MoU), the aim of which is to facilitate the free movement of Global Mobile Personal Communications by Satellite (GMPCS) terminals. This was finally implemented by the ITU in 1998[1], and as a result INMARSAT now issue a certificate of transportation for equipment for SNG (Appendix D). Anyone considering taking an INMARSAT terminal to a different country is advised to contact the telecommunications licensing authority in the country they plan to visit, to get up-to-date information on any conditions attached to use of the terminal in the visited country. Failure to secure the appropriate approvals is likely at the very least to result in confiscation of the equipment at the point of entry to the country or even, in some countries, arrest on the grounds of suspected espionage. Trans-border issues are discussed further in Chapter 6.

5.6 ISDN

To understand the use of Inmarsat-A and -B HSD services, and the use of store and forward and video-conferencing, it is necessary to look at the basic principles of the integrated services digital network (ISDN) used for public dial-up data service in many countries around the world. We will firstly look at this in the way it is used in the typical application, i.e. in a building, which we will refer to as the 'subscriber's premises'. We can then move on to examine the cross-over into the INMARSAT system.

The ISDN standards are defined by the ISO (International Standards Organisation). The standards for audio coding are set by the ITU (International Telecommunications Union), and for video coding by ISO, the IEC (International Electro-Technical Commission) and MPEG (Moving Pictures Experts Group). More details of these organizations may be found in Chapter 4 for ISO, IEC and MPEG, and Chapter 6 for the ITU.

The scope of the INMARSAT HSD 'channel' is between the INMARSAT user terminal – the satphone – and the terrestrial interface associated with the LES. From the terrestrial user's point, the INMARSAT HSD service is one of the digital networks that may be accessed via ISDN. Similarly, from the INMARSAT mobile user's point, ISDN is the usual HSD terrestrial network that is accessed via the INMARSAT LES. Hence the significance of examining ISDN in this context.

ISDN is the ITU-T (formerly CCITT) term for the digital public telecommunications network. It is offered in two packages: Basic Rate and Primary Rate. Alternatively, in the US there has traditionally been a digital data service called 'Switched 56' (referring to the 56 kbps data rate used); however, the US has now largely adopted the ISDN standard of 64 kbps, as in Europe and many other parts of the world. The Inmarsat-B HSD service, which operates at 64 kbps, therefore naturally integrates with ISDN services.

5.6.1 Classes of ISDN service

Basic Rate ISDN (BRI) channel structures comprise two 64 kbps digital 'pipes' for user data, and one signalling data pipe, collectively termed '2B+D'. It aims to provide service for individual users and business applications such as computer local area network (LAN) data links or, for broadcast applications, high-quality audio feeds. It is provided as a dial-up service by the local telephone service provider and is charged for, like a normal telephone service, based on a standing (monthly or quarterly) charge and a usage charge per minute. The usage charge per channel is similar to normal telephone rates in most areas.

Primary Rate ISDN (PRI) provides up to 30 × 64 kbps (B) data channels and 1 × 64 kbps (D) signalling channel. It is aimed at high bandwidth business applications such as video-conferencing and high capacity on-demand LAN (local area network for computers) bridge/router links. Primary rate ISDN is the large data pipe and basic rate ISDN is the small data pipe. Because the Inmarsat-B HSD service operates at only 64 kbps it is normally used with the basic rate ISDN service in broadcast applications.

5.6.2 ISDN configuration

ISDN is normally implemented using existing two-wire, twisted-pair conductors from the local telephone exchange to the subscriber premises. This can be up to a maximum distance of 5.5 km from the exchange. At an ISDN subscriber's premises, the point at which the ISDN telephone line terminates is known as the 'U-interface', as shown in Figure 5.8. The ISDN connection is terminated at the U-interface by a network termination device known as 'NT1'. In the US the subscriber is responsible for providing the NT1; in the rest of the world it is provided as part of the ISDN service. Physically the NT1 is like an oversized telephone line-box and is normally permanently wall-mounted.

The ISDN subscriber interface at the NT1 is known as the 'T point' or 'T-interface'. It converts the data from the type used between the exchange and the socket to the type used in the subscriber's premises. If a second network termination device such as an ISDN switchboard is connected to the NT1 at the T-interface, then this is designated the 'NT2' and the ISDN subscriber interface is then the 'S point' or 'S-interface'. The physical and electrical characteristics of the S-interface and T-interface are identical and they are usually referred to as the 'S/T-interface' or 'S/T bus' (also 'S0-interface' or 'S0 bus'.) Physically the S/T-interface is an RJ-45 telephone connector. It is to the S/T-interface that the subscriber equipment is connected, and we will refer to the S/T node as the S0 node from here onwards in the text. In point-to-point applications, the S0 bus can connect equipment up to 1000 m apart. When used in a passive bus configuration (i.e. connecting up to eight physical terminals) it can span a distance of up to 500 m.

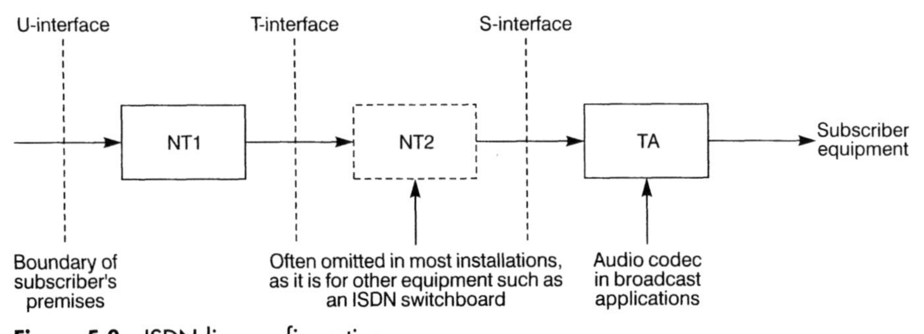

Figure 5.8 ISDN line configuration

Interface conversion between the S0 bus and the serial communications interface on the subscriber data terminal equipment (DTE) is carried out using an ISDN terminal adapter (TA.) The TA is typically packaged in a similar fashion to modems, i.e. either as a stand-alone unit or as a built-in PC card. A stand-alone TA is supplied configured with a serial communications interface such as RS.232, V.35 or X.21. The interface is normally specified by the subscriber according to the application. Nowadays most equipment capable of being used with ISDN can be supplied with an integral ISDN basic rate interface (BRI.) The DTE can be any type of data equipment such as a video-conferencing system, bridge/router or audio codec. The TA is also known as the data communications equipment (DCE.)

Stand-alone TAs can be ordered with either a dual-channel 2B+D BRI or a simple single-channel B+D BRI. As an Inmarsat-B HSD channel is equivalent to a single B-channel, it is preferable to use a single channel TA with Inmarsat-B HSD applications. The single-channel TA will normally be less expensive than a full 2B+D unit. However, a dual-channel TA will permit communication with two separate Inmarsat-B applications in different locations simultaneously, and will aid testing and fault finding. However, in the US, only very few TAs will split the ISDN to provide a single DTE connection.

There are several differing implementations of ISDN worldwide. Protocol conversion between different ISDN standards is carried out transparently by PTOs. However, ISDN BRIs differ with the various ISDN standards and so care must be taken to specify the correct country of use, and hence ISDN BRI, for the base station equipment. Fortunately, the DTE to be used with the Inmarsat-B HSD terminal uses a serial communications port (e.g. V.35, X.21, RS.232) and so is independent of any ISDN standard.

5.6.3 Channel operation

Both the ISDN B-channel and the INMARSAT HSD channel support, to all intents and purposes, a 'clear' 64 kbps digital pipe with no restriction on the bit pattern to be transferred. The ISDN offers '2B+D'– two 64 kbps digital pipes and a signalling pipe. These are 'virtual' pipes, as in reality these data channels are all multiplexed together into a single data stream.

However, there are differences in what ISDN provides against what Inmarsat-B offers. The INMARSAT HSD channel offers a single 64 kbps or 56 kbps digital pipe. It cannot directly support the same equipment that can be connected to an ISDN circuit (due to a difference related to ISDN consisting of a stream of *bytes* – 8-bit 'packets' – and Inmarsat-B HSD consisting of a stream of *bits*). The associated INMARSAT signalling, similar to the D-channel data on ISDN, is not accessible by the user. The interrelationship between INMARSAT HSD and ISDN is that the HSD channel is a 64 kbps digital pipe fully available for user data, and corresponds to one of the two 64 kbps digital pipes available to the ISDN subscriber.

As the ISDN numbering system follows the same pattern as the normal telephone system, dialling is carried out in exactly the same manner as making a normal telephone call. The subscriber number is used with the same area codes as the telephone network and international codes are also the same and used in the same way as for the telephone network.

Calls to an Inmarsat-B satphone are made in exactly the same manner as a normal international ISDN call. Dial the international access code, followed by the Ocean Region code and finally the INMARSAT mobile number (IMN) of the Inmarsat-B HSD satphone. Note that in most countries the ISDN network will route automatically to a pre-defined LES, though in some cases the LES will not support the HSD service required.

The general format is:

<<International access code + Ocean Region code +
INMARSAT mobile number>>

The Ocean Region code is:

Atlantic Ocean Region – East	871
Pacific Ocean Region	872
Indian Ocean Region	873
Atlantic Ocean Region – West	874

The Inmarsat-B mobile number (IMN) is of the form:

394 xxx xxx

where 3 = Inmarsat-B (Inmarsat-A is 1, -C is 4 and -M is 6); 9 = HSD; 4 = land mobile; xxx xxx = individual land mobile number.

For mobile-to-fixed dialling, the following parameters on the Inmarsat-B satphone need to be set prior to dialling the destination ISDN number.

- HSD channel speed: 56 or 64 kbps.
- LES to be used: ensure that the selected LES is capable of handling Inmarsat-B HSD.
- Satphone in HSD mode.

The procedures for checking and setting these parameters is given in the manufacturers' operating guide for each satphone. Once these parameters have been set the call is dialled as follows:

<<00 + destination country code + ISDN subscriber number +>>

Note that a number of LESs support access to more than one ISDN network. For example, COMSAT supports connection to the AT&T and Sprint networks. In this particular example, the user can select networks by dialling 61 or 62 instead of the normal international access code.

5.6.4 ISDN applications and Inmarsat-B HSD

All Inmarsat-B HSD satphones are fitted with synchronous serial communications ports offering similar channel capacity to a single ISDN B-channel. Several international standards have been defined for the physical, functional and electrical characteristics for different serial communications interfaces. The serial communications interfaces on Inmarsat-B satphones are all based on these international standards to greater or lesser degrees – some offering full compliance, including advanced call set-up procedures, while others offer partial compliance, i.e. data transfer functions only. In practical terms the impact of these differing levels of compliance is in the call set-up and 'clear down' and the degree to which these functions can be automated. Once the call is in progress, the actual transfer of data is common to all equipment.

5.6.5 Call set-up and clearing

For broadcast use, using a single Inmarsat-B satphone accessing a single ISDN connection, the two modes of call set-up available are as follows.

1 Handset (DTMF) dialling: A call is initiated from the satphone either by the user from the handset, the terminal console, or by DTMF tones generated automatically by a PC or other device. Upon answer by the called system the HSD port is enabled and data transfer can take place. Similarly, the call is cleared either manually by the user or by the dialling device.
2 DTR dialling (also known as 'hot dialling'): The number to be called is programmed into the Inmarsat-B satphone. When the DTE – an audio codec for example – raises the DTR line in the serial connector, the satphone automatically dials the pre-programmed number. Upon answering by the called system, the HSD port is enabled and data transfer can take place.

Call clearing is initiated either by user request or a time-out routine in either the DTE or the Inmarsat-B satphone. The DTE can be programmed to drop the line if no data transmissions have been detected within a pre-set period of time, usually set between 30 and 60 seconds, but which is entirely dependent on the user's requirements. When the Inmarsat-B satphone detects that the DTE line has dropped, it automatically clears the call. Some Inmarsat-B satphones also have a time-out function that will clear the HSD call if no data has been detected during a pre-set period of time.

5.7 Video-conferencing

Video-conferencing is the technique for allowing two-way video, audio and data signals, currently at relatively low quality, to be used for live interaction over a communications circuit. There are a number of international standards for different levels of video-conferencing. There is a low-grade video-conferencing system, designed to be used with 28.8 kbps (V.34) modems over standard telephone circuits ('plain old telephone system', or POTS) to provide a 'video-phone', utilizing the ITU-T H.324 standard[2] which was designed for this application. However, this standard does not produce anywhere near the quality required even for a breaking news story. Similarly, the use of 'web-cams' does not constitute a means of achieving viable video-conferencing on the Internet for news broadcasts due to the unpredictable fluctuations in data rates. We will therefore be concentrating on video-conferencing standards designed for use over high-speed data circuits, typically via ISDN, INMARSAT or private networks.

Video-conferencing is used either on a one-to-one basis or by groups of people who gather in a specific setting (often a conference room) to communicate with others. It may seem at first glance strange that we should be considering video-conferencing in the context of newsgathering. However, although not widely used for this purpose up until now, it is highly likely that there will be increasing use of video-conferencing techniques over the next few years. This is due to two factors: quality and cost. As time passes, the quality of video-conferencing at lower data rates is improving and the equipment cost is tumbling, as video-conferencing becomes an increasingly mass-market product. These shifts in quality and cost will make video-conferencing very attractive, particularly for breaking stories or brief updates. The advantage that video-conferencing has over normal video is that for a 'live' update using a 'head and shoulders' shot, there is very little motion. Generally, with the human face in view, the lips will be the part of the picture that has the most movement, and hence dramatic data compression is possible. Store and forward video units essentially use video-conferencing protocols for the 'live' mode of operation. In this section, we will concentrate on the video and audio aspects, ignoring the data and control details for the most part, as these are not relevant to newsgathering via Inmarsat-B.

Video-conferencing is normally achieved either as a dedicated system specifically designed for installation in a dedicated video-conferencing room or as PC hardware and software kits installed in desktop PCs. Either type will work with the Inmarsat-B HSD service. The international specification for video-conferencing over ISDN is the ITU-T Recommendation H.320[3] which is the basis for the family of standards which governs video-conferencing over narrow-band (64 kbps to 1920 kbps) networks. The recommendation was designed primarily for ISDN and has become the *de facto* global video-conferencing standard.

H.320 is a suite of specifications which define how video-conferencing systems communicate over ISDN, defining video, audio and data standards. One of the important aspects to consider in respect of using an H.320 compliant system with the Inmarsat-B HSD service is that of the audio codec to be used. H.320 encompasses three audio compression standards.

These three audio compression algorithms are designed to handle a broad range of applications. ITU-T G.711[4] uses the whole of the 64 kbps of bandwidth to provide 3 kHz telephone-quality audio; G.722[5] provides 7.5 kHz audio over a 64 kbps circuit; and G.728[6] provides 3 kHz audio using 16 kbps of bandwidth.

However, G.711 and G.722 both require 64 kbps channels and are hence unsuitable for Inmarsat-B HSD video-conferencing applications, as they would each consume all of the available bandwidth. The greater the degree of compression applied, the greater the delay introduced into the transmission, i.e. G.728 will introduce the greatest delay and G.711 the least (see Chapter 4). It is for this reason that G.711 is the standard normally adopted for ISDN telephones. However, when using a narrow band 64 kbps service like Inmarsat-B HSD it is only possible to use G.728 because of the bandwidth limitations. An H.320 compatible video-conferencing system using Inmarsat-B HSD will use 16 kbps (G.728) for audio and 1.6 kbps for H.221 framing, leaving 46.4 kbps for video. On the other hand, G.722 is used for audio-only contributions, as we will see later in this chapter.

The key component of H.320 is the H.261 video compression algorithm standard[7], which uses two principal video resolutions from the ITU-T range (Table 5.2): 352 × 288 CIF (common intermediate format) and 176 × 144 QCIF (quarter common intermediate format.) QCIF is normally used for low bit-rate channels such as Inmarsat-B HSD and the frame rate can be up to a maximum of 30 frames per second (fps). On a 384 kbps channel, an H.320 video-conferencing system will achieve 30 fps at CIF and look comparable to a VHS picture. Many of the new modem and Internet videophones are based on the 128 × 96 pixels SQCIF (sub-quarter common intermediate format) and QCIF resolutions. While this is an impressive accomplishment in the bandwidth available, it is not suitable for quality video-conferencing or, therefore, newsgathering.

On an Inmarsat-B 64 kbps HSD channel, the same system would achieve 10–15 fps at QCIF. The H.261 algorithm includes a mechanism that optimizes bandwidth usage by trading picture quality against motion so that a quickly changing picture will have a lower quality than a relatively static picture.

Video-conferencing systems need to communicate at a resolution that can be handled given several factors. These factors include available line bandwidth, system processor capacity and the capacity of the software being employed. Another factor in video-conferencing is the frame rate of the session. The H.261/263 standard allows frame rates of 7.5, 10, 15 and 30 fps. Television operates at 25 or 30 fps and cinemas at 24 fps, and video-conferencing that operates at a true speed of 15 fps or greater looks acceptable for most users. If operating at rates as low as

Table 5.2 ITU-T coded video frame transmission formats

Picture format	*Pixels*	*H.261*	*H.263*
SQCIF	128H × 96V	Optional	Required
QCIF	176H × 144V	Required	Required
CIF (Full CIF)	352H × 288V	Optional	Optional
4CIF	704H × 576V	–	Optional
16CIF	1408H × 1152V	–	Optional

10 fps, picture jitter is noticeable, and a person speaking will show an apparent lack of continuity.

Another major factor is the ability of the software and hardware to process the data both in compressing and decompressing the images. Since a great deal of compression is involved, and each frame must be processed, transmitted, stored and re-processed, then the slower the processor or the lower the efficiency of the system, the poorer the image and the slower the apparent frame rate. While a system may operate at a high frame rate, it may appear to be operating at a lower frame rate due to insufficient processing capacity. This will result in duplicate frames, or 'inferred' frames being used which can be identical to the last frame, rather than the actual data for that frame.

The development and recent implementation of the advanced compression algorithms H.263 and H.263+[8] (enhancements of H.261), a standard developed for transmitting over the Public Switched Telephone Network (PSTN), offers a higher level of quality at low data rates that are acceptable for newsgathering. The H.263 algorithm outperforms H.261 by a factor of up to 3:1 in qualitative terms, and can transmit video data at rates much less than 64 kbps. It is certainly adequate for breaking stories where there is no other way of getting reports back to the studio centre, producing results at 64 kbps comparable to a 128 kbps (2 × BRI) circuit. A 2 × BRI video-conference picture is certainly adequate for remote education, presentations etc., and could be used for a breaking news story. Rapidly changing scenes are still not very well handled, but as soon as movement within the picture diminishes, the sharpness, colour and movement quality are impressive considering the limitations of the system. However, to achieve 2 × BRI via Inmarsat-B requires two satphones to be 'bonded' together. Therefore, the use of H.263 at 64 kbps, offering equivalent subjective quality, is certainly very attractive as it can be achieved with a single Inmarsat-B satphone.

A feature that H.261 and H.263 share is the ability to vary the frame rate within a video sequence. This is important for two reasons. Firstly, the bit rate in a video-conference can be very low, so an encoder must be able to lower the frame rate to maintain reasonable visual quality. Secondly, the encoder must be able to dynamically adjust to sudden changes in video content in real time without warning. For example, at a scene change, the first compressed frame of the new scene tends to be large in terms of data because of the dissimilarity with the previous frame. With variable frame-rate encoding, the encoder can code the frame and then skip a few input frames to catch up before encoding the next frame.

Other standards specified within H.320 are H.221, H.230 and H.242, which are used respectively for framing, control and indication, and communications procedures.

Dedicated video-conferencing systems are normally supplied with a synchronous serial communications port (usually X.21, V.35 or RS.232) which will enable direct connection to the Inmarsat-B HSD terminal. On the other hand, most PC-based video-conferencing systems are supplied with an integral ISDN BRI. To use these systems with an Inmarsat-B satphone it is necessary to either fit an additional PC card with a serial interface or use a commercially available ISDN S0 bus adapter.

These convert the ISDN BRI S/T-bus interface on the video-conferencing system to a serial interface suitable for connection to an Inmarsat-B HSD terminal. In 1999, there was only one such S0 bus adapter suitable for use with Inmarsat-B on the market.

Video-conferencing systems typically do not support automated dialling or call set-up through the serial communications port. The video-conferencing system will therefore normally be configured as for a leased circuit and the call manually dialled and cleared on the Inmarsat-B HSD terminal. Once the call is connected, it typically takes 5–10 seconds for each of the video-conferencing systems to synchronize and connect. If the call is established with an audio connection but no video is present then it is highly likely that one or both of the video-conferencing systems is in G.711 or G.722 audio mode. In this case it will be necessary to clear the call down, reconfigure one or both of the terminals (both should be configured for G.728) and re-dial the call. Alternatively, the video mode could be switched from CIF to QCIF to force a 'capability block' exchange, effectively resetting the system.

5.8 Store and forward

5.8.1 Principles

A live video transmission of a quality conventionally acceptable for a news broadcast requires a communications channel (terrestrial or satellite) with a capacity of at least 4 Mbps. Clearly this capacity is not available with the Inmarsat-B HSD service and to achieve this in a remote location, away from high-capacity terrestrial circuits, normally requires the use of a full-bandwidth SNG system. However, although the Inmarsat-B HSD service is commonly used for live radio broadcasts, data rates of 56/64 kbps are not capable of providing sufficient capacity for traditionally acceptable live video. By encoding the high-quality video and audio in a process where it is digitized, compressed and then stored on a computer hard disk, the data can then be later transmitted through the Inmarsat-B HSD system. Once received at the studio, it can then be integrated into the news broadcast for transmission, and few viewers would notice the difference.

This solution is the essence of store and forward video applications and gives users the ability to transmit high-quality video at low data rates. These data rates would normally be considered poor or unacceptable for TV transmission. Although the video material is not live, it need not be more than a few hours old and the trade-off in terms of portability – 40 kg in total compared to up to 1000 kg for an SNG flyaway system – makes it a very attractive alternative. In addition, Inmarsat-B HSD channels are assigned on demand and are available over a much wider area of the globe than Ku- or C-band SNG capacity.

The fundamental principle of the system is that video is played into a store and forward unit which digitizes the video and stores it as a data file on the hard disk of a PC. Once this process has been completed, the store and forward unit can be connected to an Inmarsat-B satphone and an HSD call established to a companion unit at the destination. When the connection has been established, the video/audio data file is transferred over the link and recorded as a data file at the destination. On completion of the transfer, the data file is played out as converted real-time video and audio.

The key to successful implementation of store and forward techniques is in the coding (digitization and compression) of the video. Uncompressed digital video would create enormous data files of gigabyte proportions. However, compression of the video during encoding reduces the file size to much more manageable proportions. In fact, compression ratios of between 25:1 and 100:1 are typical with the actual compression selection depending upon the desired video quality. Choosing a high sampling rate will result in higher quality pictures than a lower sampling rate, but will increase the transmission time. Store and forward units offer a range of sampling rates, the highest of which will result in reasonable quality pictures and sound being transferred at a rate of approximately 30:1. One minute of video will therefore take approximately 30 minutes to transfer. This can vary slightly according to the video standard the original material was shot in – 625/50 transfer rates are slightly shorter. The time taken for the data transfer depends on two parameters: the size of the file and the data bandwidth of the link.

The length of the video and the sampling rate at which the video is digitized determines the size of the transmitted data file. The length of the video clip is determined largely on subject matter and editorial criteria, although it may be decided to keep it short in order to get the material back in time for the news broadcast. The amount of movement in each frame does not have a drastic effect on file size, and the sampling rate largely determines the size of the file.

The transmission times are thus fundamentally determined by three factors: video sequence length, video coding rate of the file (which can range from 512 kbps to 3 Mbps) and data channel rate (56 or 64 kbps for the Inmarsat-B HSD service). Some typical transmission times for one minute of broadcast quality video are given in Table 5.3, with times and costs based on a 64 kbps Inmarsat-B HSD channel. A similar transmission using a 56 kbps channel would increase transmission times and costs by about 10%.

Table 5.3 Transmission times and costs for 1 minute of broadcast-quality video coded as MPEG-1 on Inmarsat-B HSD*

Coding rate (Mbps)	*Picture quality*	*File size (MB; typical)*	*Tx time (min)*	*Tx cost (US$; estimated)*
1.0	Low	7.5	17	187
1.5	Medium	11.25	25	275
2.0	High	15	33	363

*Assumptions:

1 Transmission times are for 1 minute of 525/60 video – 625/50 video transfer rates are slightly faster.
2 Actual data throughput on the 64 kb/s channel is typically 60 kb/s.
3 HSD indicative costs are assumed to be an average of US$11 per minute.

The resultant pictures can have a 'filmic' look, reminiscent of 16 mm news film that was commonly seen up until the 1980s before the widespread use of ENG (electronic news gathering). The quality of the pictures is often compared to VHS, but the overall effect is not that displeasing. The quality is often much improved if the shot scenes have high light levels, so that the pictures have few dark areas. In the conversion from a light signal to an electronic signal, dark areas of pictures create video noise in the camera and on the videotape. When the frames of pictures are digitized, the digital processing is unable to differentiate between noise and picture detail in dark areas. This causes ambiguity in the digitizing process, and the dark areas have resultant distracting artifacts as the coding process attempts to treat the random noise as rapidly changing detail. Digital video also contains detail components that the human eye cannot detect. By removing these components further data reduction can be achieved (see Chapter 4).

The video is either coded as a Motion JPEG, MPEG-1 or MPEG-2 file. Each of these coding standards is described in more detail in Chapter 4. The resultant data file is transmitted according to the ITU H.261 and G.722 protocols, which form part of the H.320 video-conferencing protocol as discussed in the previous section. The H.320 protocol encompasses both H.261 for video and G.722 for audio.

H.261 as a video coding standard was designed for coding rates between 40 kbps and 2 Mbps. MPEG-1 is an ISO/IEC standard and defines a bit-stream for compressed video and audio optimized for encoding at 1.5 Mbps. However, the principles of operation of both algorithms are similar. With H.261, only the image elements that have changed need to be saved or transmitted because the other information exists in the previous frame or frames. In contrast, MPEG-1 video compression looks forward as well as backwards when estimating motion. This means that although the coding delay is slightly longer, the quality and motion handling of MPEG-1 is considered superior.

The MPEG standard has three separate parts that cover video coding, audio coding and the interleaving of the two streams. Because MPEG encompasses both video and audio, no audio and video synchronization problems exist with MPEG encoding. Transmission of the video files over the Inmarsat-B HSD is carried out using an integral file transfer system.

5.8.2 'Live' mode

Store and forward units are usually also able to offer the facility of transmitting video and audio in real time – 'live' mode. However, at 64 kbps the results are currently very disappointing, with jerky, poor definition pictures and the audio out of synchronism with the video by almost 5 seconds. This causes unacceptable lip-synch errors which are particularly noticeable on video featuring a reporter speaking to the camera – a very common news sequence.

To produce generally acceptable broadcast news pictures in real time with store and forward video units, the bandwidth of the digital pipe has to be in the region of six times the normal INMARSAT HSD rate, which is 384 kbps. Normal Inmarsat-B satphones are not able to operate in this mode, and require either multiplexing or 'bonding' with other satphones to access data rates above 64 kbps, or have to be specially modified to access leased capacity at this data rate. Since such leased capacity is not readily available on the INMARSAT system, as it is primarily configured to providing on-demand dial-up access at data rates of 64 kbps or lower, its use is not particularly viable for newsgathering.

The video codecs are now available as PC cards, and highly portable store and forward systems are commercially available weighing only about 12 kg (25 lb) and designed for use in rugged conditions. Two common systems used in newsgathering in the field are the TOKO VAST-p and the Livewire SkyLink Video II, and the principal characteristics of each of these units is outlined below.

5.8.3 TOKO VAST-p

Manufactured in Japan by TOKO, this unit is shown in Figure 5.9. VAST-p is an acronym for 'Video and Audio Storage and Transmission – portable'. It is a very rugged unit, designed with the government market in mind, and weighs only 6.5 kg (14 lb). It is supplied in a shock-resistant 'roto-molded' shipping case weighing a further 5.5 kg (12 lb), so that the whole unit, ready for shipping, weighs only 13 kg (26 lb). It has replaceable sand and dirt filters on air-intake ports for hostile environments. The system will only operate from 100 to 240 VAC, so an external inverter is needed if DC operation from a vehicle battery is required.

It provides two-way video and audio for video-conferencing as well as store and forward application, transmitting at data rates from 2.4 kbps to 2 Mbps. The video-coding algorithm is proprietary, based on MPEG1/H.261/JPEG. This means that the files are coded in a proprietary manner for transmission depending on the mode selected, so they can only be received and decoded by another TOKO VAST-p unit. The video and audio are also sent as two separate segments, so an incomplete transmission may result in a full complement of video pictures, but only part of the audio.

If there is a break in transmission in store and forward mode, the file transfer will resume from where the break in transmission occurred when the connection is re-established. However, the result will be seamless. It offers one 'live' audio channel, and composite video input and output in either NTSC or PAL. Identical units are used in the field and at the studio centre, so that it is just as easy to send material from the studio centre to the field if required (though unlikely for newsgathering use).

It is supplied with different storage options so that it can store up to 2 Gb (2000 Mb) of information for up to 2 hours of video and up to 99 files in store and forward mode. The storage is arranged to hold a number of files or 'clips'. Clip locations have to correspond on both the send and receive units. For instance, if the clip to be sent is stored in File Location 12 on the transmitting unit, then File Location 12 has to be empty and ready to receive the data in the receiving unit. This limits its use in a fully automatic mode, but with discipline it is not necessarily a significant limitation. It has simplified coding rate selections and has a user-friendly menu-driven control panel requiring few key presses for quick 'plug and go' operation.

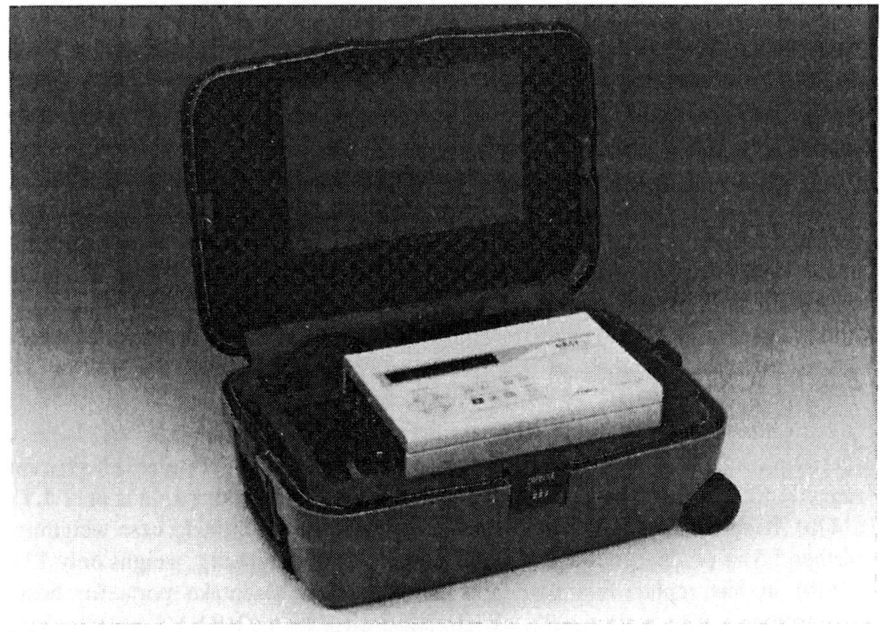

(a)

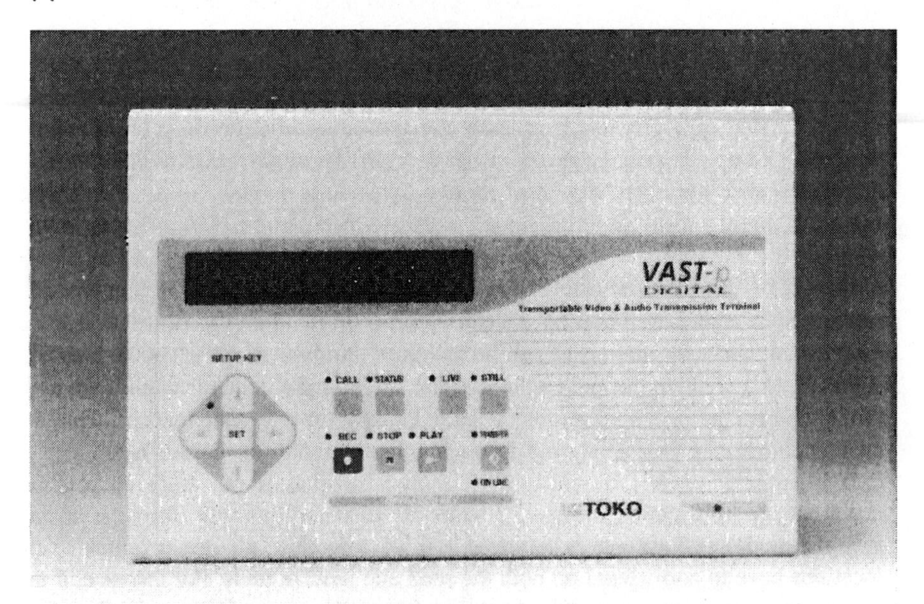

(b)

Figure 5.9 (a) Toko VAST-p; (b) Toko VAST-p operating panel. *(Photos courtesy of Toko America Inc.)*

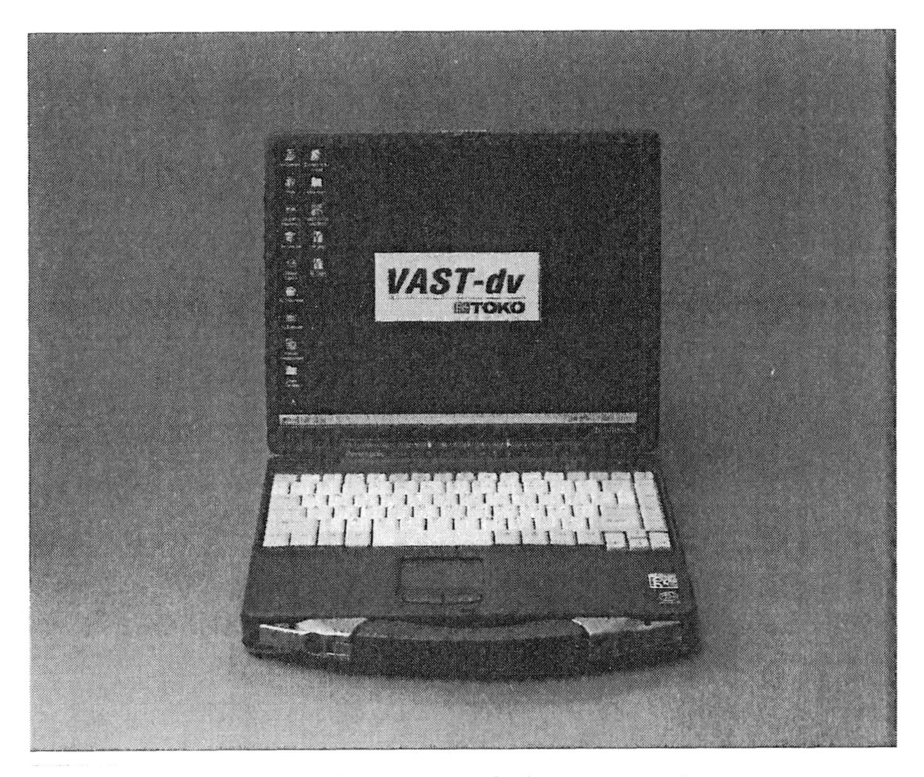

Figure 5.10 Toko VAST-dv. *(Photo courtesy of Toko America Inc.)*

Toko has now launched the VAST-dv, a notebook PC-based system (Figure 5.10) designed to deliver MPEG. This offers user-selectable MPEG1 and MPEG2 settings and coding rates from 1 to 15 Mbps, though data transmission rates as low as 64 kbps are possible.

The VAST-dv has an IEEE 1394 'Firewire' port, allowing easy connection to digital video cameras, and the VAST-dv software in a standard notebook PC allows the VAST-dv to double as the user's primary PC as well, weighing less than 2 kg (5 lb).

5.8.4 Livewire Voyager

Developed and manufactured in the UK by Livewire Digital Ltd, the Voyager (formerly marketed as the SkyLink Video II) is shown in Figure 5.11. It is a store and forward system that can handle news-quality motion video and audio, clip lists, high-quality stills and other forms of disk-based data. Any number of clips may be recorded and transmitted without entering storage locations or file names. It uses MPEG-1 compression, offering good motion handling, stereo audio and lip-synch. Because the video and audio are transmitted together, partially received files can be used, although the system will also automatically append to them if a subsequent

Figure 5.11 Livewire Voyager. *(Photo courtesy of Livewire Ltd)*

connection is made. Selection of video compression parameters is made by simply selecting the footage type, i.e. 'piece-to-camera interview' or 'long distance action'. The standard types can be modified or complemented by user-defined ones. It produces MPEG-1 format files that can be replayed by any compliant hardware or software MPEG player, and it features on-screen replay of recorded footage and simple maintenance of video clips. The system provides for rough editing of the video source, possibly reducing the need for field-editing equipment and thereby reducing the volume of equipment required. An optional non-linear editor is available.

The Voyager is designed around a rugged PC, weighing 12 kg (25 lb) and running under Windows 95 or NT. It offers access to standard software packages such as e-mail and word-processing. The system may be operated from the keyboard, mouse or optional remote control console. The system will operate on either 90–240 VAC or 10–32 VDC. Like the TOKO VAST-p, the Voyager is highly compact and ready to use quickly. The base unit, located at the studio centre, offers fully automatic operation, handling simplex or duplex INMARSAT HSD calls without user

Figure 5.12 Livewire Voyager Lite. *(Photo courtesy of Livewire Ltd)*

intervention. It offers mail-box type facilities for outgoing files and can be integrated into a computer network, with automatic call set-up, file transfer and call-clearing.

Livewire, like Toko, has also just launched a notebook PC-based store and forward system called 'Voyager Lite' (Figure 5.12). It is particularly designed to operate with the latest Inmarsat-M4 satphones, and the system may be directly connected to one or two Inmarsat-M4 satphones so that the additional bandwidth gained can deliver store and forward MPEG-2 material in the same time as MPEG-1. Voyager Lite has no analogue inputs, using the 'Firewire' standard link to the digital video camera. It uses software compression techniques to generate MPEG-1 or MPEG-2 streams in the range 128 kbps to 15 Mbps, which although cost-effective, does introduce post-processing delays. However, these are minimized by a software package that will perform the compression and transmission. The package also offers frame-accurate video and audio editing direct from the camera, though (optionally) non-linear editing can be performed with the same notebook PC.

The VAST-p and Voyager (not Voyager Lite) units have been in use with a number of newsgatherers (both broadcasters and news agencies) for a number of years, and each have their pros and cons. Costs are similar, and it is very much a matter of preference as to which units a newsgathering organization chooses to use. At the time of writing, there has not yet been an opportunity for evaluation of the new notebook PC-based units.

5.9 Radio contribution applications

The human ear hears frequencies between 20 Hz and 20 kHz, and is most sensitive in the range 2–4 kHz, while the human voice typically produces sounds in the range 400 Hz to 2 kHz. The low frequencies tend to be vowels, while the higher frequencies are consonants. These frequency ranges determine the most suitable type of audio coding and compression to use for audio broadcast purposes.

Uncompressed digital audio (as with video) can require a large amount of bandwidth to transmit. As has previously been outlined, several international standards and proprietary systems exist for the coding (digitization and compression) of audio signals in real time, all of which use different techniques to apply differing degrees of compression to the audio input – music or speech. The degree of compression applied determines the amount of bandwidth required for the coded audio data stream. The compression process also introduces a time delay that becomes greater as the degree of compression increases. Both of these factors – required bandwidth and coding delay – have to be taken into account in determining the most suitable coding algorithm for a particular application for use over the Inmarsat-B HSD 64 kbps service. For video-conferencing, G.728 was the optimum coding standard because it allowed for the maximum bandwidth in an Inmarsat-B 64 kbps channel for video. However, where only audio is to be digitized and sent over the channel, G.722 is often used. Audio compression in general is described in more detail in Chapter 4, including MPEG audio.

Given that satellite propagation delays are of the order of 250 ms the need for greater audio bandwidth (i.e. greater compression) has to be balanced against a desire to keep any coding delays to a minimum. The following describes the three internationally recognized and commonly used coding standards, suitable for use over the Inmarsat-B HSD service for audio applications.

5.9.1 Coding standards

G.711

The ITU-T standard G.711, which uses 64 kbps of bandwidth to provide 3 kHz telephone quality audio, has very little coding delay and is generally used in telephony over ISDN. The shortest possible coding delay is highly desirable for satellite transmissions, as there is an inherent propagation delay of about 250 ms. However, the relatively narrow audio bandwidth of G.711 has limited use in broadcast applications and so generally a higher level of compression is required at the expense of greater coding delays.

G.722

The ITU-T G.722, which also uses 64 kbps of bandwidth, is a common encoding method used over both ISDN and satellite circuits for audio broadcast use, based on ADPCM (see Chapter 4). A coding delay of about 6 ms makes it very suitable for live interviews and voice feeds over a satellite circuit. However, the coding method is optimized for speech (the sound is split into two bands, the lower and higher frequencies, to make the coding more effective), and although the audio bandwidth of 7.5 kHz is generally considered perfectly acceptable for radio news items, it is unsuitable for transmission of music.

MPEG Layer II

ISO/IEC MPEG Layer II is not widely used in radio news applications, although it is used in radio broadcasts of music. There are several different implementations of the MPEG Layer II standard, hence affecting the range of data rates available (see Chapter 4). However, the principal disadvantage in terms of newsgathering is the processing delay, making MPEG Layer II difficult to use for 'live' contributions which involve interaction with the studio. The actual delay is much greater than the theoretical delay figures. MPEG Layer II is often generically (but not strictly correctly) referred to as MUSICAM (see Chapter 4). MPEG Layer I, is not widely used in broadcast applications.

apt-X

This coding 'standard' is a proprietary standard developed by Audio Processing Technology (APT) in the UK and though common it is not an international ITU standard. It uses ADPCM with a very low coding delay of less than 4 ms and, like G.722, delivers 7 kHz bandwidth in a 64 kbps HSD channel. However, because it is proprietary, it does not offer the universality of G.722.

5.9.2 Choice of coding standard

The advantages of G.722 for live newsgathering can now be appreciated, and it is likely to remain the favoured coding system for the present. However, some organizations do prefer having the option of using MPEG Layer II (and Layer I) and hence tend to use multi-standard codecs.

Typically, audio codecs are fitted with a serial synchronous communications port such as RS.232, X.21 and V.35 for direct connection to the Inmarsat-B HSD satphone, although some may require interface converters. These interface converters are normally off-the-shelf items and can be ordered according to the type of interface conversion required.

On the terrestrial side, some audio codecs may also have an integral ISDN BRI, or one may be available as an option. However, many do not and will require connection to the NT1 using an ISDN TA with the corresponding serial synchronous communications interface (e.g. RS.232, X.21 or V.35).

G.722 audio codecs have no user-configurable settings and are connected directly to the Inmarsat-B satphone. In general, there is a high level of compatibility between G.722 audio codecs of different manufacturers. Nevertheless, if any problems with the link are encountered the use of different makes of audio codecs has to be a prime suspect.

All audio codecs are duplex in operation (i.e. data can be sent simultaneously in both directions). Many MPEG Layer II audio codecs also incorporate the G.722 codec and are capable of symmetric and asymmetric dual-mode operation. This enables G.722 in both directions, MPEG Layer II in both directions, or MPEG Layer II in one direction and G.722 in the other direction. Asymmetric operation offers an advantage if the studio-to-remote path is in G.722, as delay is then minimized for 'live' contributions while offering maximum audio quality with MPEG Layer II in the remote-to-studio path. Care needs to be taken to ensure that the codecs at each end of the link are similarly configured.

Unlike G.722 audio codecs, there is not a high degree of compatibility between different makes of MPEG Layer II audio codecs because there are different implementations of the standard. If in doubt about the compatibility of two codecs, check them first on an ISDN circuit, as even the satellite delay can have an adverse effect.

There are a wide variety of audio codecs that can be used with INMARSAT HSD; some are codecs alone, while others are combined terminal adapter and codec units. Of this wide range, we will briefly examine the Glensound GSGC5 (a combined terminal adapter and codec) and the Comrex DXP.1 (a codec only).

Glensound GSGC5

The UK-made Glensound GSGC5 combines a simple audio mixer, TA and codec in a very compact battery-powered unit specifically designed for radio newsgathering (Figure 5.13). It provides 7 kHz bandwidth circuit using one of a variety of

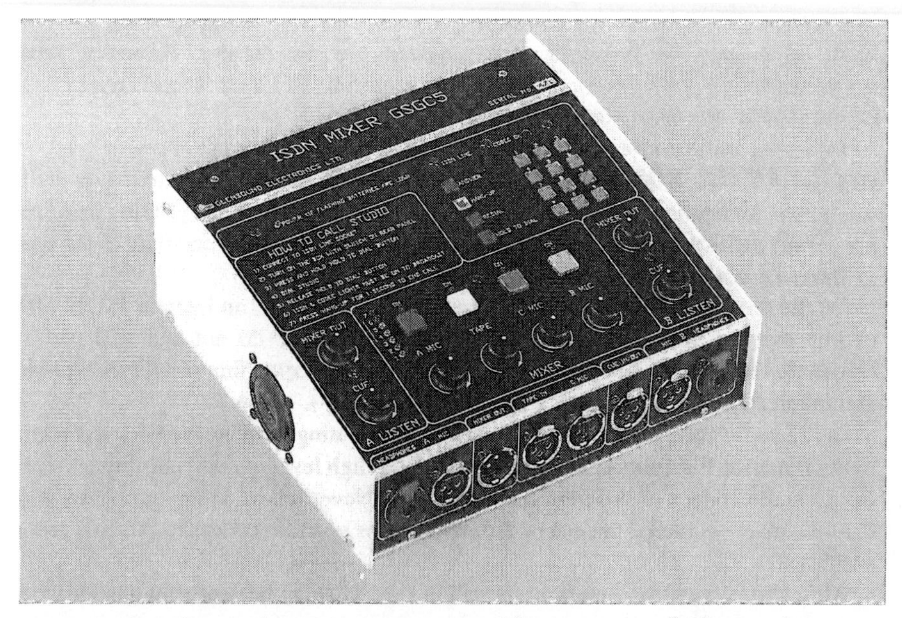

Figure 5.13 Glensound GSGC5 reporter unit. *(Photo courtesy of Glensound Electronics Ltd)*

integrated codecs – G722, apt-X, a special dual-standard codec, or MPEG Layer II – which is factory fitted. The dual codec is particularly ingenious as it automatically analyses the 'called' codec and uses the appropriate identified coding (either G722 or apt-X). The GSGC5 is compact, fully self-contained and easy to operate, making it ideal for journalists. It can be plugged into an ISDN-2 S0 bus (thus it can be used directly with an Inmarsat-M4 satphone) or, with the appropriate interface (X.21 or V.35), be connected to an Inmarsat-B satphone. Facilities are provided for two programme microphones, an 'effects' microphone and a tape machine. An internal DTMF generator is used to generate ringing, dialling, engaged and other tones. This generator can also be used, after a call has been established, to signal the studio. A DTMF detector and sounder are also fitted so that the studio can signal the reporter. Although normally battery operated, it can also be used with an external DC supply.

Comrex DXP.1

The Comrex DXP.1, made in the US, is a digital audio codec designed for broadcasters. It is based on the G.722 standard to send and receive 7 kHz audio at transmission rates of 56/64 kbps, and is particularly notable for its wide interoperability with other brands of G.722 codecs. It is compact, portable and relatively low-cost, with an adjustable headphone output and an input that is switchable between microphone and line levels. A TA is needed to interface it to an ISDN line and there is a cable available to convert the standard data output to X.21 or V.35 for INMARSAT HSD connection.

5.10 Operation

The following explanations give an idea of the differences in operation between the Inmarsat-B and Mini-M satphones, and demonstrate how units of each type are typically assembled and operated. These are currently the two INMARSAT systems most commonly used by newsgatherers. However, it is likely that if Inmarsat-M4 delivers what it promises, it will supplant Inmarsat-B very rapidly – but at the time of writing, the service has only just been launched.

5.10.1 Inmarsat-B

There is a relatively wide selection of brands of Inmarsat-B satphones, although some are re-badges of other makes. These are the principal satphones currently found in use:

- Glocom Global Phone 3000I
- MagnaPhone MX6060

- Nera Saturn Bp/Bt
- Marconi Obidos
- STS Lynxx (formerly California Microwave Lynxx)
- IN-SNEC GLOBALIS-B

The Inmarsat-B satphone typically consists of three assemblies: the antenna and its mount, the console/master handset and the transceiver unit. Some manufacturers combine the console and the transceiver units into one unit, with the antenna either fitted directly to the transceiver or connected via a remote cable.

Almost all Inmarsat-B satphones have identical antennas. These are four flat panels, each about 50 cm square, which clip together to form one large antenna panel array of approximately one square metre. The panels can be assembled in any order, and they each have a mounting clip at one corner. Each petal is positioned in turn so that it engages the horizontal and vertical mounting rails and the mounting clip locks into place on the locking tab located in the centre of the mounting plate. Note that the plastic clips commonly fail in operation.

The following leads are typically supplied.

- Mains/line input cable: The line input voltage can be between 90 and 280 VAC, so the unit can be used anywhere in the world.
- Car battery (DC) power lead: usually with crocodile clips for direct connection to a vehicle battery.
- Antenna cable: a thin cable connecting between the antenna socket on the satphone to the antenna.
- Handset cable.

Siting the satphone

To operate, the satphone needs to be able to 'see' one of the four INMARSAT satellites. When siting the antenna, a compass is used to check that there is a clear view towards the satellite – to the north if in the Southern Hemisphere and to the south if in the Northern Hemisphere. The satphone can be used inside a building if the antenna is pointing out through a window; the window does not need to be open, but beware of metallized coatings on the glass used to prevent glare or to offer privacy. The window obviously needs to offer a view of the sky in the right direction.

Don't use the compass near the satphone while it is switched on, as the electromagnetic field can cause a false reading. In addition, the use of a compass near vehicles (metal affects the reading) is not advisable, nor inside some buildings, as many modern structures have an integral steel frame that will affect a compass reading. The view towards the satellite must not be obstructed by tall buildings or trees close by. For safety reasons, the antenna should not be pointed towards areas where people may pass or gather, and ensure that the cables don't cause a trip hazard.

A few seconds after the satphone is switched on, the handset display should give various status indications, telling the user which LES is selected (usually the one last used). To make a reliable HSD call the antenna must be pointing accurately at the chosen Ocean Region satellite. This is not the case for simple voice calls which, because of a much lower data rate (2.4 kbps), do not require such critical alignment. There is usually a chart supplied with the terminal that will give an estimate of the azimuth and the elevation of the antenna. The whole satphone may be rotated until the antenna points in the direction indicated by the compass. The handset display shows some representation of signal strength. Some satphones go a step further and will give a numerical display of bit error rate (BER – a measure of the quality of the digital signal) and signal strength; you should refer to the manufacturer's manual for an explanation of the exact indication.

Safety

Inmarsat-B satphones are heavy and can be physically awkward to lift. The antenna produces significant levels of non-ionizing radiation and the manufacturers' recommendations for minimum clear distances in front of the antenna must be followed: 5 m is a typical 'safe zone'. Standing behind the antenna is perfectly safe, and it is advisable to mark out the safe zone with visual warning tape. However, in HSD mode, all Inmarsat-B satphones have an automatic transmission safety cut-out which operates within 5 seconds of an obstruction appearing in front of the antenna. This is not the case when it is operating in voice-only mode, when care must be taken that no-one stands in front of the antenna. Such is the resilience of the data link for voice transmissions that the satphone may be able to maintain a connection even with a person standing in front of the antenna.

Operating

All commands are keyed into the master telephone handset or the control panel, which usually displays all information needed. Operational controls are accessed from the handset with error messages and the progress of the call displayed on the handset. Once aligned and set up, the satphone is ready for operation. Each manufacturer's unit will vary in the keystrokes required to set the mode of operation and dialling sequence. Different codes are used in the dialling sequence to establish the type of call (voice, HSD etc.) as indicated earlier in this chapter. Any additional units such as audio codec or store and forward units will also need to be connected before an HSD call can be made.

5.10.2 Inmarsat Mini-M

The operation of the Inmarsat Mini-M satphone is very much quicker and straightforward. The satphone comes in a small case and the antenna is usually integral to the case. It is simply a matter of flipping the antenna up, pointing it approximately in the direction of the satellite, and optimizing the position in azimuth and elevation to get the strongest signal strength. Once this has been done, the handset can be used to dial the call directly (see Figure 5.14). Because the RF power from the Mini-M satphones is much lower, there need only be a 1 m area kept clear in front of the satphone.

Figure 5.14 Media satphone 'farm' in Baghdad. *(Photo courtesy of Paul Szeless)*

5.11 Conclusions

The use of INMARSAT for newsgathering is increasing, and with the new Inmarsat-M4 satphone combined with the compact video store and forward laptop systems, more newsgatherers are going to be able to afford entry into this area of technology. Of all current INMARSAT traffic, 40% is data, and this is predicted to reach 70% by 2003, so that INMARSAT are planning to rapidly develop their data services and increase capacity. The use of INMARSAT services for delivering radio news contribution is well established, and again the ability of the new Inmarsat-M4 satphones to easily interface with off-the-shelf ISDN products and applications is going to further enhance its use for audio newsgathering, as well as expanding its use for video.

The challenge of the up and coming GMPCS services is covered in Chapter 9, but at the time of writing it is clear that the INMARSAT system will remain a principal means for newsgathering from remote and difficult locations.

References

1. ITU (1998) Ninth Plenary Meeting, Resolution 1116, Implementation of the GMPCS-MoU arrangements.
2. ITU (1998) ITU-T Recommendation H.324 (2/98), Terminal for low bit-rate multimedia communication.

3. ITU (1997) ITU-T Recommendation H.320 (7/97), Narrow-band visual telephone systems and terminal equipment.
4. ITU (1988) ITU-T Recommendation G.711 (11/88), Pulse code modulation (PCM) of voice frequencies.
5. ITU (1988) ITU-T Recommendation G.722 (11/88), 7 kHz audio-coding within 64 kbit/s.
6. ITU (1992) ITU-T Recommendation G.728 (9/92), Coding of speech at 16 kbit/s using low-delay code excited linear prediction.
7. ITU (1993) ITU-T Recommendation H.261 (3/93), Video codec for audiovisual services at p × 64 kbit/s.
8. ITU (1998) ITU-T Recommendation H.263 (2/98), Video coding for low bit rate communication.

The author thanks INMARSAT for allowing him to quote extensively from *Inmarsat-B High Speed Data Reference Manual.*

6

Across the spectrum: regulation and standards

The technical practicalities of satellite newsgathering are difficult enough to master. To understand the administrative, regulatory and commercial aspects can be just as challenging, and so in this chapter, we will look at these issues and put them into context. Be warned, however, that to understand this aspect of telecommunications requires a good memory for abbreviations and acronyms!

Operating SNG uplinks virtually anywhere in the world requires some degree of administrative process to be undertaken, and any operation has to take place within the international regulation of spectrum. These individual national administrative procedures are used to regulate SNG uplinks both in terms of political and technical controls. The constant push to provide the latest news is the driver behind allowing satellite newsgathering greater freedom, particularly into places where it is not wanted and even viewed with mistrust and fear. Broadcasting and telecommunications are matters of national security and social policy in many countries, and in some countries the concept of foreign newsgatherers bringing in their own transmitting equipment is a question of sovereignty. The issues of obtaining permissions and licences in different countries will be examined later, along with the impact of impending new international agreements.

To operate an SNG uplink system anywhere in the world, there are normally three areas of administration that have to be covered:

- Permission has to be granted to the uplink operator, on either a temporary or permanent basis, to operate the uplink in the country and area where it is to be operated.
- The SNG uplink operator has to obtain authority, usually by way of registration, with satellite system operators who may be providing the space segment to be used, to access that space segment with the uplink equipment.
- Space segment has to be secured by way of a satellite transponder/channel booking.

Looking at all these issues, we will see how and where they interact – and to begin with, we will look at how the electromagnetic (or frequency) spectrum is regulated as a resource. To appreciate fully the administrative context within which satellite newsgathering operates, we need to understand the regulatory aspects of spectrum. There are a bewildering number of organizations and bodies involved in the setting of standards, regulation and administration of the spectrum. Amongst all of this, we need to understand where the process of satellite newsgathering overlaps with international and national regulation.

6.1 Spectrum

There is a constant pressure for the resources of bandwidth and power by telecommunications services providers seeking to deliver data, telephony and video services, particularly in the sectors of the frequency spectrum used for satellite communications. The desire to deliver more services to more people has never been greater nor technologically more feasible, and to meet this pressure there is a need to be ever more efficient in the allocation and use of spectrum. The drive to provide these services comes from the obvious opportunities for commercial gain, spurred on by a desire to provide increasingly sophisticated means of communicating.

In telecommunications, the electromagnetic spectrum resource is commonly referred to as 'RF' – the abbreviation for radio frequency. In this context, 'radio' does not refer to radio media, such as radio stations and their broadcasts, but is used in the scientific sense, as defined in a dictionary as 'rays' or 'radiation'. This can be confusing for non-scientists, but is in fact the original definition of 'radio'. It has also been historically referred to as 'airwaves' or as the 'ether', and a specific frequency in the spectrum can be referred to as a 'wavelength'. But all these terms refer to aspects of electromagnetic spectrum.

The known electromagnetic spectrum ranges from nothing to gamma radiation, with visible light just over half way up the spectrum, as shown in Figure 6.1 (the commercial satellite frequency bands of interest for SNG are shown expanded). Electromagnetic spectrum can be viewed as a natural element, such as water or air, but only available in a relatively controlled amount to be useful. It can be imagined that without international agreement and regulation of usage there would be chaos. Each country could use whatever spectrum and transmitter power level they pleased, but as a natural resource and no respecter of man's political boundaries, the airwaves would be crowded with transmissions interfering with each other. Anyone who has tried to listen to short-wave or medium-wave radio at night will hear this type of interference as atmospheric conditions alter at night to 'throw' the signals further afield. There is also a significant amount of interference from space affecting many parts of the frequency spectrum.

The international regulation of the spectrum deals in fact with 'normal' conditions, and cannot fully take account of random atmospheric effects caused by the onset of night or abnormal weather conditions, which can particularly affect the lower end of the spectrum as used for most radio transmissions.

There are a number of regional and global standards-setting organizations, which are linked either directly or voluntarily. There is, however, a single global body that

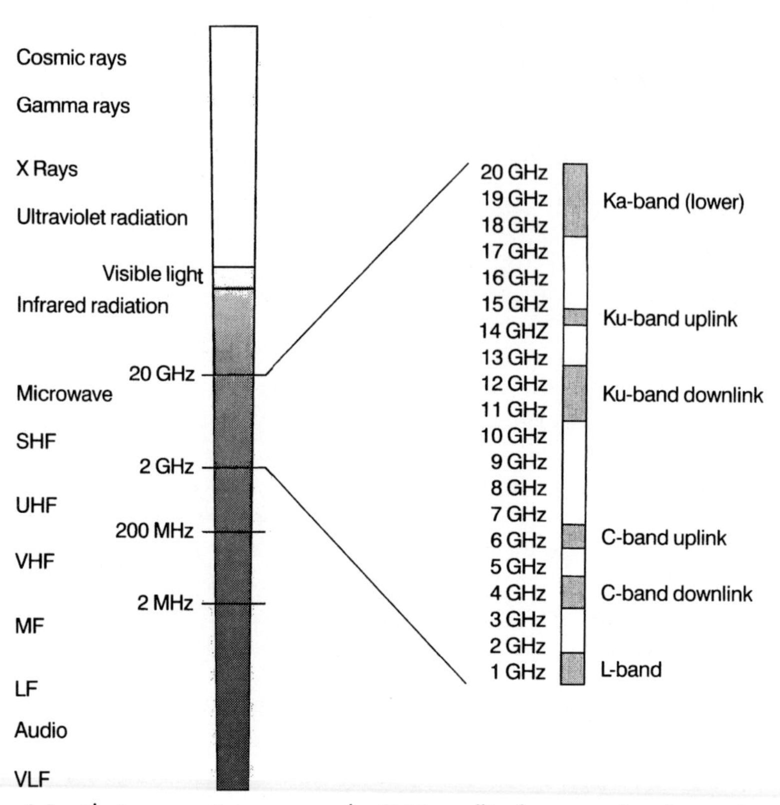

Figure 6.1 Electromagnetic spectrum: the SNG satellite frequency bands are shown expanded

oversees all aspects of the use of spectrum – the International Telecommunications Union (ITU). It encompasses all administration of spectrum, and it is worth spending a little time examining this organization.

6.2 The ITU

The International Telecommunications Union (ITU) is concerned with international co-operation in the use of telecommunications and the radio frequency spectrum, and with promoting and facilitating the development, expansion and operation of telecommunications networks in developing countries. It is the leading world body for the co-ordination and development of technical and operating standards for telecommunications and radiocommunications (including satellite) services.

The ITU, founded in Paris in 1865 as the International Telegraph Union, changed its name to the International Telecommunication Union in 1934, and became a specialized agency of the United Nations in 1947. It is an intergovernmental organization (IGO), currently made up of 185 member states and 363 members (scientific and industrial companies, public and private operators, broadcasters and regional and international organizations), within which the public and private sectors co-operate for the development of all telecommunications for the common good. It creates and defines through agreement international regulations and treaties governing all terrestrial and space uses of the frequency spectrum as well as the use of the geostationary satellite orbit. Within this framework, countries adopt their national telecommunications legislation. It also has a particular mission to promote and facilitate the development, expansion and operation of telecommunications networks in developing countries. However, it is important to realize that the ITU does not have the power to impose regulations – it is not a supranational power – and its strength is through agreement between member countries.

The ITU is therefore the ruling world body for the co-ordination and development of technical and operating standards for telecommunications, including satellite services. However, it should be noted that the ITU has no powers of enforcement, and relies on the consensual co-operation of all its members to achieve its objectives.

6.2.1 A short history of spectrum administration

The invention and subsequent patenting of wireless telegraphy (the first type of radio communication) by Guglielmo Marconi in 1896 saw the utilization of this new technology primarily for maritime purposes. The beginning of spectrum administration was prompted from an international dispute involving the Marconi Company.

Guglielmo Marconi formed a company to commercially exploit his invention in the maritime market – which was clearly the major area of development. In 1901, Marconi signed an exclusive contract with the insurers Lloyd's of London to provide Marconi operators and Marconi equipment on board insured ships to track their progress, and Marconi established a presence in all the major seaports of the world. Meanwhile, since competitors from America and Germany had appeared, the Marconi Company established its most controversial policy, known as the 'non-intercommunication rule'. Marconi operators on ship or shore could only communicate with other Marconi operators. Clients using other apparatus were excluded from the Marconi network; only in the event of a serious emergency was this rule to be suspended.

In March 1902, the German Kaiser's brother, Prince Henry of Germany, was returning home to Germany after a highly publicized visit to the United States. He was sailing aboard the German liner *Deutschland*, which was equipped with wireless equipment made by a German competitor to Marconi – Slaby-Arco. None of the Marconi stations on either side of the Atlantic would communicate with the ship because of its rival apparatus. Prince Henry, who tried to send wireless messages to both the US and Germany, was outraged.

Following this incident, Germany demanded that an international conference be called to discuss wireless telegraph ('radiotelegraph') communications and the monopoly that the Marconi Company held over maritime communications. A month before this first international wireless conference in Berlin in 1903, Telefunken was formed by two competing German firms, Slaby-Arco and Braun-Siemens-Halske, to present a united German front against Marconi, encouraged by the German government. Although the conference was supposed to address a number of wide-ranging issues, the only real issue was the Marconi Company's refusal to communicate with other systems. All the countries at the conference, with the exception of Italy and Great Britain, favoured compelling Marconi to communicate with all ships because they opposed his *de facto* monopolization of the airwaves.

The emerging problems surrounding the financing and regulation of the new technology, and the sanctity of each country's territorial airwaves, were embodied in the Marconi-German confrontation. The issue over whether a private company should gain dominance over a natural resource such as the airwaves and dictate their use was overriding. Germany, France, Spain and Austria had all assumed control of wireless in their own countries under governmental control because of its military significance. While Marconi was involved in commercial exploitation, the governments of these countries saw huge strategic value in the airwaves.

However, the issue was not resolved at this first conference, but in 1906 the International Radiotelegraph Conference was held in Berlin and the International Radiotelegraph Convention was signed. The annex to this Convention contained the first regulations governing wireless telegraphy – the Conference required any public shore station to communicate with any wireless-equipped ship, so defeating the monopoly of the Marconi Company, and included the adoption of the 'SOS' distress signal. These and other regulations, which have since been amended and revised by numerous World Radio Conferences held over the years, formed the basis of the Radio Regulations (RR).

In 1924, the Comité Consultatif International des Communications Téléphoniques à Grande Distance (International Telephone Consultative Committee, acronym CCIF) was set up ('F' stands for 'fils', a French colloquialism for cable). The Comité Consultatif International des Communications Telegraphes à Grande Distance (the International Telegraph Consultative Committee, acronym CCIT) was set up in the following year. Then, in 1927, the Comité Consultatif International des Communications Radio à Grande Distance (the International Radio Consultative Committee, acronym CCIR) was also established, and these committees were jointly made responsible for all technical studies in telecommunications. The International Consultative Committees (CCIs) thus became involved in the preparation of the Union's regulatory conferences at which international agreements governing all types of telecommunications were concluded. (In 1956, the CCIF and the CCIT merged to form the Comité Consultatif International des Communications Téléphoniques et Telegraphes à Grande Distance, acronym CCITT.)

In 1927, the Union allocated frequency bands to the various radio services existing at the time (fixed, maritime and aeronautical mobile, broadcasting, amateur and experimental). This was to ensure greater efficiency of operation, particularly in view of the increase in the number of services using frequencies and the technical requirements of each service.

In 1947, after the Second World War, the ITU held a conference with the aim of developing and modernizing the organization. Under an agreement with the United Nations, it became a specialized agency of the United Nations in October 1947, and the conference decided that the headquarters of the organization should be transferred in 1948 from Bern to Geneva. The International Frequency Registration Board (IFRB) was set up to manage the frequency spectrum, which was becoming increasingly complicated and congested, and the Table of Frequency Allocations originally introduced in 1912 became mandatory. This Table allocates to each service using radio spectrum specific frequency bands with a view to avoiding interference between stations, such as in communications between aircraft and control centres, ships at sea and coast stations, radio stations, or spacecraft and earth-based stations. As mentioned at the beginning of the chapter, freedom from interference is one of the primary reasons for international co-ordination on spectrum usage.

The Frequency Allocation Table forms Article 8 of the ITU Radio Regulations, is over 200 pages long, and is the most frequently referred to document of the ITU. It divides the world into three administrative regions: Europe, Africa and Northern Asia (Region 1); the Americas (Region 2); and Southern Asia and Australasia (Region 3). Each region has individual allocations of frequencies for some services, and common allocations for others. Frequencies are allocated for primary and secondary usage, so that some parts of the frequency bands are allocated on an exclusive basis to some services and shared for other services. National administrations are not bound to allow usage of frequencies as allocated in the Frequency Allocation Table in their country, as local priorities may dictate otherwise. On the other hand, national administrations in general do not permit usage of a particular frequency band for a purpose other than that defined for that region in the Frequency Allocation Table.

The inauguration of the space age in 1957 with the launch of the first artificial satellite, SPUTNIK 1, by the former USSR posed new demands on the ITU. In 1959, to meet the challenges of the space age, the CCIR set up a Study Group responsible for studying space radiocommunication, and in 1963 an Extraordinary Administrative Conference for Space Communications was held to allocate frequencies to the various space services. The first telecommunications satellite, 'TELSTAR 1', had been launched and put into orbit in 1962 and the ITU subsequently administered space communication uses of spectrum.

6.2.2 The role of the ITU

As well as administering frequency allocations for satellite use, the ITU also acts as the global co-ordinating body for the allocation of orbital slot positions for satellites. The ITU allocates sectors of the geostationary arc nominally to countries in the region below each arc sector. There are of course certain areas of the geostationary arc which give excellent continental and transcontinental coverage, and these are highly sought after by satellite operators as they offer significant commercial advantages.

National administrations, in co-operation with the ITU, require operators of potential satellites to lodge an application for a particular orbital slot in the geostationary arc with the relevant national administration. The national administration then co-ordinates the application with the ITU, checking for potential conflicting demands with adjacent national administrations, and thus the ITU, as well as allocating the arc sectors, also acts as a global 'clearing house' for individual orbital slot allocations. These applications are made years in advance of the actual placing of the satellite in position, and many organizations place applications and hold allocations for orbital slots that may never be used – otherwise known as 'paper' satellites. In an environment where there is an ever-increasing demand for slot allocations, there is great pressure to reduce the number of 'paper' satellites.

6.2.3 The structure of the ITU

The ITU Council governs ITU affairs between each of the ITU Plenipotentiary Conferences, which are held every four years. The ITU Council is composed of forty-six Members of the Union elected by the Plenipotentiary Conference, representing all five regions of the world (Americas, Western Europe, Eastern Europe and Northern Asia, Africa, Asia and Australasia). The Member Governments are party to the Constitution and Convention, which are the basic parts of the ITU treaty. Only the ITU's Plenipotentiary Conference can alter these basic treaties, which are augmented by the International Telecommunication Regulations and the ITU Radio Regulations (ITU-RR).

The ITU structure is based on three sectors:

- The radiocommunication sector (ITU-R) provides the forum for treaty-level agreements on international use of radio frequencies and for broadcasting, radiocommunication and satellite transmission standards. The regular summit conference for this sector is the World Administrative Radiocommunication Conference (WARC).
- The telecommunication standardization sector (ITU-T) establishes global agreements on standards ('Recommendations') for telecommunications. The regular summit conference for this sector is the World Telecommunication Standardisation Conference (WTSC).
- The telecommunication development sector (ITU-D) provides technical assistance to developing countries within a new strategic planning framework. The regular summit conference for this sector is the World Telecommunication Development Conference (WTDC).

Of these, only the WARC is at treaty level and can devise and amend the Radio Regulations. Broadcasters, telecommunications carriers and other elements of the communications industry have a close interest in the work of the ITU, and many participate in the ITU sectional work in Study Groups. Each year, there are as many as eighty ITU meetings, including the specialized Study Groups. The WARC is the relevant body which has direct control over the frequencies used for satellite newsgathering around the world.

6.2.4 Radio frequency standards

As previously mentioned, in 1927 the International Radio Consultative Committee (CCIR) was established by the ITU, one of three Consultative Committees. The CCIR undertook tests and measurements in the various fields of telecommunications, and contributed to the drawing up of international standards. It was also in 1927 that the ITU allocated frequency bands to the various radio services existing at the time (fixed, maritime and aeronautical mobile, broadcasting, amateur and experimental). The primary aim was to ensure greater efficiency of operation in view of the increase in the number of services using frequencies and the technical requirements of each service.

The CCIR was originally the standard-setting section of the ITU, and regulating the use of frequencies is an essential aspect of the work of ITU. To carry out this work, it was decided to separate the standards-setting activities of the CCIR from its management of the radio-frequency spectrum in terrestrial and space radiocommunications. The standards-setting functions have been merged with those of the former International Consultative Telegraph and Telephone Committee (CCITT) to form the telecommunication standardization sector, ITU-T. The management of the radio frequency spectrum, along with the regulatory activities formerly carried out by the International Frequency Registration Board (IFRB), was integrated into the new radiocommunication sector, ITU-R.

The role of ITU-R is to ensure the efficient and economical use of the radio spectrum, including the geostationary satellite orbit, and carry out studies from which recommendations are made. Areas covered include:

- Spectrum utilization and monitoring
- Interservice sharing and compatibility
- Science services
- Radio wave propagation
- Fixed-Satellite Service (FSS)
- Fixed services
- Mobile-Satellite Service (MSS)
- Sound broadcasting
- Television broadcasting

Of particular interest in ITU-R is the Study Group known as Working Party 4 (WP 4), which deals with fixed-satellite services, and the sub-group WP 4SNG. WP 4SNG undertakes studies specifically relating to satellite newsgathering, and in particular has issued several recommendations for SNG, the most significant being Recommendations ITU-R SNG.722–1 ‘Uniform technical standards (analogue) for satellite news gathering (SNG)’ and the subsequent ITU-R SNG.1007 ‘Uniform technical standard (digital) for satellite news gathering (SNG)’.

WP 4SNG also compiles the *SNG Users Guide*, which details not only the relevant contact points in member administrations which deal with SNG access in each member country, but also lists SNG service providers for each country. It is published by the ITU and is available in a hard copy version on a subscription basis or electronically, free of charge, via the ITU website (www.itu.int).

6.2.5 WARC

Under the control of the ITU, the World Administrative Radiocommunication Conferences are held every two years, lasting four weeks, along with a Radiocommunications Assembly. The main function of WARC is to review and revise as necessary the Radio Regulations, on the basis of an agenda adopted by the ITU Council following consultation of the membership of Member Administrations.

The general scope of this agenda is established four years in advance and the ITU Council establishes the final agenda usually two years before each conference. The agenda usually focuses on a particular aspect of spectrum management and specific services, though there is normally a WARC to consider the entire frequency spectrum every twenty years. WARC may also recommend to the ITU Council items for inclusion in the agenda of a future conference and give its views on forthcoming agendas for at least a four-year cycle of WARC.

After WARC-92, the 'Administrative' part of the title was dropped, to become the World Radicommunications Conference (WRC), following the decision to extend participation to allow observers from non-governmental organizations (NGOs) to attend. However, only Member Administrations have voting powers.

The WRCs are attended by experts and administrators from each member country to rule by international agreement on the allocation of all radio frequencies in the range from 9 kHz to 400 GHz, and to define RF power levels for each service in each region. The WRC are open to all ITU Member Administrations and to the United Nations, international organizations, regional telecommunication organizations, intergovernmental organizations (IGOs) operating satellite systems, the specialized agencies of the United Nations and the International Atomic Energy Agency (IAEA). In addition, system operators authorized by their country to participate in the work of the sector are allowed to participate.

The deliberations of WRC are long and weighty, and its reports are highly technical and not light reading. The radiocommunication agencies of each country adopt the relevant WRC agreements into national regulations for their own administration, although there may be regional amendments.

WRC-97

The most recent WRC was held between 27 October and 21 November 1997 in Geneva and, not surprisingly, it considered a number of satellite-related services. Its deliberations resulted in the production of 600 pages of completed text negotiated by nearly 1800 delegates over the four-week period. In summary, the following was decided:

- It decided to begin to re-plan the Broadcast-Satellite Service (BSS) in Regions 1 and 3.

- It made regulatory provisions to permit the future operation of non-geostationary Fixed-Satellite Service (FSS) systems in both the Ka- and Ku-bands, and the conclusion of agreements on important spectrum allocations for the operation of space systems. In particular it was concerned with the Earth-exploration satellite service and the growing demand for feeder links of non-geostationary Mobile-Satellite Service (MSS) systems (see Chapter 9).
- It also reached a compromise solution to the difficult issue of accommodating MSS interests in the Radionavigation Satellite Service in the 1.5 GHz band – this involves safety services and must be protected from harmful interference. This was achieved through the promise of the ITU-R undertaking studies on an urgent basis to determine the feasibility of a downlink allocation in a portion of the band in the future.

Not least, the conference made a number of recommendations for the agenda of the next WRC, to be held in the spring of 2000 (WRC-2000.) This is delayed by six months from the normal two-year cycle because of the burden of work in progress.

The wide range of satellite services regulated by the WRC includes Meteorological, Radio Navigation, and Radio Astronomy among many others, as well as FSS, MSS and Broadcast-Satellite Service (BSS). MSS includes Land Mobile, Aeronautical Mobile and Maritime Mobile Services; satellite newsgathering services are encompassed within FSS and MSS.

The significance of the ITU and its organization is in establishing the overall regulatory framework within which SNG functions, and the future development of SNG will depend to a large degree on the decisions of the ITU.

The ITU is a regionally orientated organization, and it is clear that regional variations are allowed for and encouraged to meet the particular needs of that region. Europe has for a number of decades sought to establish its own regulatory framework as part of the overall principle of achieving a unified European approach. Several organizations have been established.

6.3 CEPT

The Conference Europeane des Administrations des Postes et des Telecommunications (CEPT) was founded in 1959 as an official body for developing postal and telecommunications services between European administrations, and making recommendations to improve these services. It formulates and recommends telecommunications standards for member countries, which although not obligatory, are in general adopted by all members. The European Radiocommunications Committee (ERC), governed by the CEPT, administers spectrum matters for Europe through the European Radicommunications Office (ERO) in Denmark, which is staffed by experts in the field of frequency management. The ERO is focused on undertaking studies looking ahead to demands for frequency management in Europe.

6.4 ETSI

The European Technical Standards Institute (ETSI) is a non-profit-making organization set up in 1988 on the joint initiative of the EU and the CEPT to produce telecommunications standards for telecommunications across the European Union and beyond. Its aim is also to facilitate the 'single market' for telecommunications within Europe by harmonizing separate technical standards to form a set of harmonized European standards.

ETSI plays a major role in developing a wide range of standards and other technical documentation as Europe's contribution to worldwide standardization in telecommunications, broadcasting and information technology. ETSI's prime objective is to support global harmonization by providing a forum in which all the key players can contribute actively. Based in Sophia Antipolis (France), ETSI is an open forum that has almost 700 members from fifty countries representing national administrations, telecommunication network operators, manufacturers, broadcasters, service providers and users. Its members, who are responsible for approving its objectives, therefore determine the Institute's work programme. As a result, ETSI's activities are maintained in close alignment with the market needs expressed by its members.

Any European organization proving an interest in promoting European telecommunications standards has the right to join ETSI and thus to directly influence the standards-making process. ETSI produces voluntary standards – some of these may then be adopted by the EU as the technical basis for Directives or Regulations – and because those who subsequently implement these voluntary standards have a hand in defining them, the standards are practical rather than abstract.

The relationship between the standards and the Directives of the EU is often very close. Standards are formulated which may later become a 'technical basis for regulation' (TBR) and eventually may be converted into a 'common technical regulation' (CTR) of the EU. The time-scale for these standards to become CTRs is measured in years rather than months, and some national administrations may adopt these standards in advance of them becoming a common EU standard. Throughout all of this, however, is the focus on developing harmoniszed European standards. ETSI conforms to the worldwide standardization process whenever possible and the work is co-ordinated with the activities of international standardization bodies, mainly the ITU-T and the ITU-R.

ETSI consists of a General Assembly, a Board, a Technical Organization and a Secretariat. The Technical Organization produces and approves technical standards. It encompasses ETSI Projects (EPs), Technical Committees (TCs) and Special Committees. More than 3500 experts are working for ETSI in over 200 groups.

There is a very wide range of standards produced by ETSI covering telecommunications, but of particular interest for satellite newsgathering is CTR-030 (formerly ETS 300 327 and then TBR-030) and ETS 300 673, as they relate to 'type-approval'.

6.5 Type-approval

Type-approval of an uplink system may be required in a country before a licence can be granted. This may be an international standard, as in Europe, or a national standard, such as that in the US.

Type-approval is also used in a slightly different sense. As will be seen later, satellite system operators require registration of an SNG system before access is permitted to their system. In particular, the performance of the antenna is of great interest as it is the single most important component that has an effect on interference with adjacent signals on a satellite. To facilitate registration for the end-users, many antenna manufacturers have obtained 'type-approval' of their antenna products with satellite system operators and national authorities.

6.5.1 CTR-030 (ETS 300 327)

TBR-030 was originally published by ETSI as standard ETS 300 327 in 1994, issued as TBR-030 in 1997, and adopted by the EU as CTR-030 in September 1998. This standard was formulated to establish minimum performance specifications for the operation of SNG uplinks in Europe. It is limited to:

- Systems working in the Ku FSS bands.
- Systems having attended operation, i.e. operator constantly with the uplink.
- Systems that have an antenna size of less than 5 m.
- Systems that can operate to satellites with 3° spacing in the geostationary arc.

In common with ITU regulations, the standard is aimed to minimize the risk of interference, and defines parameters that the system has to meet in terms of off-axis power radiation, transmit polarization and mechanical performance to maintain pointing accuracy.

There is also another ETSI standard – ETS 300 673 – originally issued in 1997 and being revised in 1999, which deals with the electromagnetic compatibility (EMC) required of satellite earth stations, including SNG uplinks.

It is a common requirement for licensing in European countries that SNG uplinks meet both these standards. Indeed, as will be seen later, the German RTP (formerly BZT) type-approval is directly related to these standards. Proof of a system meeting both CTR-030 and ETS 300 673 may be demonstrated by having obtained German type-approval from the former Bundessamt für Zülassengun in Telekommunikation (BZT).

The BZT was privatized in early 1998 as part of the overall privatization of state-controlled German telecommunications and is now called CETECOM GmbH. 'BZT approval' (as it is still commonly referred to) is recognized in most other European countries that currently have no type-approval system of their own.

6.5.2 FCC

In the US, the Federal Communications Commission (FCC) is the regulatory body for radio spectrum. In 1927, the Federal Radio Communications (FRC) was formed to allocate radio spectrum. The Federal Communications Commission was formed in 1934 by combining the functions of the FRC and interstate regulation for common carriers.

The FCC sets the standards that all radio transmission equipment must conform to, including SNG uplinks. All SNG uplinks have to be registered and licensed with the FCC prior to use in the US, and there are technical criteria that must be met before the uplink can be granted a licence. These are set out in Title 47 (Telecommunications) of the Code of Federal Regulations (CFR), Part 25 (Satellite Communications).

The FCC, in accordance with ITU recommendation, requires that the antenna meet the requirement to be able to operate to satellites at only 2° spacing in the geostationary arc. This is defined in terms of antenna transmission sidelobe peaks, which automatically sets a lower limit to the antenna size (due to the physics relating to beam width). This means that in the US, it is not accepted that it is possible to use an SNG antenna smaller than 1.2 m.

On the other hand, EUTELSAT, the principal European regional satellite operator, also demands compliance to 2° spacing for operation onto their capacity, but will allow antennas as small as 0.9 m provided that compliance to 2° spacing can be proven. The 2° criterion is often referred to by engineers as the '$29 - 25 \log_{10}\theta$' characteristic (pronounced '29 – 25 log theta'), and this technical characteristic is described more fully in Chapter 2. The reason for the demand for 2° spacing is that, in the 1980s, it was anticipated that the number of satellites was going to steadily increase and there was going to be a need to move satellites closer together to increase capacity. Therefore the old CCIR requirement of '$32 - 25 \log_{10}\theta$', which effectively defined 3° spacing, was modified by a number of satellite operators and some national administrations, including the FCC in 1983.

Most SNG antennas registered in the US are 'type-approved' by the antenna manufacturers to FCC requirements in terms of transmit beam patterns (i.e. compliance to the $29 - 25 \log_{10}\theta$ characteristic). The registration application requires the type-approved patterns (i.e. design patterns) for the particular model of antenna to be submitted, along with other information including maximum uplink power and modulation schemes to be used in the system, as well as administrative information. If the antenna is not type-approved by the FCC, then range test patterns of the specific antenna requiring registration have to be submitted.

The FCC also requires a practical test of the SNG uplink to be completed within one year of the initial registration to demonstrate compliance. This test can either be carried out on an antenna test range or more usually on-satellite, and the results are then submitted to the FCC. The licence is then finalized if the uplink tests are satisfactory.

The registration application also requires an RF hazard study to be submitted. This deals with the potential hazard from non-ionizing radiation, and this issue is discussed fully in Chapter 8. In essence, the study has to demonstrate that the operation of the uplink will not cause any significant hazard to either the public or the uplink operator.

The FCC process is somewhat bureaucratic, but provided all the steps are carried through in the correct order and manner, then registration is a relatively straightforward process.

6.6 National control and trans-border issues

The movement of mobile satellite transmission equipment around the world is often fraught with difficulties. Some countries prohibit any use of mobile satellite communications equipment totally, while others permit it only in particular circumstances, such as for disaster relief or emergencies, or in limited geographical areas. High licence fees, taxes and customs duties cause significant difficulties for newsgatherers; in some countries additional type-approval is sometimes demanded, even though the equipment has been type-approved elsewhere or meets internationally recognized standards, as a further hindrance.

Often these regulatory barriers exist because the country does not have a policy or regulatory framework covering mobile satellite telecommunications or because they fear 'bypass' of their terrestrial network (even in regions where there is no network to bypass), so reducing their telecommunications revenues.

6.6.1 The control of SNG

The national administrations dealing with telecommunications, even in these days of increasing privatization, are still in general under government direction, and they control access to satellite capacity by SNG uplinks to varying degrees. The level of control varies from countries where the use of SNG uplinks cannot even be contemplated, to countries where there is a highly regulated control system for allowing access to space segment by SNG uplinks through licensing and frequency clearance. The latter is often used as a way of generating a lucrative income stream.

Cross-border issues are probably the largest remaining problem in the free movement of satellite newsgathering. Newsgatherers usually want to take their own facilities into a foreign country for some or all of the following reasons:

- The use of their own SNG uplink may be more cost-effective than using the local telecommunication facilities.
- There may be a feeling that greater editorial freedom is available by having their own facility.
- The story being covered may involve areas of a country that do not have any local telecommunications structure to support the rapid transmission of reports.

The technology allowing easy movement of SNG uplinks has advanced at a far more rapid rate than political and regulatory recognition in a significant number of countries. Although many national administrations allow satellite newsgathering operations by their own nationals within their borders, many still forbid foreign operators from entering their country. However, changes in global agreements should lead to the easing of restrictions with respect to allowing foreign newsgatherers to transport their own equipment into other countries.

In particular, INMARSAT has worked hard to ease the restrictions on trans-border use of some types equipment needed as part of its system (see Chapter 5). To this end, in 1997 the ITU ratified a Recommendation[1] that at least encouraged national administrations to allow the passage across their borders and the use of INMARSAT satphones for satellite newsgathering for radio. INMARSAT now issues a certificate (with the text in seven languages) for carriage with INMARSAT satphones to ease the problems of importation into a country for use by newsgatherers (see Appendix D). Furthermore, the ITU has implemented a Memorandum of Understanding on the use of equipment for Global Mobile Personal Communication by Satellite (GMPCS)[2], which is dealt with in Chapter 9.

6.6.2 World Trade Organisation

In the aftermath of the Second World War, three international bodies were set up by the UN in 1948 to promote agreements on international economic development and trade. These three organizations were the World Bank, the International Monetary Fund (IMF) and the International Trade Organisation (ITO). Although the other two have survived, the ITO never flourished and instead the General Agreement on Tariffs and Trade (GATT) was established. The aim of GATT was to encourage free trade between nations by reducing tariffs, subsidies, quotas and regulations that discriminate against imported products.

In 1995, after over seven years of negotiation in the so-called 'Uruguay Round' of discussions of the members of GATT, a new body called the World Trade Organisation (WTO) was created to administer all world trade agreements. In the same year, the General Agreement on Trade and Services (GATS) was agreed, and the WTO was henceforth charged with administering both GATT and GATS. There are 132 member nations who are signatories to the WTO.

The GATS covers virtually all services including those involved with broadcasting such as audio-visual and telecommunication services. Agreement was reached in the WTO negotiations within the Group on Basic Telecoms (GBT) in 1997.

The successful agreement comprises commitments by about seventy countries with regard to market access and rules concerning fair market practices. As a result of this Memorandum of Understanding (MoU), basic telecommunications services will be covered by the GATS, ensuring that market access is transparent and available on terms and conditions that are reasonable and non-discriminatory. The purpose is to lower customs duties on information technology and telecommunications equipment, including mobile satellite earth stations, in the context of the Information Technology Agreement, agreed by WTO ministers in December 1996.

Many countries use the UN Central Product Classification (CPC) as the basis for determining their commitment under GATS. However, the CPC is not updated at the same rate as technology advances, and it is constantly out of date. Some countries deem broadcasting to be an audio-visual service, while in other countries it is deemed a telecommunications service. Different aspects of broadcasting can be split into either of these service definitions. However, only thirteen countries have made market access commitments for audio-visual services.

The WTO agreement covers the provision of telecommunications services involving simple transmission such as voice and fax, and data transmission including the supply of international and domestic services on a facilities basis and through resale.

Countries have committed to varying degrees of market access through allowing market entry for new carriers and resellers and through foreign investment in existing carriers and resellers. The stress is on the phrase 'varying degrees' so far as satellite newsgathering is concerned. The free movement of SNG uplink systems is thought to potentially fall under this agreement, although many countries may deem it an audio-visual service. Some undertook significant market liberalization while others made commitments based on their existing regulatory regimes. Final details on the extent of the liberalization for each country are not yet fully known. Nevertheless, the WTO agreement represents an important step toward a worldwide liberal telecommunications market.

The important point for our purposes is that the WTO MoU on Telecommunications offers the prospect of increased freedom of movement for SNG uplinks across national borders within the next five years.

6.6.3 National controls

There are two main reasons for reluctance by national governments to allow free movement of SNG uplinks across national borders. Firstly, there is the fear of the power of unbounded dissemination of information that the use of satellite newsgathering can bring, and many countries have been very slow to react to the rapid changes in technology. Many governments are suspicious of the motives of newsgatherers who wish to operate uplinks, fearing open criticism of their political regimes which their own population may see or hear coming back via broadcasts from abroad. So they forbid any news organizations using satellite newsgathering within their borders as a matter of political control. They may not even allow their own broadcasters to use this equipment, and certainly would not allow foreign newsgathering agencies to bring in their own SNG equipment. Some allow foreign SNG uplinks to operate but insist on government supervision during transmissions, including 'screening' tapes before they can be fed. The ease with which foreign broadcasters are able to comment on policies and actions is a direct reflection of the fear that exists relating to the power of the operation of an SNG uplink.

It may be that a government has no legislation to cope with licensing the use of SNG uplinks, though increasingly this is changing in many countries. Governments may allow the use of satellite newsgathering by their own national entities, which may be either a single monopoly supplier or a limited and controlled number of providers.

Secondly, some national administrations bar the use of foreign SNG uplinks because of the anticipated loss of income via their own broadcast and telecommunication facilities. A 'licence' or access fee may be levied if SNG equipment is to be allowed into a country, which is estimated to equate to this anticipated loss of revenue. Some foreign newsgatherers may pay this simply to obtain the flexibility that having their own facility can give them, or they may decide not to bring in their own uplinks and use the facilities available within the country.

In general, there are two levels or types of access granted to SNG uplink operators, depending on the degree of advancement of telecommunication administration in a country. These categories are loosely 'permissions' and licences. The point at which a 'permission' can be termed a licence is a little blurred. There is a further control, called frequency co-ordination, which is related to purely technical considerations of potential interference by the SNG uplink.

6.6.4 Permissions

Where a country has no system of licensing for SNG uplinks, a formal 'permission' issued by the government of the country is the administrative route by which foreign newsgatherers obtain clearance to take uplinks into a country, usually on a temporary basis. Often this 'permission' is granted on a political basis, after reference by the national telecommunication administration or the Ministry of Information (which are sometimes one and the same) to the political leadership of the country. The permission is granted depending on a political calculation weighing up whether granting access may work to the advantage of the leadership. Therefore very often permission may be granted on one occasion and refused on another. Permission may be granted for a foreign satellite uplink if it is to cover an international sporting event taking place in the country in question, but refused if it is to cover a news event of political interest. As previously mentioned, the fact that their own population may see or hear material detrimental to the political leadership coming back via broadcasts from abroad – courtesy of CNN International or BBC World Service, for example – can be a significant fear.

There may also be a financial price associated with that permission. This may be related in some way to the perceived loss of revenue (which can, in reality, be a significant loss of foreign revenue) because a foreign broadcaster uses their own facility rather than that country's TV station and satellite earth station. Sometimes this fee can run to thousands of dollars, to the extent that it directly equates to the equivalent use of a country's facilities. On these occasions, the only benefit that newsgatherers obtain is the known quality and convenience of their own facility.

Incidentally, it is often the case that where a permission has to be obtained with a national administration which is apprehensive or obstructive to the idea of a foreign operator sending in their own uplink, local contact on the ground is by far the best way of moving matters forward. This is often a cultural issue, where the 'western' practice of making demands by telephone or fax are seen as being insulting, and 'pressing the flesh' over cups of iced tea or coffee is far more likely to produce results. Financial inducements to officials are common in many cultures and do not hold the same negative connotation as in western culture.

6.6.5 Licences

The approximate differentiation between permissions and licences is the legislative structure underpinning telecommunications that exists within a country. In countries with a statutory system regulating the use of telecommunications, including satellite newsgathering, a structure is usually in place for the issuing of licences, subject to certain provisions or restrictions. Licences can be granted on a per occasion basis or even on an annual or regularly renewable basis. Typically SNG uplink operators in their native country are granted permission to operate under the terms of a licence.

Obtaining a licence may involve a number of steps. It may involve simply filling out a form giving details of the technical parameters of the uplink. It may require details of any satellite system registrations the uplink already has, or a declaration as to whether it is going to be connected to a telecommunications network or operate as a stand-alone uplink – which is typically the case for satellite newsgathering. It may involve also submitting what are referred to as 'range patterns'. These are frequency test patterns of the performance of the antenna measured on a test range, and define the performance of the antenna, and in particular prove its 'sidelobe' performance, i.e. its directional performance. This is particularly important as it shows the integrity of the antenna, particularly demonstrating that it will not cause interference.

Upon approval of the technical parameters, and payment of the necessary fee, the licence is issued for the specified period. There may be a further fee for renewal or if alteration to the licence is required due to changed parameters of the uplink system.

Some countries do not require so much technical information or proof that the antenna meets a particular standard. They simple require the uplink to be registered with them and the requisite fees to be paid.

Here are some examples of licences available for SNG uplinks in various countries around the world (this is the type of information available from the ITU *SNG Users Guide* referred to previously).

- In the UK, licences are granted on a per occasion basis or on an annual basis by the Radiocommunications Agency (RA). Fees are typically £500 per occasion or £8500 per annum for a TES (or transportable earth station – another acronym for an SNG uplink).
- In Germany, licences can be granted for a ten-year period, with an annual fee (currently DM30), by the Reguliersrungbehorde für Telekommunikation und Post (Reg.TP), provided type-approval for the system has been obtained (see below.) This was formerly, before privatization, the Bundesamt für Post und Telekommunikation (BAPT).

- In France, the Autorité de Régulation des Télécommunications (ART), again provided type-approval for the system has been obtained, will grant a licence for up to four years after initial application, at a current cost of 40 000FF.
- In Japan, the Ministry of Posts and Telecommunications (MPT) will only allow Japanese nationals to apply for licences. Additionally, no SNG uplink can operated for or by a foreign government or its representative. However, recently the Space Communications Corporation (SCC) of Japan, a private satellite company which owns and operates the SUPERBIRD series of satellites in the Pacific Ocean Region, has been enabled to sub-license foreign operators under their own licence for SNG uplinks.
- In Hong Kong, the Office of the Telecommunications Authority (OTA) will issue a Self-Provided External Telecommunication System (SPETS) Licence to operate an SNG uplink to foreign operators, upon payment of a fee, subject to certain conditions being met.
- In the US, an uplink is registered and licensed with the FCC subject to it meeting the required technical standards. Following approval, the registration is valid for the life of that system, and no payment is required beyond minimal initial filing and legal fees. (Incidentally, as in Japan, no SNG uplink is permitted to operate if acting on behalf of a foreign government or its agent.) Foreign operators may be able to facilitate the process if they have an ally amongst one of the US networks who will be able to help them with their application. (However, it has to be recognized that with the large SNG hire market in the US, the need for any foreign operator to bring their own uplink in is questionable.)

In Europe, the EU has a Memorandum of Understanding (MoU) in place on the mutual recognition of satellite communication licences. It is the beginning of an arrangement to achieve 'one-stop shopping' for licences for SNG uplinks across Europe.

This MoU was signed in 1992 by France and Germany, and in 1993 by the Netherlands and the UK. Through the agreement, an SNG operator wishing to obtain a licence in two or more of the four countries may apply to the regulatory authority of their choice, who then becomes the 'co-ordinator'. The co-ordinator is then responsible for collecting the information necessary to process the licence application in each country. The co-ordinator contacts the regulatory authorities concerned and co-ordinates the whole process and is responsible for conveying all the licences granted to the SNG operator. The agreement guarantees a maximum waiting period of two weeks for the processing of the applications. However, this system exists only in theory, as it has not been put into practice by the national administrations involved. Complications and disputes over the mechanics of the process have prevented the MoU being put into action as yet.

6.6.6 Frequency clearance

In some countries, frequency clearance is required for each individual transmission of the SNG uplink – this is in addition to having a licence, and the licence may be withdrawn if it is discovered that frequency clearance has not been obtained. The purpose of the frequency clearance is normally to ensure that the uplink transmissions will not interfere with any other transmission. The most likely services that can be interfered with are terrestrial point-to-point microwave links operating at or near the same frequency, where the transmission beam from the uplink may cut through or be very near to the point-to-point terrestrial microwave link path. The parameters typically required for frequency clearance are the transmit frequency of the uplink, its location, the satellite to which it is going to work, its orbital position, the ITU emission code, and the times of the transmission. The ITU emission code is an internationally recognized alphanumeric code that defines the transmission characteristics of the signal, as shown in Appendix E.

The national authority will either send back an acknowledgement authorizing the transmission, or in some countries no reply can be interpreted as consent. Frequency clearances are compulsory in C-band in most developed countries, and may take weeks or months to obtain due to the high likelihood of interference with terrestrial point-to-point microwave links commonly operating in this band, which may necessitate international co-ordination. However in Ku-band, a frequency clearance is sometimes only necessary in the 14.25–14.50 GHz band as this is often shared with microwave links, again necessitating co-ordination. In the lower part of the Ku-band used for SNG uplinks (14.0–14.25 GHz), co-ordination is often not required as this is an area of the band exclusive to mobile satellite applications in many countries.

However, many countries do not require any frequency clearance at all in the Ku-band (e.g. the US), so that this is one less piece of administration for the SNG uplink operator to be concerned about.

6.7 Conclusion

It can be clearly seen in this chapter that the 'legitimate' use of SNG uplinks involves varying degrees of administration in different countries. Although these processes can be seen as irksome in the face of meeting the challenge trying to cover 'breaking' news, or even sustaining basic news coverage, it is essential in the long term for the ordered and controlled use of a scarce natural resource.

The level of administration and the financial costs associated with operating an SNG uplink in many countries is likely to increase in an effort by national administrations to control the demand for spectrum. However, on the positive side, the continuing decrease in the bandwidth and power demands of digital SNG uplinks is likely to make these types of system more attractive to national administrations. The political control of the use of SNG uplinks is likely to remain a continuing problem, though there are signs of easing of restrictions in a number of countries that have previously taken a rigid stance in opposing the use of such equipment. It is hoped that this is in part through the efforts of the WTO to open up the market. Restrictive regulations usually impede only the country's own socio-economic and political development as, in addition to socio-economic benefits, a liberalized telecommunications environment will help generate new revenues in the country. However, for almost every sovereign state, telecommunications and access to radio spectrum is invariably as much a political issue as a technical one.

References

1. ITU (1995) Recommendation ITU-R SNG.1152, Use of digital transmission techniques for satellite news gathering (SNG) (sound).
2. ITU (1998) Ninth Plenary Meeting, Resolution 1116, Implementation of the GMPCS-MoU Arrangements.

7

Fitting the pieces together: satellites, systems and operations

7.1 Introduction

Having looked at the regulatory and administrative environment in which satellite newsgathering exists, we can now move on to examine some of the principal global and regional satellite systems, focusing on those that allow access for SNG, and how these can be accessed by SNG uplinks. It is not intended to detail every satellite that is in orbit, for the information changes quite often. It is best to consult the satellite operators directly, many of whom publish their current fleets on their web sites, or use one of the published almanacs (see Appendix K and the Bibliography). It is also important to note that the information listed here for any satellite system is only current as at the time of writing, and will probably have changed by the time you read this.

Overall, there are relatively few global systems, but many regional systems. As well as carrying SNG traffic, these systems also carry services such as:

- Telephony and data, including Internet connectivity.
- Private data networks for financial institutions, corporations and a wide variety of other businesses.
- Direct-to-home (DTH)/direct broadcasting services (DBS) for radio and TV.
- Backbone video/audio distribution services.

Many satellites are in orbit to provide these other services (although there are some that cater for SNG as well as other services). The range of services is wide, and the demand in all these areas is growing, feeding a multi-billion US dollar global industry. With about US$6 billion in revenue in 1998, satellite operators constitute a dynamic business, made even more challenging by consolidation and new entrants, globalization and the search for more added value in service provision. This in turn is applying pressure on spectrum, and the result of this is that the number of geostationary satellites is increasing, with virtually all the satellite system operators expanding their fleets. There is also an increasing number of mergers and

Table 7.1 Principal satellite systems that allow access for SNG

Satellite system	*Base country*	*Coverage areas*	*Number of satellites**
Apstar	China	Asia	3
Arabsat	Saudi Arabia	North Africa and Mediterranean	5
AsiaSat	Hong Kong	Asia, Middle East, Australasia	3
Columbia Communications	US	Global	3
Deutsche Telekom	Germany	Europe	2
Eutelsat	France	Europe, Middle East, Russia	12
France Telecom	France	Europe	3
GE Americom	US	Global	12
Hispasat	Spain	Europe and Americas	2
INTELSAT	US	Global	19
Intersputnik	CIS	Asia and Europe	7
JSAT	Japan	Pacific	5
Nahuelsat	Argentina	S.America	2
New Skies Satellites	Netherlands	Global	5
Optus	Australia	Australasia and Pacific	7
Loral Skynet/Orion/Satmex	US	Europe, Americas, Asia Pacific	10
Palapya	Malaysia	Far East	2
PanAmSat	US	Global	19
Superbird	Japan	Far East	3
Telenor	Norway	Scandinavia and Europe	3
Thaicom	Thailand	Asia, Australasia, Europe, Africa	3
Turksat	Turkey	Europe, Turkey, Central Asia	2

*In orbit in early 1999, excluding planned launches.
(Note: SES Astra, the Luxembourg-based satellite system operator, does not offer any SNG access.)

strategic alliances being formed to try to dominate as much of the market as possible; a recent example of this was the merger between Hughes Galaxy and PanAmSat. Table 7.1 shows the principal global and regional satellite systems that allow access for SNG uplinks.

It is also worth noting that some organizations operate their own satellites and lease capacity on a long-term basis on other satellite systems as well, usually in order to provide as diverse an offering of services and wide area of coverage as possible. These organizations tend to be smaller in size, without the resources to economically provide this added range. They may be in the process of expanding, and sub-leasing capacity from another operator is a stepping stone to fully developing their services. For other operators, it may be a way of offering some degree of service in an area that is only going to have a limited demand for them.

All satellite system operators, whatever their size, are aiming to provide as much of a ‘one-stop shop’ for their customers as they can, whether it be purely regional service or where trans-global connectivity is required. They are also trying to offer as high a reliability as they possibly can, because of the commercial value of the traffic carried.

7.2 In the beginning, there was INTELSAT. . .

We will begin with INTELSAT, which was the first satellite system available for trans-global commercial use. It was also the first system that allowed international access for satellite newsgathering. As the first global commercial satellite system, it still holds a commanding position in the global market, but has come under intense pressure from the competition of other systems for most services in the last fifteen years. The early history of INTELSAT is closely bound up with US dominance in satellite communications.

In 1961, soon after his inauguration, President John F. Kennedy promoted the development of the US space programme as an intrinsic part of US foreign policy, including the establishment of a commercial satellite telecommunications programme. This was in response to the rapid development of the space programme of the USSR, who astonished the world by launching the first space satellite, SPUTNIK, in October 1957, closely followed by SPUTNIK 2 a month later. The US launched a number of experimental satellites in what was to become an intense race to beat the Russians; but Kennedy effectively propelled the 'space race' forward as a key strategy of US foreign policy from 1961.

In 1962, the US Congress passed the Communications Satellite Act, which called for the construction of a commercial, communal satellite system that would contribute to world peace and understanding. The Act called for the establishment of a US commercial communications organization. A private company was then established and in 1963 incorporated as COMSAT. In 1964, COMSAT entered into temporary agreements with ten other countries to establish an international telecommunications consortium to provide global satellite coverage and connectivity, forming the International Telecommunications Satellite Organisation (INTELSAT). COMSAT remains today the US Signatory to INTELSAT.

The two initial agreements were only temporary because several of the other member nations feared US domination through COMSAT. The eleven initial Signatories did not finally reach full agreement on all aspects of the new organization until negotiations were completed in 1971, and in 1973 INTELSAT officially became a legal entity. In the meantime, however, INTELSAT launched a number of satellites to establish and develop the system.

Today, INTELSAT is an intergovernmental organization (IGO) with 143 member governments (in 1999), and forming one of INTELSAT's principal governing bodies is the Assembly of Parties. Its original principal aim of providing global telecommunications links for all countries on a non-discriminatory basis is still the dominant principle, and the organization serves over 200 nations.

INTELSAT remains the primary provider of services to developing nations and remote areas of the globe. This is because of its mandate and commitment to provide universal access on non-discriminatory terms and conditions, though still mindful of some commercial viability. As we shall see, this commitment, while it is still important, may change due to the privatization of INTELSAT.

INTELSAT's activities are governed by the two separate but interrelated agreements mentioned earlier. The first, the INTELSAT Agreement, was completed by Member Nations (the Parties) and sets forth the prime objective of the organization as well as its structure, rules and procedures. The second, the INTELSAT Operating Agreement, sets forth the rights and obligations of INTELSAT Signatories and investors.

The Assembly of Parties meets every two years to consider issues of general policy and long-term objectives of special interest to governments. INTELSAT's operations are governed by its Meeting of Signatories – the investors in the INTELSAT system – and managed by its Board of Governors, which has principal responsibility for the design, development, operation and maintenance of the INTELSAT system. Signatories are designated by the member governments and include many national telecommunications agencies and companies, some of which are still owned in part by national governments. Each national PTT is usually the representative for each country in operational terms, and national administrations are represented at meetings at the set level. Under the Operating Agreement, Signatories are responsible for financing INTELSAT, with each Signatory owning a share in the organization and contributing capital in proportion to its use of the satellite system. Capital contributions support INTELSAT's operations, as well as the direct and indirect costs of designing, developing and operating the system. Signatories receive a return on capital based on the success of INTELSAT operations.

INTELSAT's satellites allow over 200 member and non-member nations alike access to global interconnection. The organization's primary focus has been the provision of international 'fixed' (e.g. telephone and broadcast) public telecommunications services, but it now provides a range of products including Internet and corporate network services. INTELSAT was the first provider of television transmission links between continents. Since its creation, there has been rapid growth in traffic, improvements in quality, and reductions in the cost of global telecommunications services.

The first satellite in the INTELSAT system was the INTELSAT I (EARLY BIRD) satellite, launched into geostationary orbit over the Atlantic Ocean in 1965. It was primarily used for telephony traffic, but was occasionally used for television transmissions including President Lyndon B. Johnson's address to the UN, the Pope's visit to New York and a concert by the Beatles.

The INTELSAT II satellites were launched in 1966/67, and INTELSAT III satellites covered all three major ocean regions – Atlantic, Pacific and Indian – by 1969, which established the first true global coverage by a commercial geosynchronous communications satellite system. The INTELSAT IV series of satellites were launched through the 1970s. With each generation of spacecraft the sophistication increased, with more accurate pointing towards the Earth's surface, higher power and improved coverage. The constellation of satellites that was established provided 'lifeline' communications and other commercial routes for international telecommunication services.

INTELSAT currently has a fleet of seventeen high-powered spacecraft in geostationary orbit: the INTELSAT V, INTELSAT VI, INTELSAT VII and INTELSAT VIII series (see Table 7.2). The newest generation of INTELSAT

Table 7.2 INTELSAT satellite fleet

Series	*Satellite*	*Orbital location*	*Launch date*	*Ku-band Txp*	*C-band Txp*	*Coverage area*
V	511	330.5°E	1985	4	21	Atlantic Ocean Region
VI	601	325.5°E	1991	10	38	Atlantic Ocean Region
VI	602	62°E	1989	10	38	Indian Ocean Region
VI	603	335.5°E	1990	10	38	Atlantic Ocean Region
VI	604	60°E	1990	10	38	Indian Ocean Region
VI	605	332.5°E	1991	10	38	Atlantic Ocean Region
VII	701	180°E	1993	10	26	Pacific Ocean Region
VII	702	177°E	1994	10	26	Pacific Ocean Region
VII	704	66°E	1995	10	26	Indian Ocean Region
VII	705	342°E	1995	10	26	Atlantic Ocean Region
VII	706	307°E	1995	10	26	Atlantic Ocean Region
VII	707	359°E	1996	10	26	Atlantic Ocean Region
VII	708	Lost on launch				
VII	709	310°E	1996	10	26	Atlantic Ocean Region
VIII	801	328.5°E	1997	6	38	Atlantic Ocean Region
VIII	802	174°E	1997	6	38	Pacific Ocean Region
VIII	804	64°E	1997	6	38	Indian Ocean Region
VIII-A	805	304.5°E	1998	3	28	Atlantic Ocean Region
IX	901	60°E	2000*	TBC	TBC	Indian Ocean Region
IX	902	62°E	2001*	TBC	TBC	Indian Ocean Region
IX	903	335.5°E	2001*	TBC	TBC	Atlantic Ocean Region
IX	904	325.5°E	2001*	TBC	TBC	Atlantic Ocean Region
IX	905	TBD	2002*	TBC	TBC	TBD

Excludes satellites recently transferred to New Skies Satellites N.V.: see Table 7.3.
*Projected launch date.

spacecraft, the INTELSAT IX series, is in production with five spacecraft on order.

INTELSAT in recent years has begun directly to deal commercially with end-users as national regulatory regimes were liberalized, under 'direct access' arrangements. INTELSAT also has direct operational contact with end-users.

7.2.1 INTELSAT registration and type-approval

The registration is a process whereby INTELSAT assigns a unique earth station code to each uplink planning to use their system. It is normally conducted with the help of the INTELSAT Signatory of the home country of the SNG uplink, who submits to INTELSAT the necessary forms. These are contained in the INTELSAT Satellite Systems Operations Guide (SSOG) 200 dealing with earth station registration[1]. SNG uplinks are classed by INTELSAT as Standard G earth stations, and the technical standards that Standard G earth stations must meet are contained in INTELSAT Earth Station Standard (IESS) 601[2].

The SNG terminal has to be type-approved by INTELSAT to operate within the INTELSAT system. INTELSAT has two types of type-approval applicable to SNG uplinks, which applies to antennas and systems pre-approved.

1 Type-approved antenna models: This approval is for the antenna and 'feed' system only. Transmission and frequency stability testing will be required after the antenna and feed have been integrated with the transmission equipment. INTELSAT will test samples of the manufacturer's production of the antenna and, subject to the necessary performance criteria being met consistently across the test batch, will issue type-approval certification that a particular model of antenna meets the necessary technical standards. A number of antenna manufacturers' models have already been type-approved by INTELSAT, who encourage the use of manufacturer type-approved antennas because it is mutually beneficial to INTELSAT and the end-user, and it allows users to begin service more rapidly.
2 Type-approved SNG uplinks: This is the most all-encompassing type of approval because it is for the entire uplink system. Using type-approved equipment eliminates the need for verification testing of each individual SNG uplink, and thus no further verification testing is necessary prior to the uplink being brought into service.

The production by a manufacturer of a 'type-approved' antenna or system means that the manufacturer's equipment meets INTELSAT's stringent performance requirements and that each unit of a particular model closely reproduces the performance of every other unit of this model. This consistency of performance is the result of advanced manufacturing and assembly technology which makes it possible very accurately to duplicate performance in production.

Where there is no manufacturer's type-approval on either the antenna or the transmission system, type-approval has to be obtained on a case-by-case basis where the individual antenna and system is tested by either the manufacturer or the end-user, and the requisite forms as contained in SSOG 200 are submitted. In the case of an antenna not being type-approved by INTELSAT, range test patterns of the specific antenna have to be submitted. This is a complicated process and it is always best to either use type-approved models of equipment or ensure that the type-approval has been obtained from the vendor of the system upon delivery to the customer.

Typically, INTELSAT will require an 'on-satellite' test as the final part of the registration procedure, where the SNG uplink undertakes a number of tests with a satellite in the INTELSAT system as directed by INTELSAT, which are monitored at an INTELSAT ground earth station.

INTELSAT, having issued a registration number to the SNG uplink, will then require a complicated bureaucratic process to be completed for each and every booking. INTELSAT can only accept bookings from one of three types of customer, as listed below.

- A Signatory.
- A Duly Authorized Telecommunications Entity (DATE): A country which is not party to the INTELSAT Agreement may designate an entity.
- A Direct Access Customer (DAC): This is a customer who has been authorized by a Signatory or DATE to directly access the INTELSAT space segment.

It has to be remembered that INTELSAT is effectively a global co-operative, and bureaucracy is an inevitable result where the first priority of an organization is seeking to be fair to all 143 parties. This means that anyone operating an SNG uplink who is not one of the above types of customer (termed a 'requesting authorized customer') has to approach INTELSAT through one of the above to place a booking.

A further example of this is the requirement for the SNG uplink operator to file a transmission plan application through a requesting authorized customer for each transmission from a location. Once this has been approved, then a booking for the required space segment can be made – again through a requesting authorized customer.

7.2.2 SSOG 600 transmission plan approval

The SSOG 600[3] is an application form for approval of a 'transmission plan' that has to be completed on behalf of an SNG uplink operator by an INTELSAT requesting authorized customer (usually of the home country of the SNG uplink operator) before INTELSAT can process a booking for the uplink. The application requires the technical parameters of the SNG transmission to be specified, along with full details of the downlink. INTELSAT will not accept an SNG booking until this form has been submitted and processed, and a link budget calculated. INTELSAT will then reply to the SNG uplink operator through the INTELSAT requesting authorized customer with either acceptance of the booking or declining the booking because the link budget is inadequate. INTELSAT will also require a copy of the permission from the country where the SNG uplink was to operate (the 'country of origin') confirming that the SNG uplink has permission to operate. This whole procedure can take hours or even days – hardly an ideal way to respond to fast-breaking news stories.

In addition, a 'matching order' is required from the Signatory of the country of origin of the transmission (i.e. where the SNG uplink is operating), verifying that it has knowledge of the transmission and accepts responsibility of ensuring that the uplink is operated in accordance with INTELSAT regulations. This can take some time if the 'originating' Signatory is in a different territory to the requesting authorized customer. There is, however, a process of 'self-matching', allowing at least the space segment to be confirmed before the originating Signatory gives their assent. This is where the SNG uplink operator (through the requesting authorized customer booking the capacity) effectively guarantees to pay all costs whether the transmission goes ahead or not (e.g. in the situation where permission to operate is not granted in time for the transmission. For the transmission to proceed, the 'originating' Signatory has to agree that the SNG uplink is permitted to operate, as they are effectively responsible for guaranteeing that they will bear the financial responsibility if the SNG uplink operating in their country interferes with third party traffic or damages the satellite (though the latter is unlikely). Due to the poor communication with some Signatories, this can lead to nail-biting delays, as there is doubt as to whether the agreement will be secured in time for the transmission.

Recently, however, in response to pressure from INTELSAT SNG users, INTELSAT will now complete the SSOG 600 on behalf of the SNG uplink operator, via the requesting authorized customer, based on the information supplied, and the requirement for a matching order has been relaxed in many cases.

This whole process is viewed as over-bureaucratic by SNG uplink operators, who regard it as their own responsibility to ensure that:

- their uplink will deliver the required power level to the satellite (all other satellite parameters being satisfied);
- they have the requisite authority to operate;
- their downlink facility is adequate to receive the signal.

On the other hand, with any commercial satellite carrier (who is not bound by any international treaty obligations) bookings can be accepted and confirmed within minutes (assuming capacity is available) directly with the SNG uplink operator. It is not the concern of the satellite operator as to whether the country where the uplink is operating agrees to the transmission, and it is the responsibility of the SNG uplink operator to ensure that the transmission parameters are adequate. However, satellite system operators will produce 'link budgets' (see Chapter 2 and Appendix B) on request if required.

7.2.3 Privatization of INTELSAT

INTELSAT remains the principal provider of SNG capacity trans-globally, though increasingly it is in competition with other satellite systems. In order to meet this challenge, and because of pressure from the WTO and other commercial imperatives, the monolithic (and often bureaucratic) INTELSAT has to change; so it has been split in two. It is important to describe this process, as it is probably the most significant change to the global telecommunications structure in the last twenty-five years.

In 1998 INTELSAT began the process of restructuring to become partially privatized. The governing bodies of INTELSAT recognized the potential conflict of interest between treaty obligations and the requirement to ensure the commercial

Table 7.3 New Skies Satellites fleet

Series	*Satellite*	*Orbital location*	*Launch date*	*Ku-band Txp*	*C-band Txp*	*Coverage area*
V	513	183°E	1988	6	32	Pacific Ocean Region
VII	703	57°E	1994	10	28	Indian Ocean Region
VIII	803	338.5°E	1997	6	38	Atlantic Ocean Region
VIII	806	319.5°E	1998	6	38	Atlantic Ocean Region
K		338.5°E	1992	16		Atlantic Ocean Region
K-TV*		95°E	2000**			

*Under construction.
**Launch delayed from 1999.

survival of their system. The first step in the privatization process was the decision by the Assembly to establish a spin-off company.

The new company has been established as New Skies Satellites N.V. (NSS), incorporated in the Netherlands in April 1998, and is now fully operational. It is currently partially owned by INTELSAT, which holds a 10% holding in the new company through an independent trust. While NSS becomes established INTELSAT is providing operational management during the transition phase. INTELSAT officially transferred five operational satellites, plus a sixth which was under construction (Table 7.3), to New Skies Satellites N.V. on 30 November 1998, and NSS plans to launch further satellites in the next few years.

The major motivation for this radical step – effectively changing the basis of what was the cornerstone of global telecommunications – was to ensure INTELSAT's future market responsiveness. Private corporations take strategic commercial decisions rapidly, whereas an international treaty organization such as INTELSAT has to secure agreement with over 140 parties in what tends to be (necessarily) an extremely cumbersome (and therefore slow) process. It can literally take years to respond to major commercial challenges in the market, and commercial decisions may directly conflict with individual national administration interests, to the potential detriment of the whole organization. As previously pointed out, bureaucracy is an inevitable result where the first priority of an organization is seeking to be fair to all 143 parties. This is one of the principal reasons why INTELSAT is undergoing privatization.

The initial partial privatization with the creation of New Skies is to lead to the complete privatization of INTELSAT, and this is scheduled to be completed by around March 2001, when INTELSAT will become a completely privatized entity. This has to be achieved without conflicting with one of the principal founding principles of INTELSAT, namely providing 'lifeline' services across the globe to all countries irrespective of wealth. There are still large areas of the world which depend on the INTELSAT system to provide satellite communications either in volume or with any permanence. The question of how to provide basic services at an affordable cost for these countries and at the same time operate on a commercial basis is not easily resolved, and the Intelsat Board is reviewing several corporate models in the organization's restructuring efforts as it moves closer to privatization. A final decision with member governments on how to complete the privatization of Intelsat is expected by the end of 2000[4], followed by the implementation of the privatization by mid-2001. Currently it is proposed that the essential lifeline function will be preserved by the fully commercialized INTELSAT.

In 1998, a serious threat to the privatization of INTELSAT emerged due to opposition in the US legislature. Because INTELSAT is based in Washington DC, and the US was one of the powerful parties behind the establishment of INTELSAT, the privatization has attracted interest from the US government. It supports the privatization in general, but naturally considers that it has a significant 'stakeholder' interest in the future of INTELSAT. This interest in the privatization of INTELSAT has caused the US government to seek to dictate the terms and the timetable for the privatization process. While INTELSAT is an international organization, and not within the jurisdiction of the US, the proposed actions by the US Congress could potentially harm US users of the INTELSAT system or those wishing to transmit signals into the US via INTELSAT. It may be that access would in some way be restricted, or that INTELSAT is limited in how it can operate in the US. At the time of writing, this is still an ongoing issue and unresolved.

7.3 PanAmSat

PanAmSat is the world's largest commercial satellite system, with nineteen satellites currently in orbit. Originally the first privately owned international satellite operator, it merged with the Hughes Communications 'Galaxy' domestic US satellite system in 1996, to form a global fleet to challenge INTELSAT. Hughes is the majority shareholder, and it brought a fleet of eleven domestic satellites to the merger, combining with the four international satellites operated by PanAmSat at the time. Since the time of the merger, another four satellites have been brought into service and a further seven are planned to be operational by the end of 2000 (see Table 7.4).

The Pan American Satellite Corporation (PanAmSat) was founded in 1984 by the late René Anselmo, a multi-billionaire businessman who had worked for a number of years in the Spanish-speaking television industry, establishing the Spanish International Network. He was acutely aware of the monopoly power of INTELSAT in providing international satellite telecommunication services and wanted to create a commercial challenge to this monopoly. Anselmo's aim was to launch the world's first privately owned international satellite, with a particular focus on providing services between Ibero-America and Spain.

Consequently his company PanAmSat launched PAS-1 in 1988, and positioned it over the Atlantic Ocean, making it the first private-sector international satellite service operator. CNN became its first customer the following year, using PAS-1 for distributing its programming to Latin America. In 1991, PanAmSat announced plans to become a fully global service operator with the construction and launch of

Table 7.4 PanAmSat satellite fleet

Satellite	*Orbital location*	*Launch date*	*Ku-band Txp*	*C-band Txp*	*Coverage area*
Galaxy I-R	133°W	1994		24	US
Galaxy III-R	95°W	1995	16	24	US/Latin America
Galaxy III-C	95°W	2001*	24	24	US/Latin America
Galaxy IV-R	99°W	2000*	24	24	US
Galaxy V	125°W	1992		24	US
Galaxy VI1	99°W	1990		24	US
Galaxy VII	91°W	1992	24	24	US
Galaxy VIII-i	95°W	1997	32		Latin America
Galaxy IX	127°W	1996		24	US
Galaxy XR	123°W	2000*	24	24	US
Galaxy XI	99°W	1999	40	24	North America
SBS 4	77°W	1984		10	US
SBS 5	123°W	1988		14	US
SBS 6	77°W	1990		19	US
PAS-1	45°W	1988	18	6	Americas, Caribbean, Europe
PAS-1R	45°W	2000*	36	36	Americas, Caribbean, Europe
PAS-2	169°E	1994	16	16	Asia-Pacific
PAS-3	43°W	1996	16	16	Americas, Caribbean, Europe
PAS-4	68.5°E	1995	24	16	Europe, Africa, Middle East, Asia
PAS-5	58°W	1997	24	24	Americas, Caribbean, Europe
PAS-6	43°W	1997	36		South America
PAS-6B	43°W	1998	32		South America
PAS-7	68.5°E	1998	30	14	Europe, Africa
PAS-8	166°E	1998	24	24	Asia-Pacific
PAS-9	58°W	2000*	TBA	TBA	Americas, Caribbean, Europe
PAS-10	TBA	2000*	TBA	TBA	TBA

*Projected launch date.

another three satellites, which were all in place by 1995. Alongside the normal broadcast distribution and business television that trans-global satellite systems rely on for the backbone of their business, PanAmSat made a positive effort to become established in the SNG market by offering Ku-band capacity specifically targeted at satellite newsgathering.

The history of Hughes Galaxy goes back a little further. In 1979 Hughes Communications Inc. (HCI) was created with five employees. The parent company was Hughes Electronics Corporation, a subsidiary of General Motors Corporation. The objective of the company was to foster a commercial market for satellite communications technology that previously had been reserved almost exclusively for scientific and military purposes.

HCI introduced many pioneering concepts into the US domestic satellite market, such as:

- transponder leasing;
- cable TV-dedicated satellites;

- in-orbit spare satellites;
- digital compression technology for DTH services in the US and Latin America;
- high-power satellite-based mobile telecommunications by American Mobile Satellite Corporation (AMSC);
- global satellite service for business and consumer markets featuring video and high-speed data communications (Galaxy Spaceway).

HCI pioneered the concept of marketing satellites dedicated exclusively to cable programming when it launched its first commercial spacecraft, Galaxy I, in 1983. Home Box Office (HBO) became the first Galaxy I customer, signing a contract for six transponders.

By 1984, Galaxy II and III had been launched to expand the fleet, and Hughes Galaxy had become a significant player in the US domestic satellite (domsat) market. In 1985 HCI entered its first international joint venture with the formation of the Japan Communications Satellite Company (JCSAT) – the first private telecommunications firm to operate in Japan.

In 1989, HCI acquired four of the Western Union 'Westar' satellites to become the largest commercial C-band satellite operator in the US. The Westar satellites were renamed as the 'Galaxy' fleet of satellites. In 1990 HCI acquired the SBS series satellites from the Satellite Transponder Leasing Corporation (STLC), a subsidiary of IBM. In 1993 HCI sold its share in JCSAT, enabling JCSAT's merger with a newly licensed satellite competitor in Japan, now called Japan Satellite Systems.

In 1996, Hughes Electronics announced the proposed merger of its Hughes Communications Galaxy business with PanAmSat to form a new publicly held company, a move that was completed after US government approval. The newly merged company operates under the PanAmSat name, with a 72% stake held by HCI.

7.4 Regional satellite systems

Having looked in some detail at the global satellite systems, it is worth having a brief look at the regional systems to gain an overview of the development of satellite communications.

7.4.1 North America

Since the early development of satellite communications largely occurred in the US, the growth of US domsat and some of the companies involved is of considerable interest. These companies spurred on the commercial development of satellite and spacecraft technology, which has in turn benefited the rest of the world. As well as Hughes, other companies such as AT&T, Western Union (who owned the Westar satellites mentioned above), GTE Corporation and Loral, have all played a part in establishing the multi-billion dollar US domsat market.

The nature of the US market is such that companies have often been merged and then further merged again, forming an almost bewildering variety of alliances and changes of name. The following is a brief summary of the principal features of the North American market, and all the systems discussed below cater for SNG use as well as programme distribution.

Loral Skynet

Loral Skynet is a US satellite service operator that operates the Telstar satellite fleet. Initially part of AT&T Communications, the Skynet division was one of the pioneers in the development of the US domsat system.

The launch of Telstar 1 in July 1962 was AT&T's first venture into satellite technology. Telstar 1 was a joint experimental operation between AT&T's Bell Laboratories, which designed and built the satellite, and NASA, which launched it into an elliptical orbit. The Telstar satellite experiments demonstrated the feasibility of transmitting multiple simultaneous telephone calls, data and video signals across the Atlantic – though the satellite only covered the Atlantic Ocean for a maximum of 102 minutes per day – and Telstar lasted for six months before it expired. AT&T continued to provide satellite services during the 1970s through a joint venture with GTE, leasing four COMSTAR satellites from the COMSAT Corporation and providing voice communications to the United States from seven earth stations.

The COMSTAR satellites reached the end of their life during the 1980s, and AT&T Skynet re-established the Telstar series, offering services to the broadcasting market. The first of the Telstar 3 series of satellites was launched in 1983, with Telstar 303 still in service in 1998. AT&T Skynet was acquired by Loral Space & Communications in 1997 to become Loral Skynet.

Loral Skynet has continued the Telstar series, with Telstar series 4 and 5 satellites currently serving the US domsat market. Loral Skynet also plans to launch Telstar 7, 8 and 9 series satellites. With the recent privatization of Mexico's satellite system, Loral Space & Communication acquired a controlling interest in Satelites Mexicanos (SatMex), and Loral Skynet now manages the SatMex fleet.

GTE SPACENET

The GTE Satellite Corporation was formed in 1972 to provide an earth station network for telecommunications within the US. By 1975, it had formed a partnership with AT&T to further develop its system, and by 1980 it had commissioned the construction of its own GSTAR series of Ku-band satellites. In 1983, GTE SPACENET was formed with the acquisition of the Southern Pacific Satellite Company, which had also commissioned the construction and launch of its own SPACENET series of satellites. Seven GSTAR and SPACENET satellites were launched into orbit between 1984 and 1990. By the time GTE SPACENET had been acquired by GE Americom in 1994 it had become the largest volume provider of US domsat space segment.

GE Americom/GTE SPACENET currently operates eleven satellites providing coverage of North America, with the GE-1E satellite covering Europe. In South America, GE Americom is a partner in the Argentinean-based NahuelSat satellite system.

Anik

Telesat is Canada's primary satellite system and was formed in 1969. It launched Anik A1 in 1972, and was the world's first commercial domsat in geostationary orbit.

Telesat currently operates two Anik E series satellites, both launched in 1991. Anik E1 is used primarily for business telecommunications services, while Anik E2 is used for broadcast purposes, carrying the bulk of Canada's television signals. The first of the Anik F series of satellites is due for launch in 2000. Canadian broadcasters obviously also have access to the US-based domsat systems.

7.4.2 European regional satellite systems

The European market is smaller than the North American market, with SES Astra and EUTELSAT being the primary operators for the region (though SES Astra does not provide any access for SNG). France Telecom operates the 'Telecom' series of satellites, Deustche Telekom operates 'Kopernikus', Hispasat operates the 'Hispasat' series and Telenor operates the 'Thor' series. Some of these companies also operate on INTELSAT and EUTELSAT capacity, but all these companies are important providers for SNG capacity within Europe. However, without doubt the principal SNG space segment provider is EUTELSAT.

EUTELSAT

The European Telecommunications Satellite Organisation (EUTELSAT) was founded in 1977 by the CEPT to develop a European regional satellite system for satellite telephony distribution. Membership criteria for countries are that they are a European sovereign state, and a member of the ITU. EUTELSAT was formally established as an organization in 1985.

In association with the European Space Agency (ESA), EUTELSAT launched the OTS-2 (Orbital Test Satellite) in 1978 to experiment with and develop the use of Ku-band communications over Europe (OTS-1 was lost on launch in 1977). EUTELSAT 1 Flight 1 was launched in 1983, followed by a further three satellites in the EUTELSAT 1 series and six satellites in the EUTELSAT 2 series. ESA still controls the EUTELSAT 1 satellites on behalf of EUTELSAT, although EUTELSAT itself controls the rest of its fleet. Table 7.5 shows the current EUTELSAT fleet of satellites.

EUTELSAT's Signatories use the satellites both as part of their domestic and international networks and also to lease capacity to other users. Their shareholding is determined annually according to use of capacity. The initial membership of seventeen countries has currently grown to forty-seven (in 1999), with more countries waiting to join.

Table 7.5 EUTELSAT satellite fleet*

Series	*Satellite*	*Orbital location*	*Launch date*	*Ku-band Txp*	*Coverage area*
I	F4	25.5°E	1987	10	Europe, Mediterranean, Russia
I	F5	12.5°E	1988	10	Europe, Mediterranean, Russia
II	F1	13°E	1990	16	Europe, Mediterranean
II	F2	10°E	1991	16	Europe, Mediterranean
II	F3	16°E	1991	16	Europe, Mediterranean
II	F4	7°E	1992	16	Europe, Mediterranean
HOTBIRD	1**	13°E	1995	16	Europe, Mediterranean
HOTBIRD	2	13°E	1996	16	Europe, Mediterranean
HOTBIRD	3	13°E	1997	16	Europe, Mediterranean
HOTBIRD	4	13°E	1998	16	Europe, Mediterranean
HOTBIRD	5	13°E	1998	16	Europe, Mediterranean
W	1	Lost on launch			
W	2	7°E	1998	24	Europe, Middle East, North Africa
W	3	7°E	1999	24	Europe, Middle East, North Africa
W	4	TBD	2000*	24	Europe, Middle East, North Africa
W	1R	TBD	2000*	24	Europe, Middle East, North Africa
SESAT		36°E	2000*	18	Russia (Siberian Europe)
ATLANTIC BIRD	1	12.5°W	2001*	20	Europe, Americas, North Africa, Middle East
EUROPESAT		29°E	2000*	36	TBA
DFS Kopernikus		28°E	Taken over		Europe (ex-Deutsche Telekom)
Telecom 2A		8°W	Taken over		Europe (ex-France Telecom)

*Projected launch date.
**Originally called EUTELSAT II F6.

Non-Signatory service providers in some EUTELSAT member countries have also been licensed to operate national and international networks, with Member Government authorization to access EUTELSAT capacity.

In January 1996, EUTELSAT's Assembly of Parties approved an amendment to the founding treaty, the Convention, permitting new entities such as telecommunications operators or broadcasters to invest in the EUTELSAT satellite system and have direct access to it, in the same way as current Signatories.

Eutelsat is planned to become a private company in 2001. The privatized company, Eutelsat S.A., will be based in Paris and comprise a Directorate and a Supervisory Board. It will initially consist of shareholders that are current Eutelsat Signatories.

EUTELSAT acts as a wholesaler of its satellite capacity, and provided the SNG uplink is EUTELSAT-registered, it is possible to buy space on EUTELSAT capacity via a number of service providers/resellers.

As with INTELSAT, EUTELSAT demands that transmitting earth stations wishing to operate onto their capacity fulfil stringent performance criteria. EUTELSAT classes SNG uplinks within their Standard L. The purpose of this Standard is to define the minimum technical and operational requirements under which approval to access leased space segment capacity may be granted to an applicant by EUTELSAT.

EUTELSAT offers a type-approval procedure for manufacturers in the same way as INTELSAT for two configurations, i.e. the antenna system including the feed, and the complete earth station. Operational standards are contained within EUTELSAT System Operations Guides (ESOG) and technical standards in the EUTELSAT Earth Station Standards (EESS). The Standard L specification is contained in EESS 400.

7.4.3 Other regional systems

In the Middle East/Africa region, the principal providers are Arabsat and Turksat. In the Asia/Pacific region, the principal providers are AsiaSat, Apstar and JCSAT. The market is changing so rapidly, particularly in the expanding markets in the Asia and Pacific regions, that it is difficult to cover adequately even the current situation in a book. Therefore, for the latest information see the reference to source materials in Appendices G and H. Within any region, space segment for SNG is always available from either a regional or global satellite operator.

7.5 The use of capacity

We can now turn to examining the different ways of accessing satellites, and look at the different types of satellite capacity available, the difficult process of determining which is the best capacity for an application and how satellite capacity is actually accessed. SNG can be used across all types of capacity but we will look at occasional capacity in particular.

The price structure of capacity, whether it is occasional, long- or short-term lease, pre-emptible or not so, is determined on at least the following four parameters:

- period required
- bandwidth
- power
- geographical coverage

Satellite capacity can be bought for periods from 10 minutes to 15 years, and this describes the range of services from occasional to long-term lease.

Bandwidth can be bought from 100 kHz to 150 MHz. For analogue video, the smallest bandwidth channel is 17 MHz and the largest is 36 MHz, with 27 MHz being a typical channel size. For digital video, the data rate of the channel is usually between 5 and 34 Mbps, with typically 5–7 Mbps for video distribution to cable head-ends, 8 Mbps for news contributions and 17–34 Mbps for sports and events.

Most audio is carried either as an analogue sub-carrier on an analogue video (and an analogue video can have several audio sub-carriers associated with it) or, more typically nowadays, as a digital signal. The data rate can be as low as 64 kbps for a mono speech signal and 128 kbps for a stereo speech signal or better quality mono signal; but for a high-quality music circuit, the data rate is typically up to 2 Mbps.

There is also the issue of the power required from the satellite, as this is a factor which is one of the key parameters in determining the life of a satellite. If a large amount of transmitted power is required from the satellite to the ground, this will affect the pricing. Similarly (and this is particularly relevant to SNG), if the satellite channel has to have increased gain in the receive stage (to be more 'sensitive'), this will also affect the pricing. Adjustments to the power either in transmit or receive are referred to as 'gain-steps', and any alteration to the nominal gain-step of a transponder to accommodate a customer's requirements will have an interactive effect on all the other channels in the transponder. These effects need to be calculated by the satellite operator and the impact on other customers assessed and minimized.

Finally, capacity is priced according to the 'beam coverage' required. There are different beam patterns depending on the frequency band, and the precise geographical coverage is defined by the pattern chosen.

In the Ku-band, there are 'spotbeams' and 'widebeams' (also called 'broad-beams', 'superbeams' or in Europe 'Eurobeams'); while in the C-band there are 'global', 'hemi-' and 'zone' beams. Ku-band spotbeams and C-band zone beams are often 'shaped' to cover landmasses rather than oceans, as obviously the landmass is where signals need to be concentrated. Sometimes island groups are included if there are significant population numbers present to be served.

Spotbeams offer the highest concentrated power coverage in the Ku-band, either in terms of the uplink or the downlink, and therefore command the highest prices. A spotbeam will usually cover an area approximately 3000 km (2000 miles) in diameter, though this can vary from satellite to satellite, and is determined by the design of the antennas of the satellite. Sometimes two spotbeams may be combined to provide a larger service area, as in the US where 'CONUS' is sometimes used to describe the coverage of a particular transponder where it covers both the eastern and western halves of the US (CONUS being an acronym for CONtiguous United States). On some US domestic satellites ('domsats'), beams are referred to as half-Conus; for example as in 'west-half CONUS'.

Ku-band widebeams offer broader coverage, perhaps 8000 km (5000 miles) in diameter, but are only available at lower power levels because the power is spread over a wider area.

In the C-band, a global beam describes a wide area low-power beam that covers a significant amount of the Earth's surface – although plainly it is impossible for a beam to literally cover the whole of the globe. In fact, a global beam symmetrically covers just less than half the hemisphere, with the centre of the beam often centred on the Equator.

A hemi-beam is a C-band beam that covers approximately half the area of a global beam, and might typically cover a continent or straddle parts of two continents to provide inter-continental connectivity, and is therefore 'shaped'. It is a higher power beam than the global beam because the power is more focused onto a specific area.

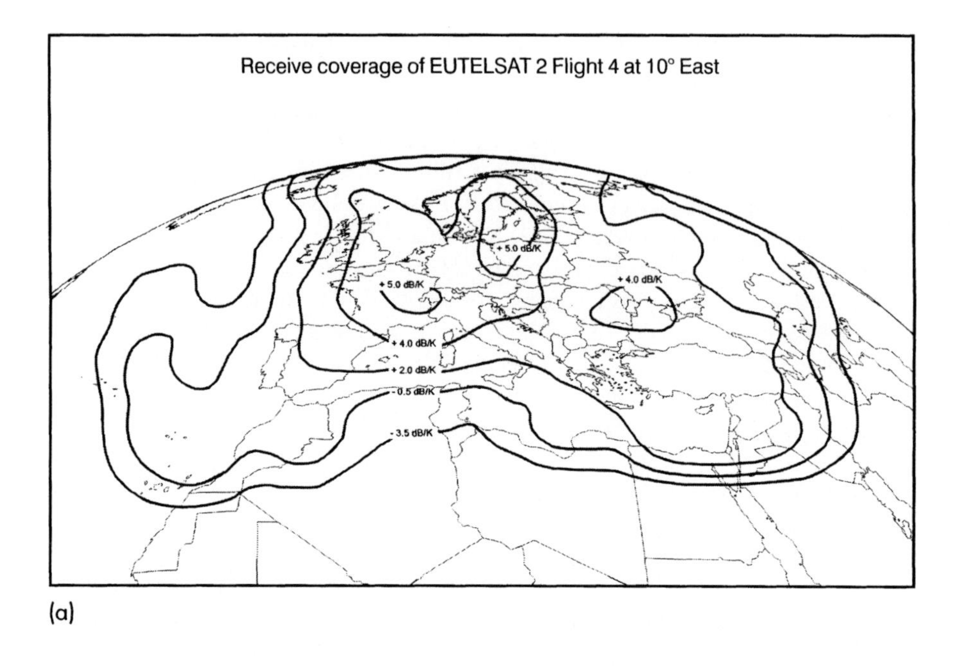

(a)

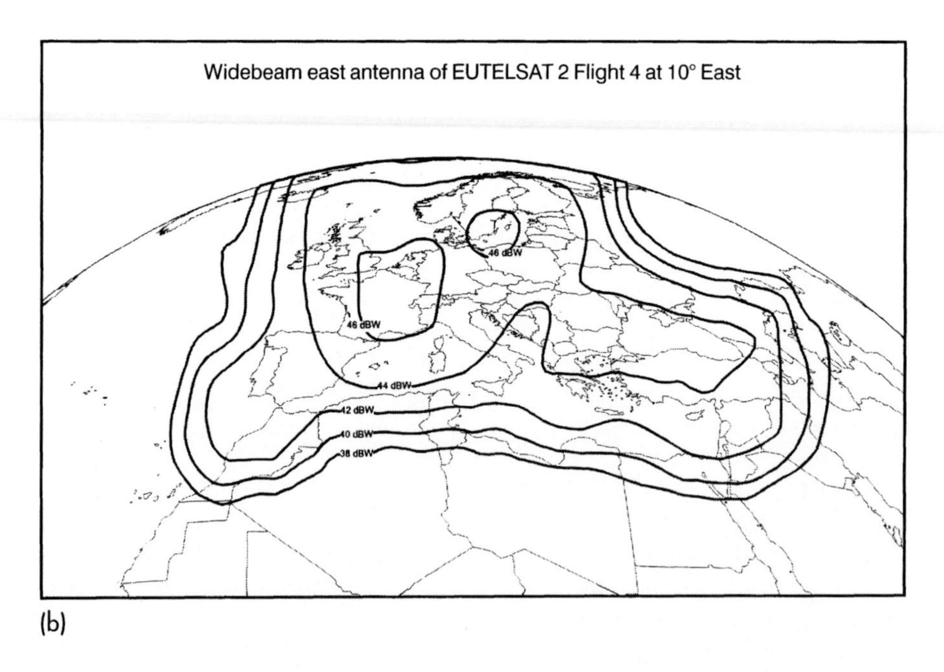

(b)

Figure 7.1 (a) EUTELSAT 2 Flight 4 uplink footprint map; (b) EUTELSAT 2 Flight 4 downlink footprint map. *(Courtesy of EUTELSAT)*

Finally, zone beams can be thought of as being the C-band equivalent of the Ku-band widebeam, with high gain covering a reduced area.

Each satellite has a defined area of coverage, both for the uplink and the downlink, and the satellite operator defines these in planning a service from a satellite before it has even been launched. The coverage is published by the satellite operator in the form of 'footprints', which show the geographical coverage of each of the uplink and downlink beams from the satellite; Figure 7.1 shows a typical footprint map.

7.6 Types of capacity

Satellite capacity is broadly divided into 'distribution' and 'contribution'. Distribution capacity is used to deliver signals to cable 'head-ends' or transmitters for terrestrial distribution, or for direct-to-home (DTH) satellite delivery. Contribution capacity provides routes for transmitting material from point-to-point (generally) for inclusion in distributed programmes after (usually) some production process.

There are several different types of satellite capacity that can be accessed by SNG uplinks, and satellite system operators broadly class these as follows:

- Dedicated/non-dedicated
- Pre-emptible/non pre-emptible
- Long-/short-term leasing
- Occasional
- Inclined orbit

The type of capacity determines, in a number of ways, the method of working for an SNG uplink (this is true whether it is a truck or flyaway system). An uplink can access capacity which covers a number of the above types; for instance, it could have a booking for 'occasional', 'dedicated' capacity, which might be in 'inclined orbit'.

7.6.1 Dedicated and non-dedicated capacity

The term used for capacity on a satellite that has been specifically set aside for a particular purpose is 'dedicated'. So for instance, on INTELSAT and EUTELSAT satellites, there is dedicated capacity allocated individually for services such as business data, video (multi-lateral video distribution, DTH and SNG), audio and radio services, public telephony, paging and messaging, and Internet, amongst others.

The reason why there is a separation of capacity for dedicated uses is that this allows the satellite operator to optimize the technical parameters of a transponder (or group of transponders) providing a particular service. For example, a satellite operator may concentrate DTH services on one particular satellite, so that consumers can have their satellite dishes pointed at a single satellite only for a wide range of channels. For instance, by the end of 1997, EUTELSAT maintained that their Hot Bird series of satellites at 13°E was reaching almost 70 million households via cable and satellite across Europe, North Africa and the Middle East – 16 million directly via satellite distribution alone.

The characteristic for data tends towards lower power services but great in number, and therefore if they are grouped together, the technical performance can be optimized for a large number of similar signals. It is usually preferable to keep large-power analogue signals together, away from lower power digital signals.

Similarly, SNG places particular demands on the satellite operator. SNG signals tend to be made up of a number of relatively short-duration signals of differing power levels, with transmissions starting and ending in an apparently random sequence. Therefore, it too tends to be grouped onto a particular transponder or group of transponders on a satellite. This allows the perturbations of transmissions frequently beginning and ending from possibly interfering with other services. The delicate power–bandwidth balance of a transponder can be upset by a pattern of transmissions separated by no activity, and therefore it is preferable to keep such transmissions away from continuous services.

There is an increasing amount of capacity allocated for SNG purposes, which is kept available for 'occasional' traffic. Often the capacity allocated to occasional SNG traffic is capacity not normally required for any other purpose except for 'restoration' – a service we will look at later.

Non-dedicated capacity is spare capacity that can be used for SNG or any other temporary use on occasions if it is available. Every satellite operator always has some spare capacity available to accommodate in-service failures on satellites, and if the circumstances are right, this capacity may become available for SNG. For instance, if an important news story breaks in a particular area, satellite operators may be prepared to offer capacity not committed to some other service for SNG, provided that the acceptance of SNG traffic does not cause problems with adjacent services. It has even been known for a satellite operator to 'shuffle' services around to create suitable capacity for SNG, if they perceive that there may be a significant financial advantage (due to demand) in offering SNG short-term capacity on a story.

Capacity may also change from 'dedicated' to 'non-dedicated' at various times. This may be a transponder, or a part-transponder, that is vacant between the ending of one long-term lease (also sometimes called a 'full-time' lease) and the beginning of the next.

7.6.2 Pre-emptible or not?

Capacity is also classed as being either 'pre-emptible' or 'non pre-emptible'. These terms refer to the level of guarantee of service.

Pre-emptible service is where, under certain conditions (and these can vary according to the conditions of the lease), service can be interrupted or discontinued. The pricing of the lease reflects this potential loss of service, and the reason for offering capacity is that it may suit both the satellite operator and the customer to take a lease on this basis. For instance, it may allow the satellite operator to offer capacity that is normally reserved for 'restoration' service or capacity that is currently available but cannot be absolutely guaranteed for either commercial or technical reasons to be available over the whole term of the lease. For the customer, it can be a lower cost alternative to purchasing guaranteed service, perhaps because the service being bought is either not essential to their business or where it fulfils an interim solution until capacity that is more reliable becomes available. So it is a mutually beneficial arrangement, and such capacity tends to be offered over shorter periods than non pre-emptible services. Short-term leases are typically on pre-emptible capacity, and customers can generally expect little or no notice of termination of service if the capacity is required for 'restoration' service.

Non pre-emptible capacity is, as the term suggests, the provision of a service that is generally absolutely guaranteed. In case of failure, either due to spacecraft or transponder faults interrupting service, the satellite operator is absolutely bound to provide equivalent alternative service within a strictly defined time-scale as defined in the contract with the satellite operator. Failure to do so can result in severe financial penalties being placed on the satellite operator. There is a very high premium attached to such a service, and large multi-national corporations that provide telecommunication, financial or other such high value, high dependency services typically buy it. In the world of video and audio services, it is bought primarily for essential distribution services, although some large broadcast networks buy some of this high-cost capacity for important contribution routes.

Some satellite operators offer different degrees of non pre-emptibility, with associated pricing advantages.

7.6.3 Leases

Leased capacity is where full-time capacity has been purchased for a period of time on a particular satellite, generally for a period of at least a year. Short-term leases are available but leasing is more generally considered as a long-term option.

The advantages of leasing are as follows.

- Price fixed under contract: like all long-term trading of a commodity, this could be both an advantage and a disadvantage.
- Guarantee of dedicated capacity: a competitive advantage.
- Contigency for failure may be built-in – 'restoration' service.

Leases are bought and sold on the basis of four principal factors:

- Length of lease
- Bandwidth/data rate
- Coverage
- Level of service guaranteed

Leases are deemed as long term where they are of a duration of at least a year, while short-term leases provide the capacity over days, weeks or months.

7.6.4 Long-term leases

Satellite operators may offer leases on different terms and conditions, so for the purposes of illustration we will examine in outline how INTELSAT offer long-term leased capacity.

Long-term leases are generally available on one, two, three, five, seven and ten year terms, and because of the size of the investment, the taking up of a lease is usually planned well in advance of the commencement date. A lease booked in advance is termed a 'reservation' and is offered as one of two types.

1 Guaranteed reservation (GR): Where a reservation will be accepted up to one year in advance on currently operational satellite capacity, and up to three years in advance for planned satellite capacity.
2 First right of refusal (FRR): Where INTELSAT will offer first refusal on specific capacity (subject to a fee being paid in advance). If INTELSAT receives a guaranteed reservation for capacity held under a first right of refusal, the FRR holder could choose either to upgrade the reservation to GR or relinquish the capacity. FRR reservations can be made up to three years in advance for operational capacity or five years for planned capacity.

Once a long-term lease has been entered into, it can be extended once at any time during the lease term at the prevailing lease rate, and there is a minimum length of extension of 1 month. However, there are penalties:

- If the lease is a pre-emptible service and is cancelled during the term, there is a penalty payment equivalent to the first two years' service and 25% payable on the remaining term of the lease.
- If the lease is on non pre-emptible capacity and is cancelled, then full payment on the entire lease is due.

There are other special discounts available on other types of leased capacity and on renewal of an existing non pre-emptible lease.

7.6.5 Short-term leases

As previously mentioned, these are available for short periods and the same pricing formulae are applied, taking into account bandwidth, power and coverage required. Short-term leases are particularly popular where a major news or sporting event is taking place in an area where interested broadcasters and newsgatherers do not already have leased capacity available. A number of satellite operators made capacity especially available for the Winter Olympic Games in Nagano, Japan in 1996, the 'handover' of Hong Kong in 1997, the Gulf crisis in early 1998, the 1998 Football World Cup and the Sydney Olympics in 2000.

To give an idea of scale of cost, a short-term lease for a single 9 MHz digital channel (which can carry an 8 Mbps signal) would cost in the order of $US30 000 per week in 1999. For a longer term, a five year lease on a 36 MHz Ku-band spotbeam (suitable for analogue or digital traffic) providing European coverage on INTELSAT 705 (342°E) would cost US$2.2 million per annum (US$11 million over the five year term).

Particularly on major breaking news stories, satellite operators will go to the lengths of 'steering' satellite beams to provide coverage in areas where coverage is not normally available. Satellites have also been brought into service that have only just been launched, and are still in their pre-service test phase (which every satellite goes through), or satellites that are in-orbit spares have been moved temporarily to a different orbital position.

Short-term leases offered on this type of capacity are usually offered for periods of one to four weeks, and can be highly lucrative for satellite operators as the major US and European newsgatherers scramble to secure capacity for their SNG uplinks to operate on. Newsgatherers from the Far East are able to route through the US to cover these events.

Short-term leases are charged at a higher pro rata rate than long-term leases, and can provide a useful source of business on satellite capacity that becomes available between long-term leases. Some satellite operators keep a certain amount of capacity available for short-term leasing because of the high returns that can be gained.

7.6.6 Occasional capacity

Occasional (ad hoc) capacity is scarce in some areas of the world and relatively plentiful in others. For most newsgatherers, the availability of occasional 'on-demand' capacity is critical. Even though a newsgathering organization may have leased capacity, a story can break in an area out of the geographical service area of their leased capacity, or create such a demand for feeds that extra capacity needs to be bought in. Smaller newsgatherers may depend totally on being able to access occasional capacity, whether they are global or regional newsgatherers, and this is an important market for satellite operators.

However, whenever there is a big news story in a poorly served area, it is often a struggle for satellite operators to meet the demand. This is certainly true of Ku-band, although the situation is improving in some areas of the world where previously there had been little or no Ku-band capacity, e.g. Southern Africa and the Far East.

The relative cost of occasional capacity is high in comparison to leased capacity, but it is reasonable when considering the extra work that a satellite operator has to undertake to meet occasional service requests.

To offer occasional service, a satellite operator will have to:

- Offer a 24-hour, 365-day per annum booking operation, with multi-lingual staff.
- Provide adequate technical back-up rapidly to process and execute service orders.
- Make available capacity on transponders, allowing for constant fluctuation of transmissions.
- Operate a sophisticated billing process.

Such an operation tends to be relatively labour intensive, with time-critical systems in place, and that in turn makes for an expensive operation. Bookings are usually of a minimum 10 minute duration, extendible in 1 or 5 minute increments. It is not unusual for a booking to be placed at less than 15 minutes notice.

There are high cancellation charges, which are factored depending on how much notice is given before the booking is cancelled. Virtually all satellite operators will demand 100% payment with less than 24 hours notice of cancellation, with varying penalty charges depending on the length of notice given up to 14 or 28 days. However, discounts are given for regular periodic bookings or for a commitment to use a given number of hours of capacity in a month. Some satellite operators offer discount thresholds as the accumulated total of bought occasional capacity increases through a year.

7.6.7 Inclined orbit

As described in Chapter 2, satellites of interest for newsgathering purposes are in a geostationary orbit, but the process of sustaining a perfectly maintained geostationary orbit cannot be continued indefinitely. Station-keeping manoeuvres consume energy on board the satellite that cannot be replenished, as the thruster motors that correct the position and attitude of the satellite consume liquid fuel rather than electrical power. At a certain point in the life of a satellite, a decision is taken to abandon North–South station-keeping manoeuvres, as they are not critical to sustaining the correct orbit. East–West station-keeping is considered important, as satellites can be spaced as close as 2° and if this axis of station-keeping is not maintained satellites may move in too close together. A significant proportion of fuel is used to sustain North–South station-keeping, so the serviceable life of the satellite can be considerably extended if these manoeuvres are abandoned.

When North–South station-keeping is abandoned, the satellite is said to be in 'inclined orbit' – this state is more fully described in Chapter 2. Essentially as the satellite is 'observed' from the Earth's surface, it no longer appears to be stationary in relation to that point on the Earth's surface, and requires tracking by both the SNG uplink as well as the downlink earth station. When there is insufficient fuel to even sustain East–West station-keeping, the satellite is deemed to be finally terminated, and is moved out to the 'graveyard' orbit (see Chapter 2).

What has this to do with the commercial issues of selling space segment? Capacity that is completely station-kept (i.e. stable) commands a higher premium than capacity deemed to be in inclined orbit. The disadvantage of inclined orbit capacity is that both the uplink and the downlink have constantly to track the path of inclination. At the downlink, this is commonly performed by auto-tracking equipment, which not only gradually moves the antenna to keep the satellite on-beam but also 'learns' the daily cycle so that it is able to predict the direction to move the antenna. In terms of SNG uplink, this tracking has to be performed manually, and although this cycle can also be learnt by the uplink operator, it is often beneficial to make use of daily computer predictions to aid the uplink operator.

Hence this capacity is available at lower cost on satellites which are in the final phase of their operational life. Typically capacity is available at discounts of 50% or more of the stable orbit cost, and is obviously very attractive if the disadvantages can be accommodated. In fact, inclined orbit capacity is attractive for SNG, as the majority of transmissions are point-to-point and therefore the characteristics of an inclined-orbit satellite can be coped with.

With the range of types of capacity, it is a constant juggling act for satellite operators to maximize the usage of as much of their capacity as possible, with the right mix of capacity usage to be able to cope with the occasional satellite anomaly or failure. It may be thought that it is desirable for a satellite operator to have as much capacity on non pre-emptible use as possible, but in fact, it is essential to have capacity that can be 'pre-empted' so that anomalies can be accommodated. Which brings us on to 'restoration' service, which we have mentioned several times in this chapter.

7.6.8 Restoration

'Restoration' is the term used by satellite operators to describe the process of providing alternative capacity for use by a particular service due to either a temporary or permanent in-flight failure. As can be imagined, the provision of a satellite service on a permanent lease basis can become a commercial and business lifeline upon which a large or small company depends. For instance, the loss of a high-capacity data feed from the regional headquarters of a large financial institution such as a bank or an insurance company to its head office, which may carry all the transactions and financial information for the previous 24 hours, could be catastrophic for the institution. It pays for the reliability of the service, and if service is lost, it expects back-up service to be provided within minutes or certainly within hours.

The provision of a service with 99.99% reliability commands a very large premium from the satellite provider, and in return, when service is lost for whatever reason, the satellite operator has to be able to provide an alternative quickly. Therefore satellite operators always keep some capacity available for such an occurrence, and have extensive contingency plans to cope with in-orbit failures. In the event of a serious failure, the satellite operator will try to use whatever capacity it has available on its own satellites, and if necessary will utilize capacity from another satellite service operator until the problem can be solved. Any significant problem is termed by satellite operators as an 'anomaly'.

It has to be said that the majority of failures occur on launch, which by its very nature is the time of greatest risk in the life of a satellite. Such failures are usually caused by some problem with the launch vehicle, and each year there are always a number of failures. These affect not only commercial geostationary satellites, but also scientific and military satellites destined for both geostationary and non-geostationary orbits. Almost every year, one organization will suffer an unfortunate number of failures. For instance, in 1998, PanAmSat had a particularly bad year, suffering the total loss of Galaxy IV (terminal failure after five years) and Galaxy

X (lost on launch 1 minute 20 seconds after leaving the pad), and also anomalies on PAS-4, PAS-5, PAS-8, Galaxy VII and Galaxy VIII-i. Other organizations suffered as well, of course, and in other years others have suffered as badly as PanAmSat did in 1998.

Let's look at two recent examples of major in-orbit failures which are worth examining as each instance shows how satellite operators are able to respond to such crises.

INTELSAT 605

At 15:13 GMT on 11 September 1997, the INTELSAT 605 satellite, located at 335.5°E, suffered an anomaly which could potentially cause a major disruption to traffic on the satellite. The central telemetry unit (CTU) had failed – this is the central on-board system feeding back confirmation signals confirming that ground control instructions have been acted on. Note that this was the back-up CTU unit: all mission-critical components on-board a satellite are duplicated, and the original CTU had failed in 1992, forcing INTELSAT engineers to use the back-up unit (for five years, though, before this anomaly).

The result of this failure was that the satellite was not responding to interrogation from the ground. However, INTELSAT was able to observe that transponder control instructions sent to the satellite were having an effect, and that the transponders all seemed to be working normally. What they could not be sure of was for how long this situation would last. Tests continued over the following hours, and INTELSAT engineers were able to verify operation by observing slight changes in the 'beacon' signal transmitted from the satellite, indicating that instructions were being executed. (The beacon signal serves a number of functions – not only does it act as a distinctive frequency marker for the satellite, making it easier for earth stations to identify the satellite, but also certain basic 'health' monitoring signals from the satellite are modulated onto it.)

INTELSAT's contingency plan for this eventuality on the satellite was to 'pre-empt' all traffic on INTELSAT 601 at 332.5°E, move it from this orbit position to INTELSAT 605's position at 335.5°E (3° of movement), and resume the traffic, which included a significant number of PSTN (Public Switched Telephone Network) services. However, INTELSAT 601 carried a significant amount of video traffic and INTELSAT was conscious of the loss that would be suffered by fixed ground stations that accessed video services at the 332.5°E slot. Since INTELSAT 605 appeared to be operating normally for the moment, INTELSAT decided to formulate a different plan that would not cause such a major disruption to video services at 332.5°E.

The enormity and complexity of the undertaking, and the ability of INTELSAT staff to plan the relocation of hundreds of carriers and leases within about four or so hours was amazing. The actual transitions took several weeks to accomplish because the satellites were so heavily loaded. This included the utilization of occasional-use TV capacity at 335.5°E and 325.5°E as bridging capacity until all carriers and leases could be relocated. The principal activity was completed in phases over about six weeks without a hitch.

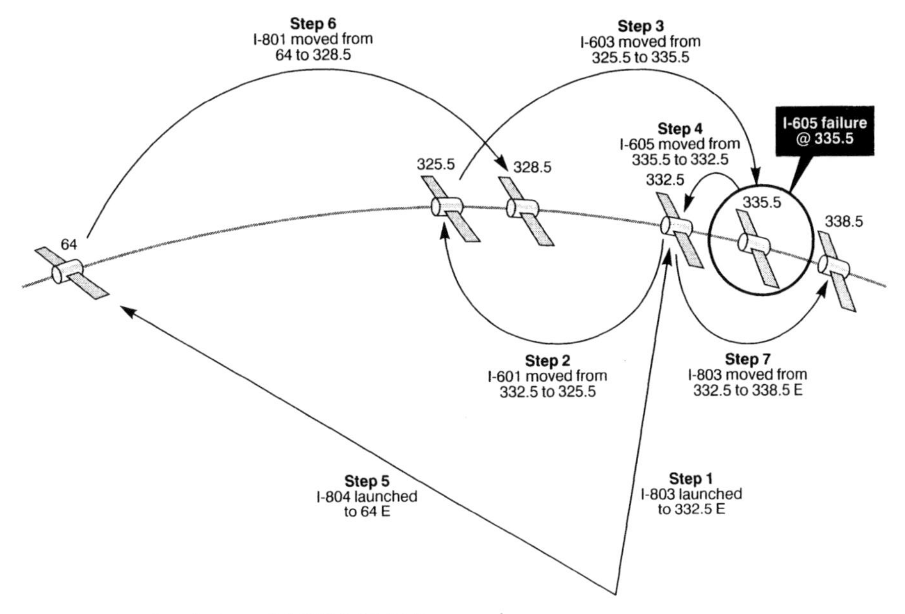

Figure 7.2 Resolution of INTELSAT-605 anomaly

This alternative plan depended firstly on the success of the planned launch of INTELSAT 803 on 23 September (12 days after the anomaly occurred), and secondly the successful launch of INTELSAT 804 in December 1997. The INTELSAT Board of Governors approved this plan on the afternoon after the anomaly (12 September). What was to occur over the following weeks and months was a carefully orchestrated set of manoeuvres involving six satellites (shown in Figure 7.2), finally completed in May 1998.

Why such a complicated set of manoeuvres over such an extended period? Fundamentally, the problem lay in the careful transfer of PSTN traffic in a large number of phases, mirrored by re-alignment of ground earth stations. To do this effectively, the global beams normally allocated for use for video services had to be used for PSTN traffic, and video services were suspended while these alignments were occurring.

It should be noted that INTELSAT VIII series satellites were not suitable for PSTN services due to differences in beam design from the INTELSAT VI series, and therefore could not be used as a long-term replacement for any INTELSAT VI series service.

The result was that INTELSAT was able to maintain all its existing services (though it did delay the implementation of a number of new services intended for the new satellites until they were finally parked in their original orbit positions). INTELSAT – and their customers – were fortunate in three ways. Firstly, INTELSAT 605 did not suffer a total failure, only the monitoring was affected. Secondly, the launch of INTELSAT 803 had been planned for only 12 days after the anomaly occurred, and the launch was successful and on time. Thirdly, INTELSAT was able to gain the co-operation of two separate groups of customers – the PSTN services and the video services – to maintain as high a level of service as possible to both groups. There was disruption for both groups of customers, and INTELSAT had to be very imaginative in its deployment planning. However, this incident is just one example of how a satellite provider can cope with a potentially very damaging anomaly.

Telstar 401

A far more serious anomaly occurred with Telstar 401, an AT&T satellite serving North America at orbital slot 97°W. On Saturday 11 January 1997 at 11:15 GMT, the AT&T ground control station in Pennsylvania, US lost all contact and control of Telstar 401. The satellite was still relatively young – coming into service in early 1994 – and total loss of contact with a satellite is extremely unusual.

For the next six days, AT&T engineers tried to re-establish contact, but the satellite was finally declared permanently lost on Friday 17 January 1997.

The services of forty transponders were suddenly lost (24 × 36 MHz C-band, 16 × 54 MHz Ku-band) and amongst the services carried were ABC, PBS (Public Broadcasting Service) and other US network TV feeds. Some of these feeds had contracts that guaranteed capacity and these services were immediately transferred to Telstar 401-R with no discernible interruption to service. Other customers who did not have such contracts had to wait until AT&T could find alternative capacity. Some traffic was also moved to SBS-6, an HCI satellite. Telstar 401 also carried a considerable amount of SNG traffic within North America, and the ripples of this loss spread out for some time afterwards. With such a catastrophic loss, it was extremely difficult to respond in the way that INTELSAT was able to later in the same year with INTELSAT 605, where the organization suffered a controllable anomaly.

AT&T and Lockheed Martin (the satellite manufacturer) spent some time trying to analyse what occurred, as such a loss was very unusual so early in a mission. The cause of the loss was finally attributed to a severe magnetic storm, which had been detected approaching the Earth five days earlier and which reached the Earth in the early hours of 11 January.

Unfortunately for AT&T, this occurred during the transition of the AT&T Skynet system (of which Telstar 401 was a valuable part) to Loral Space & Communications (see Chapter 6), and this seriously devalued AT&T Skynet, as the satellite was valued somewhere in excess of US$130 million.

It is a fact that in the last few years the number of failures either on launch or in orbit has been increasing, and there is a view in the industry that standards of construction of both launch vehicles and satellites has deteriorated, and hence resulted in these failures. It is not clear why this may have happened, and to try to analyse it here would be pure speculation, but insurance premiums have been increasing as a direct result of these failures. The setting of insurance premiums in any business is a statistical analysis process, and the statistics on losses in the satellite industry have been on a downturn.

7.6.9 Leases and newsgathering

Plainly one of the biggest factors when considering leasing is cost. It will cost several millions of US dollars per annum to lease a transponder on a satellite that will give guaranteed capacity over even a limited geographical area. For even a large corporation, this considerable investment has to show a guaranteed return. Leased capacity for purely newsgathering is unlikely to show this return, as few news organizations can find enough raw news to gather to fill even a single channel on a satellite transponder for 24 hours a day, 365 days a year. The capacity will have to be used for other purposes such as sports traffic or providing connectivity between studio centres. Additionally, non-broadcast data traffic (transmission plans allowing) – intra-company telephony, Internet/intranet access – can be used to improve the usage of the capacity. Sub-leasing of capacity to other users including newsgatherers may be considered if editorial competitive advantage is not lost.

In these days when every news operation is under severe financial pressures, long-term leased capacity can be both liberating and restricting. It is liberating in that there is the ability to get material back from anywhere within the footprint of the lease and this relieves the burden of finding capacity to cover a news story, which may in itself be creating such a demand that occasional capacity becomes scarce and/or expensive. On the other hand, it can be restrictive for it can indirectly apply subtle editorial pressure in deciding whether to cover a story out of the footprint of the transponder if it means buying occasional capacity. This is considered a form of 'double-spend' as additional capacity is being bought in at a higher price, while capacity already paid for is lying 'idle'.

There is also the pressure that the leased capacity is costing money constantly, and every minute and every hour it is not being used for any traffic, money is being spent 'unnecessarily'.

The taking up of a long-term lease involves a difficult decision process unless the level of traffic is guaranteed. There has to be a careful calculation and a degree of gambling in trying to weigh up whether the traffic levels are sufficient to tip the balance in favour of the expenditure on a lease. Careful statistical analysis of past, current and predicted traffic levels has to be carried out for the calculation. For many users it is a very complex decision, and if there is a choice to be made the different deals available from each satellite system can make direct comparisons even more tortuous.

Transponder loadings need balancing, and the inclusion of any ancillary traffic (such as data) also has to be taken into account. Some users – the US networks are an example – have transatlantic leases with US downlink beams and spotbeams on Europe for the uplink, which for special events can be steered to cover a different part of the Western Hemisphere. This is not always possible if the lease is a half-transponder, as the satellite operator has to take into account the users on the other half of the transponder. However, being able to steer the capacity in this way is a very cost-effective way of utilizing capacity. A significant amount of the traffic for the US networks is from Europe (usually London) into the US (usually New York), and so London forms a European 'gateway'. For example, if the transponder can be steered away from centring on Europe to cover a story in Africa, then the London traffic can be booked on occasional capacity – if the 'big story' is not generating much traffic from London it may not be a significant issue.

This ability to steer beams is, as previously mentioned, not a decision that is taken lightly, and requires close liaison and co-operation with the satellite operator. But this scenario is becoming more common, and latest generation spacecraft currently being launched have more of this type of flexibility designed in, including greater transponder switching ability as well as steering of beams. Satellite operators have adapted to a market that is demanding more of this flexibility.

Leased capacity allows the use of both fixed earth stations at foreign bureaux as well as SNG uplinks to readily access guaranteed paths into the contribution hub – the master control room (MCR) at the network headquarters.

It is usually a half- or full transponder (typically 36 MHz or 72 MHz respectively). If it is a half-transponder, then any changes have to be negotiated with the occupant of the other half of the transponder. This is because what occurs in one half of the transponder has an effect (possibly detrimental) to the occupant of the other half of the transponder. Therefore, power limits have to be calculated and set to achieve the correct power sharing arrangement. If one side 'saturates' its side of the transponder, then this will affect the other occupant, possibly causing interference. This is certainly the case where large power analogue signals are involved, but less of a problem in the digital domain where lower powers are used.

The satellite operator has to carefully balance up the requirements of each occupant of a transponder, and then each transponder as part of the whole payload of the spacecraft. Operators want to maximize the capacity and revenue while minimizing prime electrical power consumption, which will impact the overall level of service of the satellite. Therefore digital carriers are preferred, as they are more spectrally efficient and operate at lower power levels.

Sometimes deals can be struck which allow a user to buy capacity in addition to the long-term leased capacity at a discounted rate and even on a different satellite belonging to the satellite operator.

Leased capacity also gives operational freedom in terms of how earth stations – fixed or SNG uplinks – can operate on that capacity. Once the satellite operator is assured that the user is competent, a good deal of freedom is allowed the user on how they power up and down on the transponder, provided it accords with the transmission plan submitted to the satellite operator. The satellite operator control centre often leaves the management of a leased segment of the transponder very much up to the lessee, and they can power up ('come up') and power down ('come down') at will. So long as the other occupants of the transponder and the satellite itself are not subject to interference from the uplink, then the long-term leaseholder can enjoy far greater freedom than the occasional user.

7.6.10 C-/Ku-band 'cross-strapping'

Unrelated to the commercial basis on which the capacity is being bought, some satellite operators offer a facility called 'cross-strapping'. This is where the uplink transmits in one frequency band and the satellite receives and 'cross-straps' (frequency translates) this signal to a different frequency band to the downlink. It is also particularly useful in parts of the world where there is extensive C-band coverage, but the customer wants to receive the signal in the Ku-band. A satellite can receive a transmission from an uplink in the C-band, the signal undergoes an on-board frequency-shifting process to Ku-band and is then transmitted to the downlink. This process is controlled from the satellite operator's TT&C (tracking, telemetry and control) ground station.

7.7 Uplink operations

In this section, we will methodically progress through all the necessary steps to carry out a successful SNG transmission. Some of these steps are a once-only process, others are repeated on every transmission.

7.7.1 Practical steps in registering an SNG uplink with a satellite system

It was mentioned in Chapter 6 that almost all satellite systems require registration of an SNG uplink before bookings can be accepted. The purpose is to ensure that the SNG system operates within the prescribed parameters of the satellite system, and will not cause any interference or disruption to any other service on a satellite.

The procedure for this is in some ways similar to obtaining an operating licence for a particular country. Technical details and range patterns are always required, and the satellite system operator may require an 'on-satellite' test to be carried out. The satellite system operator will nominate a test time, and the uplink initiates a transmission under the co-ordination of the satellite system control centre. Such tests last typically thirty minutes, as the satellite system operator makes various checks as the power level from the uplink is gradually increased, and then various checks on performance are made at full power. The main point of these tests is to verify that the uplink cannot cause interference with other traffic on a satellite, and to confirm the integrity of the system. Once the tests have been completed, the satellite system operator issues a registration number for the uplink which is normally the registration mark for that system for life.

INTELSAT and EUTELSAT have a reciprocity agreement that allows an uplink registered on one system to be registered on the other system with the minimum amount of paperwork and usually without any tests. This is because the technical requirements of both satellite systems are very similar, and it is not in the interests of either INTELSAT or EUTELSAT to make customers undertake virtually identical test procedures. The regulations and tests are there to protect their systems, but SNG systems sold by reputable manufacturers will meet these requirements unless the uplink antenna is faulty or out-of-specification due to some damage. Other reciprocity agreements exist between some of the smaller satellite systems. Some satellite systems will accept prior registration with INTELSAT or EUTELSAT as adequate proof of uplink system integrity and allow operation on their system without any further tests, or they may require the SNG system to be tested individually on their system.

The registration number given by INTELSAT or EUTELSAT becomes, in general, the primary identifying mark for that system, and any changes made to the system need to be notified to the satellite system operator. Other registration identifiers for other satellite systems may be given, but the INTELSAT system registration is the most widely recognized globally.

Sometimes a minimum antenna diameter for the SNG uplink may be specified for certain satellites that are located close together because of potential interference from 'sidelobes' of the uplinked signal. For instance, INTELSAT currently has satellites located at 60°E, 62°E and 64°E, and INTELSAT has imposed a minimum antenna size of 2.4 m at C-band and 1.2 m at Ku-band in order that the integrity of the 2° spacing can be maintained. There are SNG systems with 2.2 m antennas for use in C-band, and INTELSAT will consider SNG access to this capacity using smaller antennas on a case-by-case basis. Occasionally other satellite operators impose other such restrictions.

7.7.2 Operating an SNG uplink

As an aside, we have not discussed what qualifications or training a person operating an SNG uplink needs to have. Amazingly, considering the potential damage that an uplink can cause both to the satellite and to people in the vicinity of the SNG uplink (as we shall see in Chapter 8), there is no qualification required to operate a satellite uplink in many countries. A few countries may require an individual to hold a general radio operator's licence, but in most countries no qualifications or minimum statutory training is required. In the US, job advertisements for SNG operators sometimes request preference for an FCC general class licence or SBE certification (Society of Broadcast Engineers). The FCC requires a trained operator to be in attendance on the uplink; but what is 'trained'? Most satellite operators demand that the uplink has to be under control, but do not specify the level of skill required. But in general, neither national administrations nor satellite systems operators demand that operators should have undertaken specific satellite operations training or reached a defined level of competence before being allowed to switch an uplink into transmit.

The emphasis is, and always has been, on the equipment meeting very stringent technical requirements, but no parallel requirements for the operators. Consequently, the majority of training is 'on the job', and the company operating the uplink allows an individual to operate an uplink when they are satisfied that the operator is competent. There are only a few specific training courses in SNG uplink operations (listed in Appendix K).

7.7.3 Placing a booking for occasional space segment

Having obtained satellite system registration for the uplink, deployed the uplink to the location of the story with skilled operators (!), and secured whatever licence or permission is required to operate, the process of securing satellite capacity (assuming you are not working onto your own lease) should have already begun.

To find and book occasional capacity segment there is a process that has to be followed, step by step. In many parts of the world, finding and booking space segment is like shopping around for any other commonly traded commodity, particularly where there is a thriving competitive market to provide space segment.

Before seeking capacity, the operator will need to obtain a suitable licence or permission from the national administration of the country of operation. Having obtained this, or at least begun the process, there are a number of basic parameters that need to be established:

- The location where the uplink is going to operate.
- The registration identity of the SNG uplink for the relevant satellite systems.
- Certain basic technical parameters of the SNG uplink, such as power, antenna size/gain, Ku- or C-band, analogue or digital.
- The date and times for transmission.
- The downlink parameters, including if it is self-provided, or whether downlink facilities are also required. (Most satellite operators can also supply downlink and connectivity to the studio centre.)

A typical service order is shown in Appendix H. As previously mentioned, capacity is normally bought on the basis of a minimum booking of 10 minutes of transmission, with extra time being bought in 1 or 5 minute increments. Often a 5 minute allowance is given prior to the transmission for technical line-up, provided there is time available on that particular channel/transponder.

For analogue transmissions, it is common to speak of a 'transponder' booking, as the amount of bandwidth required (27 MHz, for example) would use most of a 36 MHz transponder, or a half-transponder in the case of a 72 MHz transponder. However, for digital SNG transmissions, where it is common for a data rate of 8 Mbps to be used, the transponder can be sub-divided into a number of channels, typically 9 MHz. This is true of both Ku-band and C-band transponders. As a very rough rule of thumb, in 1998 analogue occasional capacity cost typically US$20 per minute, and occasional digital capacity around US$10 per minute.

Certainly in the US and Europe, the process of purchasing capacity is essentially very simple. Armed with the required information, as above, it is a matter of telephoning service providers, whether they be system operators or satellite segment brokers, and placing the enquiry. In highly developed markets, the whole process of seeking offers and deciding as to which suits best in terms of either price or service can be achieved in under an hour, virtually around the clock.

Before finally committing to using the capacity, there will probably be a requirement to provide the national radiocommunications administration with the details of the transmission for frequency clearance or as a condition of the terms of the licence or permission.

7.7.4 Locating an uplink

The uplink plainly has to be located where the story is, and the logistical and safety considerations involved in the decision where to place the uplink are covered in Chapter 8. Because of the wide range of uplinks available – both trucks and flyaways – it is not possible in this chapter to give detailed instructions on how to set up an uplink, and the uplink operator must rely on manufacturers' training courses and instruction manuals. Figure 7.3 shows an example of an uplink being rigged up.

Hopefully it has become clear from earlier discussion in Chapter 2 that both satellite uplinks and downlinks have to be accurately positioned and pointed at the satellite. For SNG uplinks, essentially transient in terms of their location, this is an important consideration in deciding where an uplink can be placed.

The uplink has to be stable and, therefore, whether mounted on a vehicle or as a flyaway sitting on the ground or the roof of a building, it must not be in a position where it could be knocked or blown off alignment. If the uplink is to be used on a ship, then it will require some form of additional stabilization such as a 'gyro-stabilized' platform, even if the ship is in dock. Ships move slightly even when tied up in harbour, and therefore it cannot be assumed that an uplink will work sitting on

Figure 7.3 Rigging the uplink. *(Photo courtesy of Paul Szeless)*

the deck of a ship in harbour. Gyro-stabilized platforms are large, complex structures, and are not appropriate for temporary uplink transmissions. Even INMARSAT system antennas when mounted on vessels have to be a gyro-stabilized, despite the wide angle of the beam from the antenna.

Having said that, the BBC did achieve a 'first' in 1998 when it achieved 'live' news transmissions from a flyaway on a battleship while on 'active-service'. In the spring of 1998, the British aircraft carrier HMS *Invincible* was patrolling the Persian Gulf during the tense situation with Iraq, and the BBC were invited by the UK Ministry of Defence to provide a UK 'pool' transmission facility on-board *Invincible* for UK broadcasters. The BBC achieved this by using a 90 cm digital SNG flyaway, placed on the flight deck of the carrier. Several days of experimentation (in between flight operations) with the enthusiastic co-operation of the ship's crew followed. It was found that if the ship made a particular heading into the wind at a certain speed (around 15 knots), thus minimizing the pitch and roll of the ship, and by manually tracking the flyaway antenna to maintain correct pointing at the satellite, successful transmissions could be made. Of course, the size of the ship (20 000 tonnes) and the relative calmness of the Gulf (which is a shallow sea), allied to the good weather conditions and the wide beamwidth of the small antenna (1.2°), were all vital factors that enabled this operation to be completely successful (not forgetting the highly skilled uplink engineers).

Having found a stable position, the uplink has to be sited where it has a look-angle that gives it a clear 'get-away' to the satellite. This means that it has to be sited so that there are no obstructions such as buildings, trees or passing high-sided vehicles that can block the transmission path to the satellite. In urban areas, in particular, this can be quite difficult, as even a roof-mounted antenna on a truck can be obscured by a moderately tall building close by. It is of particular concern where the uplink may be located on streets running on an East–West parallel which may obstruct the path to a satellite to the North or South (depending on the hemisphere). If the satellite is on a particular low elevation angle (typically below 20°) this will increase difficulties in built-up areas.

7.7.5 Co-ordination and line-up procedures

Each time an uplink is to undertake a transmission, there is always co-ordination via a co-ordination telephone line, which can be a land-line, cellular or satellite phone (often colloquially referred to as the business phone or 'bizphone') (see Figure 7.4). Co-ordination describes the process by which the 'line-up', start of transmission and end of transmission are discussed and agreed by telephone between the two points. This will occur either between the uplink and the final destination of the transmission (such as the destination downlink control centre, as usually occurs with long-term leases and some short-term leases) or between the uplink and the satellite operator's control centre (for occasional transmissions). Booking information is checked and confirmed, as well as the technical parameters. If there are any problems at either point during the transmission, this is dealt with via the co-ordination line. The number of the co-ordination line must be known to the control

Figure 7.4 'Lining-up' with the satellite control centre. *(Photo courtesy of Simon Atkinson)*

centre, and must be kept clear of other traffic during the transmission in case of a problem that the control centre needs to alert the uplink operator about.

The satellite operator's control centre is usually in its home country: INTELSAT Operations Centre (IOC) is in Washington DC, US; EUTELSAT Satellite Control Centre (CSC) is in Paris, France; PanAmSat Control Center is in Homestead, Florida, US; and Loral Orion Operations Center (OOC) is in Rockville, Maryland, US.

Line-up is the term used for the initial establishment of the signal from the uplink, and it is the period during which the signal is checked for technical compliance by the reception point. Normally, after no longer than five minutes, line-up is terminated by the commencement of the transmission. When the transmission is terminated, the end time (known as the 'good-night') is agreed by the uplink and the satellite operator's control centre. This is particularly important for occasional transmissions where the chargeable period has to be accurately agreed and recorded for billing purposes.

Once the uplink is rigged and aligned to the correct satellite (commonly referred to as being 'panned-up'), a typical sequence for an occasional booking is as follows (the time of the start of transmission is often referred to as 'Z'):

- **Z – 10 minutes:** Uplink calls control centre. On identifying its uplink registration, the booking details are confirmed, including uplink frequency and polarization.

- **Z – 5 minutes:** Line-up usually begins. With clearance from control centre, the uplink starts to transmit ('brings up') an unmodulated ('clean') carrier to 5–10% of full nominal power. For an analogue transmission, the control centre may ask for energy dispersal (ED) to be switched off with clean carrier. The control centre checks the carrier is at the correct frequency, polarization and nominal power with no 'out-of-band' harmonics (intermodulation products, or IPs).
- **Z – 3 minutes:** With the carrier still at less than 10% full power, the control centre instructs modulation to be switched on. For analogue operation, the modulating signal is normally a video test pattern with an identifying caption from the uplink. For a digital signal, it will simply be typical modulation as the control centre may not be able to decode the digital modulation. Once control centre agrees the modulated signal looks correct, power is raised to full nominal power. Control centre finally checks the signal and agrees transmission can commence.
- **Z:** Chargeable period commences. Test signal is removed and programme signal now switched to transmission.
- **Z + *n* minutes:** End of transmission. Uplink 'brings down' signal and agrees end time of transmission with control centre.

7.7.6 Over-runs

During the transmission, the uplink operator may need to extend the booking; for instance, if the 'live' slot has been moved in the news programme running order, or the tape has arrived at the last minute and there is more material to be fed than had been anticipated. A last-minute booking extension is termed an 'over-run'. Although it will be in the interests of the satellite operator's control centre to grant an over-run (as it is increased income), it may not be possible because there is another booking on the same channel immediately after the booked time. Obtaining an over-run may not be possible if the uplink is unable to contact the satellite operator control centre before the end of the booked time – the control centres can be quite busy and not always able to answer all phone calls. However, it is strictly forbidden by all satellite operators to continue to transmit beyond the end of the booked time unless prior authorization has been obtained.

7.8 Multi- and unilateral operations

When a satellite transmission is from point-to-point, it is described as a 'unilateral' transmission. This is the typical SNG transmission from an uplink in the field back to a single destination such as a broadcaster's studio. Where a transmission is from one origin to a number of destinations simultaneously, this is called a 'multilateral' transmission. In the television news environment, examples of multilaterals are the daily news exchanges between members of broadcasting unions, for example Eurovison (EBU) and Asiavison (ABU), where news material shot by one member is offered up for sharing with other members – a type of news co-operative.

Another type of multilateral is where an uplink is providing 'pool' material to a number of clients. A 'pool' is where a group of newsgatherers agree to share the pictures on a story for common use, and a pool may be set up for a number of reasons. A number of broadcasters and news agencies have standing arrangements to share material with each other, and some similar arrangements exist between individual broadcasters. The purpose is usually either to save cost or because access to a particular news event has been granted on the basis that material will be shot by one camera crew and pooled with other interested parties.

7.9 The Inter-Union Satellite Operations Group

Finally, mention must be made of the World Broadcasting Union (WBU) Inter-Union Satellite Operations Group (ISOG). ISOG is an industry lobby group, made up of broadcasters from the eight broadcasting unions of the world, which seeks to influence satellite operators and the satellite technology industry, particularly in the area of SNG.

In 1985, the North American National Broadcasters Association (NANBA – but now NABA after dropping the 'National' in their title) instigated a series of meetings with INTELSAT and the North American Signatories to discuss areas of common interest. These initial discussions set the tone for the establishment of a permanent structure for dealing with the system-wide operational problems.

In 1986 the eight broadcasting unions of the world, under the umbrella of the WBU, created the Inter-Union Satellite Operations Group (ISOG). Its members are drawn from the editorial and operational membership of the broadcasting unions (see Table 7.6). Their directive was to maximize satellite resources for radio and television broadcasters by making satellite systems sensitive and responsive to operational needs, and to ensure the streamlining of processes. SNG has always been a particular focus of the group and will continue to be so.

Table 7.6 World Broadcasting Union (WBU) members

Broadcasting union	*Abbreviation*	*Headquarters*
Arab States Broadcasting Union	ASBU	Tunisia
Asia-Pacific Broadcasting Union	ABU	Malaysia
Associacion Internacional de Radiodiffusion	AIR	Uruguay
Caribbean Broadcasting Union	CBU	Barbados
European Broadcasting Union	EBU	Switzerland
North American National Broadcasters Association	NABA*	Canada
Organizacion de la Television Iberoamericana	OTI	Mexico
Union des Radiodiffusions et Televisions d'Afrique	URTA	Senegal

*Formerly NANBA.

Since 1986, the ISOG has hosted meetings twice a year. One meeting is generally held in the Americas and the other in another region of the world. Local signatories and broadcasters are encouraged to attend these meetings so that they may keep up to date with the latest developments and with each other's ever-changing requirements. A different venue is chosen for each meeting, allowing the group to focus on different regional concerns from meeting to meeting. While overall attendance remains fairly constant at about 200, the geographical mix of delegates changes with the venue. The purpose of the ISOG meetings is to provide a forum for broadcasters and satellite service operators to exchange information, outline requirements and resolve common problems.

An area in which ISOG has been particularly active in the last few years, in association with INTELSAT, is in establishing digital SNG standards for services and testing different manufacturers' MPEG-2 compression equipment for interoperability. This has led to a valuable focus on the importance of interoperability to ensure that no matter where a signal is originated, on whichever MPEG-2 compression equipment, it can be received anywhere else with different MPEG-2 compression equipment. Further details on these tests can be found in Chapter 4.

7.10 Conclusion

Two of the organizations dealt with in this chapter – INTELSAT and EUTELSAT – which had been established as non-governmental organizations (NGOs) are in the process of being privatized. The privatization of INTELSAT (and INMARSAT – see Chapter 5) is part of a push by the WTO with the Signatories to remove national monopolies operated by signatories of the respective organizations. This in turn reflects the commercial pressure on the global telecommunications market, which has also seen the privatization through the 1990s of a number of national telecommunications providers (PTTs). The steady transition of the control of spectrum from state control to licensed agencies is a feature seen in Europe, and this change goes hand in hand with a more commercial attitude to the use of spectrum. No doubt, this change will slowly spread to other parts of the world, helped by the developments in the WTO GATS initiative in telecommunication.

The processes examined in this chapter, and others, should give you a reasonable understanding of the way satellite systems are used for SNG, and the steps necessary for an uplink operator to verify that their system is compliant with the satellite system operator, and why these steps are necessary. The regulatory issues in each country vary widely, but only in rare situations (such as war) will an uplink operator be able to avoid dealing with national administrations at some level or other.

As SNG increases in use around the world, so the procedures will become less arduous compared to, say, ten years ago in the late 1980s, when the use of SNG was still relatively new to satellite system operators such as INTELSAT, and completely foreign to a large number of governments. The impact of coverage of globally significant events by CNN and the BBC has made many more individuals and organizations aware of the existence of SNG.

References

1. INTELSAT (1998) Satellite Systems Operations Guide (SSOG) 200, Earth station registration instructions and forms for earth station registration.
2. INTELSAT (1998) Intelsat Earth Station Standards (IESS) 601, Standard G – Performance characteristics for earth stations accessing the Intelsat space segment for international and domestic services not covered by other earth station standards.
3. INTELSAT (1999) Satellite Systems Operations Guide (SSOG) 600, Transmission plan approval.
4. Phillips Business Information Inc. (1999) *Satellite Today*, **2** (118); www.satellitetoday.com.

8

Get there, be safe: safety and logistics

8.1 Introduction

The operation of SNG uplinks unfortunately brings both the operator and occasionally the public into contact with a number of potential hazards. In this chapter, we will examine the range of hazards and the measures that can be taken to minimize the risks. There is health and safety legislation in many countries covering these hazards, so for illustrative purposes we will look at how these are addressed by safety agencies in the UK and in the US.

In the UK, the primary safety agency is the Health and Safety Executive (HSE), and in the US, the Occupational Safety and Health Administration (OSHA). These two safety agencies have adopted best practice in formulating their safety policies, and both of these primary safety agencies defer to specialist agencies where there are specific hazards of a highly technical nature.

Secondly, we will look at the various issues of planning and logistics when deploying SNG uplinks, both domestically and internationally. This will involve looking at the use of road transport, the transportation of SNG systems on aircraft, and the influence of time pressures on achieving a safe, speedy and cost-effective deployment. We will also look at the intertwined issues of safety and logistics when operating in hostile environments, particularly war zones.

8.2 Safety

The most important consideration when operating an SNG uplink is safety, and there is a saying in the business, 'no story is worth a life'. To consider the impact on the operation it is necessary to identify the specific risks. As we are going to spend some time looking at risks and hazards, it is perhaps worth reminding ourselves of the definition of 'risk' and 'hazard', as occasionally there is confusion in their usage.

A 'hazard' is anything that can cause harm; for example, electricity is a hazard associated with SNG equipment. 'Risk' is the chance, whether high or low, that somebody will be harmed by the hazard; for example, how much of a chance is there that someone will be electrocuted in either operating or being close to an SNG uplink. It is important that the hazards and risks are clearly identified for two reasons. Firstly, a human injury or life may depend on the correct action being taken, and secondly, the owner and/or operator of the SNG equipment may be liable to prosecution for failing to identify the hazards and take suitable steps to minimize the risks.

8.2.1 Outline of hazards

There are two levels of potential hazard encountered when operating SNG uplinks – for our purposes, we will classify them as primary and secondary. The primary hazards are:

- Non-ionizing radiation
- Electrical hazards
- Manual handling issues

The secondary potential hazards are:

- Driving of vehicles
- Operating in hostile environments, including war zones

8.2.2 Non-ionizing radiation

The mere mention of the word 'radiation' usually imbues people with fear, and one of the first lessons anyone has to learn when being involved with the use of SNG uplinks is the hazard of radiation – but it is non-ionizing radiation, which is significantly different to ionizing radiation. So to dispel some myths, let us be clear about the differences between ionizing and non-ionizing radiation.

Ionizing radiation is that emitted by X-rays, gamma rays, neutrons and alpha particles that has sufficient energy to knock electrons out of atoms and thus ionize them. When this radiation passes through the tissues of a living body, in amounts above a safe level, then there is sufficient energy to permanently alter cell structures and damage DNA. This in turn can have dramatic and potentially catastrophic effects on living tissue, including, of course, human beings. However, used in controlled doses, ionizing radiation is widely used as a medical diagnostic and treatment tool, and provided the doses are within acceptable limits and there are adequate precautions to minimize the risk, there is little to fear from such use.

This is not the type of radiation associated with microwave transmitting equipment. Microwave transmitting equipment, which of course includes INMARSAT satphones and SNG uplinks, emits non-ionizing radiation. In the following discussion on non-ionizing radiation, the term SNG uplink can be taken to include the whole range of INMARSAT satphones as these also use microwave frequencies in the 1.5 GHz band.

The whole issue of the safety of exposure to non-ionizing radiation has risen to prominence in the last few years, specifically in one area of public concern – the use of mobile phones. There is currently a public debate in a number of countries about the potential biological effects of using mobile phones, but there is no conclusive scientific evidence at the time of writing demonstrating that the use of mobile phones – even prolonged use – has any abnormal biological effect on human beings. However, there are studies ongoing in the US, the UK, Scandinavia, Australia, and a number of other countries to try to determine conclusively the answer to this issue.

The definition of non-ionizing radiation is electromagnetic radiation, which encompasses the spectrum of ultraviolet radiation, light, infrared radiation and radio frequency radiation (including radio waves and microwaves). This is of much lower energy than ionizing radiation and therefore is unable to knock electrons out of atoms. When this type of radiation passes through the tissues of the body it does not have sufficient energy to ionize biologically important atoms, and therefore to alter cell structures or damage DNA. However, it does have thermal effects, and frequencies in the range 30–300 MHz have the greatest effect as the human body can more easily absorb them. At frequencies above this range, the body absorption is less, but may still be significant if the power levels are high enough.

The primary health risk of non-ionizing radiation is considered to be the thermal effect. The absorption of RF energy varies with frequency. Microwave frequencies produce a skin effect – you can literally sense your skin starting to feel warm if you are exposed to high power levels at microwave frequencies. After all, this is the principle on which microwave ovens operate, and you can cook human tissue just as easily with this type of power as animal tissue. RF radiation may penetrate the body and be absorbed in deep body organs without the skin effect which warns an individual of danger. This is called deep burning, and there are certain parts of the human body that are particularly sensitive to these deep heating effects, for example the eyes and, additionally in males, the testicles. Therefore, power levels around microwave transmitting equipment need to be kept below a certain level to minimize the risk to people. Preferably, people need to be kept away from the most dangerous parts of this equipment, especially the antenna, which is designed to focus all this energy in a particular direction, i.e. towards the satellite.

The biological effects can be expressed in terms of the rate of energy absorption per unit mass of tissue, and this is often referred to as the specific energy absorption rate (SAR). The SAR, measured in W/kg, is generally used in the frequency range 100 kHz to 10 GHz and may be either expressed as 'whole body' or specified for specific areas of the body – the limbs or trunk, for example. It is accepted that exposure to non-ionizing radiation that induces even a body temperature rise of 1–2°C can cause the (temporary) adverse effects of heat exhaustion and heat stroke. The safety limits can also be expressed as power density levels, in W/m^2 (or mW/cm^2), usually for frequencies above 10 GHz.

The study of the effects of non-ionizing radiation is highly complex and a significant body of research has turned up other non-thermal effects. All the internationally recognized standards have so far based their exposure limits solely on preventing thermal problems, though research continues.

8.2.3 Hazards with SNG equipment

The most hazardous area when operating SNG uplinks is in front of the antenna. This is a particular problem with SNG flyaways and transportable INMARSAT satphones as they are often placed on the ground, or operated so that the beam is at a height through which people may pass. It is less of an issue with SNG trucks, where the antenna is mounted either on the roof or at least high up on the vehicle structure, and hence out of harm's way.

It is not necessarily a lower risk with a smaller rather than a larger antenna, as the potential hazard is directly related to the output power delivered from the antenna, which is related to the size of the HPA as well as the antenna. However, it is recognized that, as an SNG antenna is directional, the area of risk can be clearly defined (see Figure 8.1). This makes the task of reducing the risk to manageable proportions much more viable, even in a public area. However, the amount of power is effectively focused into a high-power beam and significantly high power can be measured even a considerable distance away from the antenna in the direction of the satellite. Incidentally, this creates a secondary risk relating to aircraft, as we shall see later.

There are two areas in the path of the antenna where non-ionizing radiation can be measured: the 'near-field' and the 'far-field' zones. The definition of these zones is complex and engineers who wish to examine and calculate these should refer to Appendix J. For the rest of this chapter, the limits quoted and measures taken in relation to non-ionizing radiation apply in general to the area around the SNG uplink.

SNG antennas are directional, and therefore the likelihood of significant human exposure to non-ionizing radiation is considerably reduced. The power densities in areas where people may be typically exposed are substantially less than 'on-axis' power densities, as the antenna radiation pattern should indicate rapidly diminishing power density the further off-axis the measurement is taken.

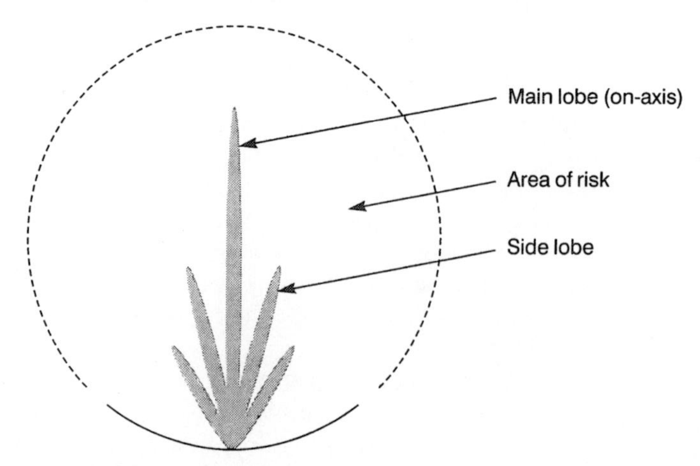

Figure 8.1 Parabolic antenna radiation pattern

Nevertheless, the potential for exposure must take into account the following:

- The direction (azimuth and elevation) the SNG antenna is pointing in.
- Its height above ground.
- Its location relative to where both the public and the uplink operators may pass.
- The operational procedures followed to minimize the risks.

When calculating and applying the limits, consideration needs to be taken of the potential time of exposure. The limits set by both US and UK safety agencies are given within a limited time-frame of exposure, but in practice, it is wise to apply these limits as if they related to an instantaneous exposure. By doing so, any uplink operator is going to be erring on the side of safety.

The International Non-Ionizing Radiation Committee (INIRC), which was set up by the International Radiation Protection Association (IRPA) in 1977, defines international standards. The INIRC has promoted research in this area and in association with the World Health Organisation (WHO) has published a number of health and safety documents covering the full spectrum of non-ionizing radiation. The national standards that follow have been developed from this international body of knowledge.

8.2.4 The position in the US

The standards-setting body in the US originally active in this area was the American National Standards Institute (ANSI) which first issued guidelines in 1982, but the work has now been taken on by the Institute of Electrical and Electronics Engineers (IEEE) in the US.

The Federal Communications Commission (FCC) first issued guidelines relating to human exposure to RF emissions in 1985, and these were subsequently updated in 1996[1] and are concerned with RF emissions in the range 300 kHz to 100 GHz.

The FCC defines maximum permissible exposure (MPE) limits, based on exposure limits recommended by the US National Council on Radiation Protection and Measurements (NCRP). Over the whole range of frequencies, the exposure limits were developed by the IEEE and adopted by the American National Standards Institute (ANSI) to replace the 1982 ANSI guidelines. There are two limits: one set for members of the general public ('general population') and the other for workers involved in operating and maintaining RF transmitting equipment ('occupational'), as shown in Figure 8.2.

The MPE for general population exposure for the frequency range 1.5–100 GHz, which encompasses all SNG frequencies, is 1 mW/cm^2, averaged over a 30-minute period. The FCC deems that general population (i.e. uncontrolled) exposures apply where the general public may be exposed, or where workers may not be fully aware of the potential for exposure or cannot exercise control over their exposure.

The MPE for occupational exposure, which allows a higher exposure limit for the same frequency range, is 5 mW/cm^2, averaged over a 6-minute period. Occupational

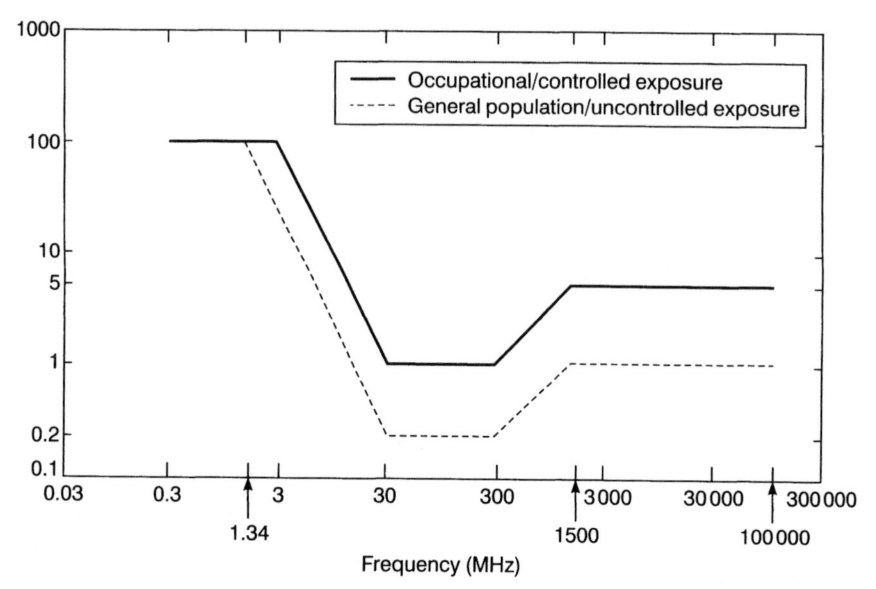

Figure 8.2 FCC limits for maximum permissible exposure (MPE) (in terms of plane-wave equivalent power density)

(i.e. controlled) limits apply where workers are exposed, provided they are fully aware of the potential for exposure and can exercise control over the degree of their exposure. Occupational exposure limits are also relevant in situations when an individual is passing through a location where occupational/controlled limits apply, provided they are made aware of the potential for exposure.

As discussed in Chapter 6, the FCC requires an RF hazard study to be submitted with a registration application for an SNG uplink, and this is based on the calculations found in FCC OET Bulletin 65[1].

8.2.5 The position in the UK

In the UK, the National Radiological Protection Board (NRPB) is responsible for research and advising best practice in the area of radiation hazards, including non-ionizing radiation. It was established in 1970 and the current recommendations were revised in 1993.

Its policy differs quite markedly from that found in the US. Unlike the US, the NRPB makes no distinction in exposure levels between the general public and workers involved in RF transmission. It also does not define absolute limits, but sets 'investigation levels', below which adverse biological effects are not likely to occur. Exceeding these investigation levels does not automatically infer biological damage will occur, but that further biological investigation may be necessary and is recommended.

It also uses different methods of measurement depending on the frequency band concerned. Within the frequency bands of interest to us, it has an exposure level for 100 kHz to 10 GHz, and a different level for 10–300 GHz. To further complicate matters, the limit for the 100 kHz to 10 GHz band has a limit expressed in SAR (W/kg) over one time period, while for the upper band it is expressed in power density (W/m^2 or mW/cm^2) over two different time periods, depending on the frequency:

- 100 kHz to 10 GHz: 0.4W/kg SAR averaged over the whole body over a 15-minute period.
- 10 GHz to 300 GHz: 100W/m^2 (10mW/cm^2) – power density on any part of the body – averaged over any 10-second period above 20 GHz.

In the band 10–20 GHz, the limit can be averaged over a period that is frequency dependent:

$$\text{Averaged time period} = 68/f^{1.05} \text{ minutes}$$

where f is the frequency in GHz. Therefore, for an uplink frequency of 14.00 GHz, with the limit at 10 mW/cm^2, the averaging period for exposure is:

$$\frac{68}{14^{1.05}} = 4.25 \text{ minutes}$$

Using this calculation, at 14.25 GHz, the averaging period decreases to 4.17 minutes, and at 14.5 GHz, it is 4.1 minutes. Therefore for a Ku-band uplink the averaging period approximates to just over 4 minutes.

Since C-band SNG is not permitted in the UK, we need not concern ourselves with exposure in the 6 GHz band and the problems with measuring SAR. There is no easy way to measure SAR directly, though equivalent measurements can be made.

The NRPB limits are more complicated than the FCC MPEs, and in practice, there is not a great deal of difference between the two limits as far as the SNG operator is concerned in terms of their own safety. Unlike in the US, there is no requirement for an RF hazard study to be submitted to the UK RA on applying for a licence for an SNG uplink.

8.2.6 Practical steps to minimize risks

Having examined the recommended maximum limits, we need to see the practical steps that can be taken to minimize the risk to both operator and general public. Assuming that the antenna and HPA equipment are in an area to which either the uplink operator or a member of the public has access, i.e. it is not mounted on a vehicle roof, then there are essentially simple measures that can usually be taken which will give reasonable protection.

Figure 8.3 Safe area cordon around an SNG antenna

In order of importance, they are:

- Rigging an uplink in an area of restricted access.
- Cordoning-off of a 'safety zone' and using warning signage (see Figure 8.3).
- Checking measurements around the perimeter of the safety zone.
- Restricting access.
- Supervising and checking during transmissions.

Obviously, the risk of exposure, particularly for members of the general public, can be greatly minimized if the uplink can be rigged in a position which has secure access while also meeting the requirements of the operation. This is not always possible, but if it can be achieved then the risk of exposure is limited to the uplink operator. Unfortunately time and logistics, as we shall see later, can conspire against achieving this aim, particularly in a breaking news story situation.

The idea of the safety zone is to protect anyone from entering an area where there is a risk of exposure over any part of his or her body. In essence this means that there must be no risk of dangerous exposure from head height down to the ground, assuming 2 m as head height (see Figure 8.4), which gives an approximate indication of the area that people need to be excluded from. A zone should be cordoned off around the front and side areas of the antenna. The limit of this safety zone depends on three factors: the angle of elevation of the antenna, the operating band (C or Ku) of the uplink, and the maximum output power capable of being

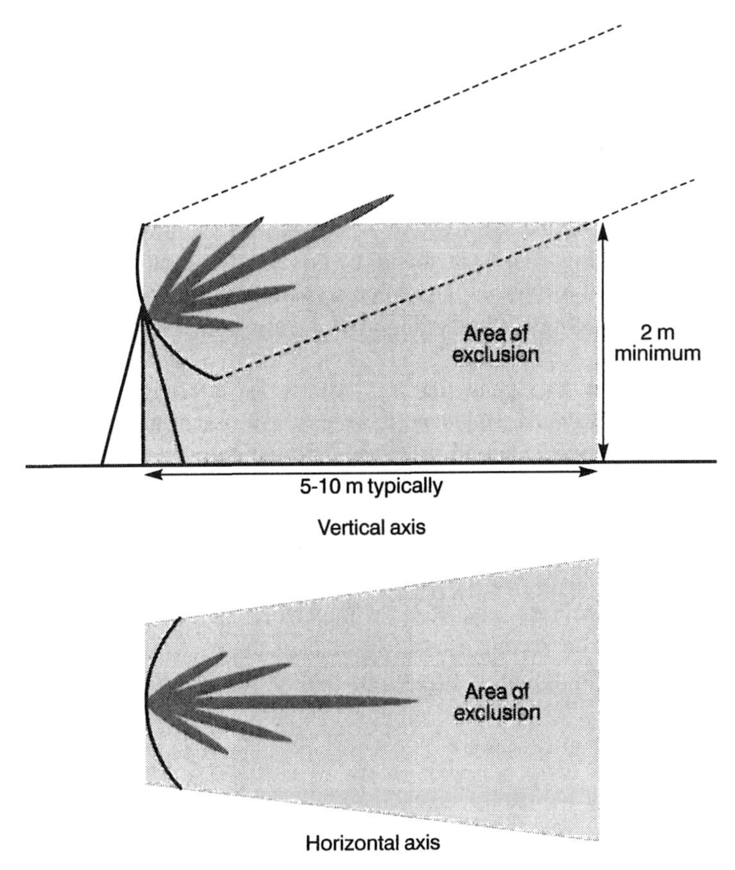

Figure 8.4 SNG antenna exclusion zone

delivered from the antenna. Generally, the nearer the equator the uplink operating location is, and therefore the greater the elevation angle, the smaller the safety zone need be, as there is a reducing likelihood of non-ionizing radiation 'spillage' from the antenna.

There are two methods of determining this area: either by theoretical calculation or by practical experiment with a non-ionizing radiation field strength meter. These meters are available from a number of manufacturers in a variety of different forms, from personal protection 'badges' with an audible and visual alarm, up to laboratory-grade precision measurement meters. The two most well known brands are the Lockheed-Martin 'Narda' and General Microwave 'Raham', and their personal protection products are available at a relatively moderate cost, considering the degree of protection they can offer.

Of the two methods of assessing the safe zone for operation, the practical experiment is perhaps the safest and easiest in a fast-developing operational environment. Having set the elevation angle of the antenna, a generous distance should be allowed and cordoned off – up to 10 m if possible. Once a satellite uplink is transmitting, the operator can then check that at the perimeter of this safety zone the exposure level is below the limit defined. If the level is below the limit, the operator can then move in slowly towards the antenna in a scanning arc path, checking the reading on the field strength meter, until the operator is near to the maximum level of exposure. This defines the minimum limit of the safety zone, and ideally the zone should be set slightly further out from the point of near-maximum exposure, remembering that there should be no risk of exposure from ground level to 2 m. Obviously, if a flyaway is sited on a rooftop, pointing out over the side of the building, then there is little likelihood of anyone being able to stray in front of the antenna.

It should noted at this point that we have so far concentrated on the risk of exposure from the antenna. However, there is also a risk of non-ionizing radiation leaking from the waveguide and its connection between the output flange of the HPA and the input flange of the antenna feed. This is particularly true of flexible waveguides, which although flexible, deteriorate over time with the flexing of the waveguide from rigging and de-rigging the flyaway equipment. The waveguide should be regularly inspected for physical signs of wear, cuts or burn marks ('hot-spots'), which are an early indication of impending failure (these marks may be noticed as a series of evenly spaced transverse marks along the length of the waveguide). Once the uplink is rigged, and before the first transmission, the HPA should be run up to full power into a 'dummy' load. The flanges and the waveguide should then be carefully checked with the field strength meter. It is a wise precaution to carry a spare length of waveguide, for obvious reasons. Leakage from flanges may be due to incorrect tightening of the flange bolts, damage due to corrosion or mishandling on the faces of the flanges, any of which would prevent a snug fit.

The operator should keep an eye on any movements of people near the perimeter and try to be vigilant in ensuring that nobody attempts to stray inside the safety zone perimeter. Signs warning of a risk should also be placed at the perimeter, and international pictogram signs are available that depict the hazard.

At regular intervals throughout an operation, the zone perimeter, the wave-guides and flange joints should be checked to ensure that the operation continues to be safe, both for the operator and the general public.

8.2.7 Operations near or in sensitive zones, including aircraft flight paths

We mentioned earlier that a potential hazard existed thus creating a risk relating to aircraft flight paths. In the UK the civil aviation governing body, the Civil Aviation Authority (CAA), considers that the high-energy beam from an SNG antenna operating in the Ku-band could cause a serious malfunction in an aircraft's instrumentation if it is in close proximity. This is considered to apply if an aircraft, on approach or take-off, was to fly through the beam from an uplink that is only a few kilometres away. It is a requirement of the UK Radiocommunications Agency (RA)[2], therefore, that if an uplink is being operated within 20 km of an airport with ILS (instrument landing systems), the operator has to inform the airport. Confirmation that this requirement has been met must be shown on the frequency clearance application made to the RA. In addition, there is a 'cone of protection' that the uplink signal must not infringe on both the approach and take-off flight paths of the airport[3] (see Figure 8.5).

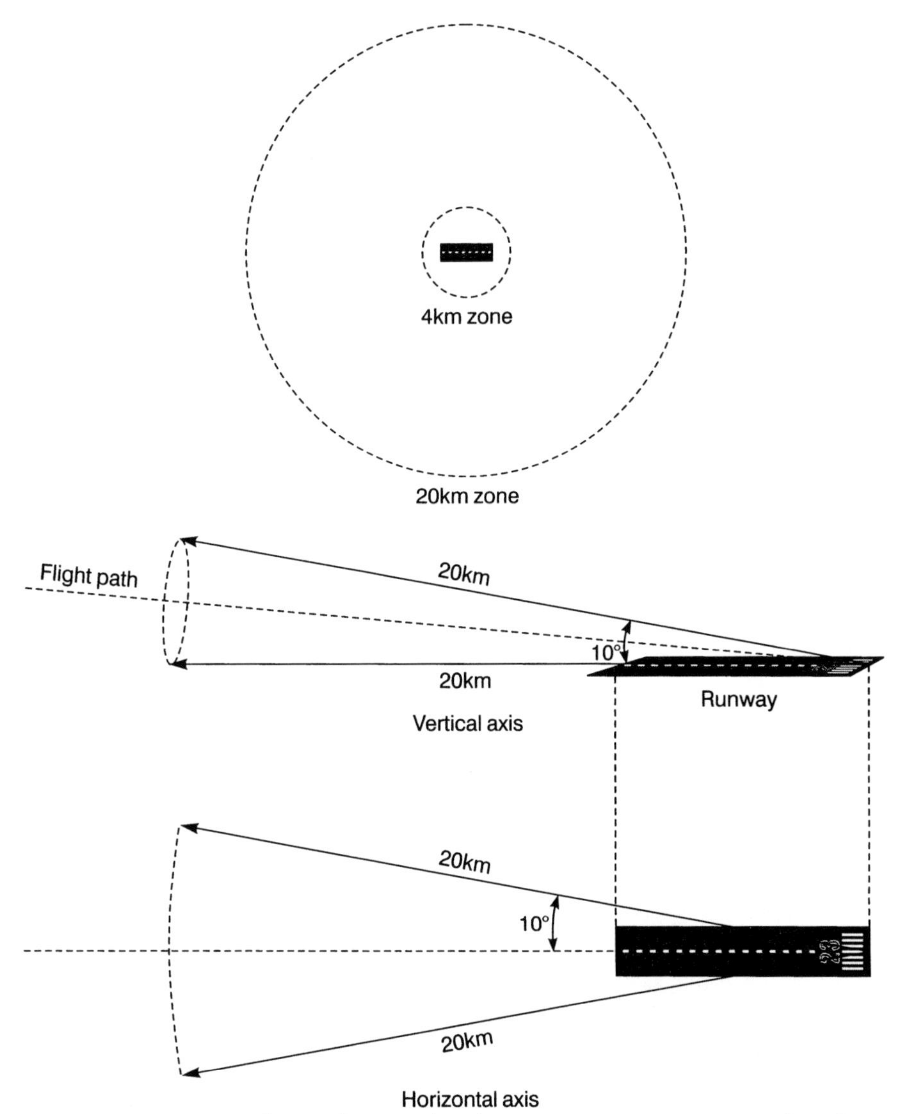

Figure 8.5 UK RA notification limits and exclusion zone on airport approach and departure paths

8.2.8 Military installations

In the UK and many other countries, operation near or at military installations is restricted, although special clearance may be obtained. This may be justified on the grounds of either potential electromagnetic interference or security. However, other countries (the US, for example) do not restrict operations beyond the perimeter of the installation. Incidentally, in deploying SNG to military installations it is paramount that a full dialogue takes place between SNG technicians and military technicians to determine that there will be no effect from the SNG uplink transmissions on military communications or weapons systems. There are strong electromagnetic fields from military systems and it is not unusual for this to have an effect on the camera rather than the SNG uplink. Radar in particular (as it produces very large but short pulses of RF) can affect the pictures, resulting in 'black line flashing' on analogue cameras or, on the now more common digital cameras, strange disturbances in broad bands in the colour and composition of the picture.

That concludes our consideration of the hazard of non-ionizing radiation. However, before leaving this aspect of SNG safety, it is perhaps worth relating a story, which although probably apocryphal, illustrates the importance of vigilance in restricting access.

In the winter of 1989/90, at the fall of the Berlin Wall, there were a large number of uplinks operating on the ground in Berlin covering the story. It was winter, bitterly cold, and many of the flyaway uplinks were being operated with almost all the uplink equipment (and the operators) enclosed in a tent, with just the antenna

Figure 8.6 The unfortunate US network antenna in Berlin

protruding out through the tent flaps. One of the US network uplinks was in transmission, with the operators huddled inside the tent behind the antenna, when a call came from the US on the co-ordination phone that the signal had suddenly been lost. The operators checked their signal monitoring equipment and were very puzzled, as everything seemed fine. Finally, one of the technicians happened to venture out into the cold, to find a passer-by warming their hands cupped around the end of the feed arm, blocking the signal to the antenna reflector! Of course, if this story is true, then the 'passer-by' must have known something about non-ionizing radiation, and it would have been a stupid prank. But, more importantly, it perhaps illustrates the importance of always trying to operate this type of equipment with your wits about you, as it is a potentially dangerous piece of equipment if not treated with respect.

8.2.9 Electrical hazards

As with any electrical equipment, electrical power has to be supplied to make the equipment function. It would be preferable if this power could be provided from low-power batteries, hence minimizing hazards associated with electrical power. Unfortunately, the demand for 'primary' supply power for an SNG uplink and all the associated equipment is quite high, even for a small system. Typically, an SNG uplink requires at least 3 kW, and large SNG systems require at least 6 kW. This can only be derived from AC power sources, and therefore there are hazards in operating SNG equipment related not only to the integrity of the power source itself, but also to the condition of the equipment.

The rugged life the equipment leads also has to be borne in mind, as well as the exposure to widely varying weather conditions, which over time can cause a deterioration in the electrical integrity of the equipment itself unless it is regularly checked by competent technicians. SNG uplinks can be operated in desert conditions, with temperatures reaching over 40°C; in tropical conditions, with relative humidity over 90%; or in Arctic conditions, with temperatures as low as –30°C. The equipment is expected to be able to perform consistently in all these conditions, and this can take its toll on the electrical integrity of the equipment over time.

There is an added hazard in that the amplifying element typically used in most SNG HPAs – the 'travelling wave tube' (TWT) – requires a very high voltage supply, which although generated internally in the HPA power supply unit (PSU), nevertheless is a potential hazard. Some of the monitoring equipment also has high voltage supplies, so all the equipment in a system is vulnerable.

Regular care of the equipment, as well as respect in its use, is vital to minimize the risks. But care also has to be taken in sourcing the supply of primary power. Generally, the uplink operator can either obtain power from a local supply, or more likely, transport a generator (usually petrol) along with the system. With a truck-based operation using vehicles built for the purpose, this is relatively straightforward and very professional and safe power systems are installed in these vehicles. They usually incorporate various safety features to ensure minimal risk to people and equipment.

With a flyaway operation, the situations are usually much more difficult. It is common to transport a system with its own generator, as one never knows if there is an adequate power supply at the destination – if there is one at all. Having arrived at the location, assuming the local supply is not reliable or safe enough (often the case in areas where flyaways are typically deployed), the generator has to be placed in a convenient position. Some thought also has to be given to refuelling arrangements (including the safe storage of fuel) as well as the safe routing of cable runs from the generator to the uplink equipment.

Finally, it is worth considering the use of a residual current device (RCD) in the supply to the uplink equipment. An RCD is a device that monitors the power supply, constantly checking for a fault. It measures the amount of current flowing in the live and neutral (line and ground) wires, and if there is a difference, then a fault may be present and the supply is broken to protect the personnel and the equipment (see Chapter 3). The assumption is that there is some current flowing out of the circuit and that it may be flowing through a person. It does not act as an over-current circuit breaker or fuse, which protects the circuit from drawing too much current beyond the capacity of the cable and the equipment, but is an additional safety device.

The advantage of the RCD is that it does not require a good earth connection, or any earth connection at all for that matter, to operate. It will protect people even when using the most basic electrical source, and is compact and cheap. Also, the risk from 'nuisance tripping' is now very low compared to early models.

8.2.10 Manual handling

The issue of manual handling in the workplace – which is about minimizing the risk of musculoskeletal disorders – is pertinent when looking at the operation of flyaway SNG uplinks in particular. As discussed in Chapter 3, flyaway systems are typically made up of a large number of boxes, and some of these cases are particularly heavy – up to 75 kg (165 lb). Although back injuries through lifting awkward or heavy cases do not account for work-related deaths, they do account for a significant amount of human suffering, loss of productivity, and economic burden on compensation systems. Back disorders are one of the leading causes of disability for people in their working years, and SNG operators quite commonly suffer some degree of discomfort from back pain caused by lifting during their work.

Back problems can either be caused by a single traumatic incident or more often through a gradual process of repetitive strain over a period. Because of the slow and progressive onset of this internal injury, the condition is often ignored until the symptoms become acute, often resulting in disabling injury. Acute back injuries can be the immediate result of improper lifting techniques and/or lifting loads that are too heavy for the back to support. While the acute injury may seem to be caused by a single well-defined incident, the real cause is often a combined interaction of strain coupled with years of weakening of the muscles, ligaments, vertebrae and discs in the back, either singly or in combination.

Typical actions that occur when handling the SNG systems and can lead to back injury are:

- heavy lifting;
- poor posture;
- reaching, twisting or bending while lifting;
- lifting with forceful movement;
- poor footing.

These are further exacerbated if the operator has not been trained in correct body mechanics – how one lifts, pushes, pulls or carries objects. Additionally, poor physical condition results in losing the strength and endurance to perform physical tasks without strain.

Working on news stories almost invariably involves very long hours, varying from periods of relative inactivity interspersed with bursts of intense activity, to periods of sustained intense activity, all of which cause fatigue. Often meals are missed, sleep patterns disrupted and dehydration can occur due to the pattern of work. If working in a different country, the operator may have to cope with changes in diet, drinking water and perhaps effects from working across a different time zone. Commonly, operators talk of 'landing on the ground, feet running', as their work begins immediately the moment they disembark from the aircraft. They will be directly involved in getting a very large amount of equipment through customs and loaded into vehicles, then quickly moving to the location and setting up the equipment as rapidly as possible. It is the nature of the business that an operation is expected to be up and running within hours of arrival, as the story is probably 'moving' and material is required for news broadcasts as quickly as possible. Once the uplink is working, of course, the work continues as the transmissions get underway.

But what does this have to do with back injury? Quite simply, it is all too easy to forget about the weight of what you are lifting if all the pressure is on you to get those cases moved into position as quickly as possible. Local porters may be available, but invariably one gets involved in moving cases as well. If you are feeling tired due to jet-lag or just generally unwell because of the change in environment, combined with the pressure by producers to get the uplink working as quickly as possible, it is hardly surprising that injuries can occur. Very experienced operators, of course, have various methods of minimizing the effects of all these factors, but nonetheless back injuries are suffered by even the most experienced.

We have dealt with the primary safety issues and can now turn to the secondary potential hazards as identified at the beginning of the chapter.

8.2.11 Driving of vehicles

The two principal methods of transportation for SNG equipment are by air and by road; the flexibility of both these methods is vital for newsgathering. In countries with a well-developed road system, the use of vehicles for domestic SNG is often the most effective and cost-efficient method to cover news. Even if a system is transported by air, a truck is used to carry the equipment on the final leg of the journey to the location. Obviously, in metropolitan areas, road travel is the only option.

SNG operators are therefore regularly required to drive vehicles carrying SNG systems. This can be a dedicated SNG vehicle or a general-purpose load-carrying van or truck, either of which can generally range in weight from 2 to 12 tonnes (4500–26 000 lb). The driving of larger vehicles in this range is in itself a safety issue.

In a number of countries, including those in Europe and North America, there are driving regulations that apply to some sizes of vehicles in this range. The regulations are designed to regulate the way that vehicles are operated with safety uppermost in mind.

Europe

Within the member states of the European Union, there are rules that apply to many vehicles in the weight range commonly used for SNG relating to:

- categories of drivers' licences relating to maximum gross vehicle weights;
- the number of hours spent driving and total hours worked by the driver;
- record-keeping;
- speed limits.

Drivers' licences generally fall into three categories with regard to the maximum gross vehicle weights (GVW) that can be driven. These categories are:

- under 3500 kg
- 3500–7500 kg
- over 7500 kg

There are also some age limits attached to these categories. In the range 3500–7500 kg, it is necessary to have a light goods vehicle endorsement (C1 class) on a standard driving licence, while to drive vehicles above 7500 kg it is necessary for a heavy goods vehicle (HGV) licence to be held.

Within the EU, there is free movement of traffic as far as vehicles from other member European states are concerned, with no customs procedures, duties or tolls. However, each country has national regulations for driving, called National Rules, and in addition, there is an EU-wide set of regulations, termed European Community (EC) Rules. Both sets of rules apply to commercial vehicles over 3500 kg, with additional restrictions for vehicles over 7500 kg. Below 3500 kg most vehicles (including SNG vehicles) are treated as cars and are therefore not subject to either National or EC Rules.

Within any country, commercial vehicles can normally be operated under one set of rules or the other, but not a mixture of both. Where vehicles are engaged in some specific non-commercial activities, including broadcasting (into which category

SNG vehicles are usually classed), then they are said to be 'exempt' from certain parts of National and EC Rules where the SNG vehicle has permanently installed 'fixed-plant' equipment. Where load-carrying vehicles happen to be transporting boxed SNG equipment, then they are not usually classed as exempt and are subject to all the rules that normal commercial vehicles operate under.

For vehicles over 3500 kg, under either sets of rules, there are restrictions on the maximum number of hours of driving each day, the number of hours of work in a day and in the week, the timing of rest breaks, and the minimum periods of rest. A driver is not permitted to drive for more than 4 hours without at least 45 minutes break time, and there is a maximum of 10 hours driving in an 11-hour duty period. There are also weekly rest requirements which limit the maximum number of hours for both driving and working.

In addition, some form of record-keeping of hours is required even for exempt vehicles, usually by the use of a tachograph. Exempt vehicles may use a driver's logbook instead of a tachograph if so desired.

For all vehicles above 3500 kg, there are specific speed limits according to the type of road, with a maximum of 100 kph (60 mph) on autoroutes. In some countries, particular speed limits apply to vehicles over 2800 kg. All countries of the EU excluding the UK and Eire ban vehicles over 7500 kg from driving through built-up areas on Sundays and public holidays, but this does not affect typical SNG vehicles as they are exempt.

North America

As in the nation states of Europe, each of the states in the US has their own state rules, although unlike the EU, there is not an over-arching set of rules which can be adopted instead of state rules. However, all states are required to set minimum standards based on federal law, administered by the US DOT (Department of Transportation). There is no distinction as found in Europe between broadcasting (and therefore SNG) vehicles and other commercial vehicles.

There are various weight classes, which differ somewhat to those found in the EU. The gross vehicle weight (GVW) thresholds are at 10 000, 18 000 and 26 000 lb (4500, 8200 and 11 800 kg respectively). Above 26 000 lb, a commercial driver's licence (CDL) is required. There are rules on drivers' hours relating to driving commercial vehicles requiring a CDL, and these are broadly similar to EU Rules, with a maximum of 10 hours driving in a 15-hour duty period. However, a CDL is not necessarily required for SNG trucks in most states as the majority are below 26 000 lb, allowing drivers with standard licences to drive such vehicles. Nevertheless, some stations still require their uplink operators to have a CDL as they are operating commercially, and if they travel interstate, it saves any argument if the other state requires a CDL for a particular vehicle size. In addition, if a CDL is held, a driver's medical certificate is required.

Below the 26 000 lb weight class, vehicles may be driven on a standard driver's licence. Most states deem any vehicle over 10 000 lb as 'goods carrying', and although a CDL is not required, drivers need to have regular medical certification and the use of a driver's logbook is a statutory requirement.

In Canada, the requirement for a different type of licence for trucks applies to vehicles with a GVW of greater than 4500 kg.

Australia

Each state has its own driving regulations, but these are now integrated into a common National Driver Licensing Scheme. There are no particular restrictions below 4500 kg, but above this limit a commercial driver's licence is required. There are different licence categories for 4500–8000 kg and over 8000 kg.

Time pressure and fatigue

Without doubt, however, no matter which country the SNG truck is being driven in, the greatest hazard the SNG operator faces is time pressure. Particularly on a breaking story, there is an inherent pressure on the operator to get to the location as quickly as possible to get on-air. This pressure may or may not be directly applied – but it is always there. SNG truck operators need to exercise considerable self-discipline in order that this pressure does not affect their driving. Stress and fatigue are the two most likely causes of a vehicle accident, both of which are common when working in newsgathering.

8.3 Logistics

8.3.1 Background

Logistics is historically a military science, though it has now extended into a wide range of commercial activities all over the world. In this section, we will look at how logistics play a part in newsgathering, primarily in the international arena, though we will spend some time looking at domestic markets.

Logistics in the context of SNG can be defined as the science of moving equipment and personnel from A to B in the quickest time at the lowest cost. It is probably the key strategy for success in satellite newsgathering, rising above how technologically advanced the equipment is, or how skilled the operators are. There is no merit in having highly skilled staff with the very latest technology if they are not where the story is. Yet the challenge of moving often quite large volumes of equipment even to relatively accessible areas of the world in the shortest time can frustrate even the most experienced newsgatherers. To do this cost-effectively is an additional pressure on all news operations.

As has previously been said, the two principal methods of transportation of SNG equipment are by air and by road, each having advantages in particular areas and offering the flexibility vital for newsgathering. That is not to imply that the use of other modes of transport is not an option. Other methods, such as by sea or rail (or even pack mule!), have been used but are more suited to operations where there has been a fair degree of pre-planning or where these methods of transportation offer particular logistical advantages.

It cannot be denied, however, that for international newsgathering, air transport offers the greatest flexibility. In 1994, the world's airlines had a total fleet of about

15 000 aircraft operating over a route network of approximately 15 000 000 km and serving nearly 10 000 airports.

Various factors have to be weighed up before choosing a particular method of routing both equipment and personnel to a location – and they do not necessarily travel by the same method on pre-planned operations. When covering stories in hostile environments such as war zones, there may not be any choice to be exercised if, for example, facilities for the press are being provided by organizations such as the UN.

8.3.2 Critical factors

There are five fundamental factors to be considered when deploying an SNG uplink to a news story. These are:

- The location
- How quickly
- The costs
- The availability of transportation
- The availability of space segment

Although they can be classified neatly into these five areas, there is a complex inter-reaction that means they have to be considered in parallel, rather than in sequence as this list might suggest at first glance. These are also not necessarily the only issues, but they are the ones that have to be considered on all deployments. It would be neat to be able to represent these as a logical flow diagram, but unfortunately the processes involved do not fit into an ordered, sequential model. Instead, Figure 8.7 shows a 'cloud' diagram which although not an exhaustive model indicates these processes and their inter-relationship.

The following discussion therefore will attempt to deal with all five of these factors. It is not possible to consider them one by one in isolation, because of the inter-relationship that exists in the decision-making process.

Getting to the location is, however, the first factor to be considered. For a local story, the answer is probably perfectly obvious, and a SNG truck can be sent. The story can be on the air within the hour and no thought to the process is necessary by the newsroom dispatcher. The location becomes an issue when it is more remote from the broadcasting centre or news agency headquarters.

It is an inevitable truism that many of the world's worst disasters, which involve the natural forces of weather and the earth, tend to occur in regions which are often the most remote and therefore relatively inaccessible. As we have all seen on the television screen (due to the power of SNG), very large numbers of people are often killed or injured in these disasters, and the emergency services and relief agencies struggle to access these areas. The news media, and television in particular, are also in a race to bring the story of the events to the wider world, and are hot on the heels of the relief services, obviously using the same methods of access. The earthquake disaster in Turkey in August 1999 provided an illustration of this, for even though

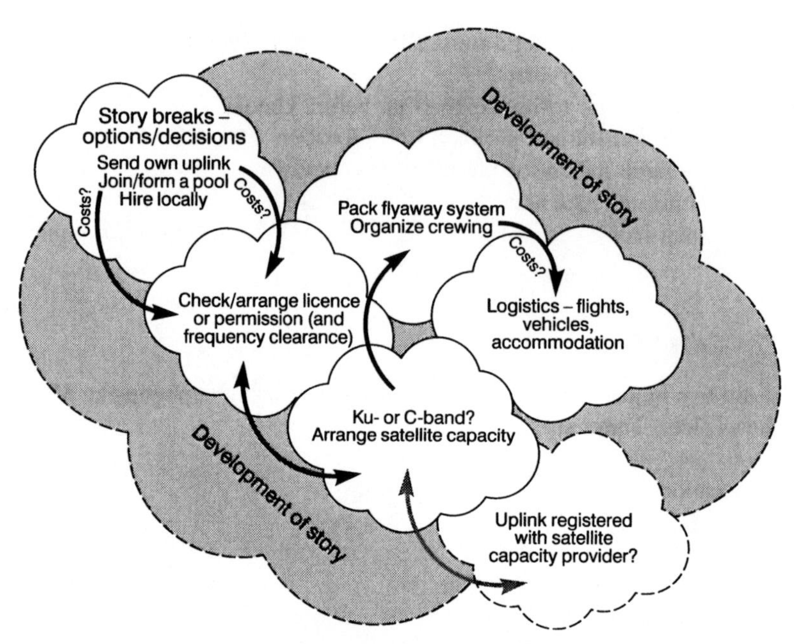

Figure 8.7 Operational planning processes on deploying SNG for a news story

Turkey is a developed country, the scramble to get there when the news broke was intense – though fortunately, on this story, the international airport in Istanbul was not affected by the disaster.

Tragic events have occurred in relatively developed regions, but where the particular location is in an inaccessible part of the region, frustrating even highly developed emergency services in their efforts to get to the scene. For example, huge areas of North America are relatively remote, and when there have been significant disasters in these remote areas, it has taken a considerable time for the emergency services to arrive and start their work. Even where the location is in a very highly developed region, the scale of the disaster can completely overwhelm a country's resources. An example of this was the huge earthquake in Kobe, Japan, in 1995, where within a few minutes, 6000 people were killed, 30 000 injured, and over 100 000 buildings were destroyed – and the whole infrastructure of the region was destroyed.

For any newsgathering team to cover these types of event, the problems of the location can dictate how the story is covered. Many of the large news networks have been forced to pool their SNG resources, as the problems of access, transportation, cost and local politics may conspire against them each having their own facilities on the ground. Often these 'pools' are organized on a national or regional basis; for instance, the US networks may form a pool serving US interests, the Japanese networks form a pool for their interests, and the EBU may provide facilities for its members. At other times, where strategic alliances exist between certain newsgatherers, joint operations may be established.

Assuming the location is in another country or continent, the choices facing the newsgatherer in terms of getting the equipment and personnel to the location of the story fall into the category of deciding how to get there by air. The equipment can either be sent on a scheduled flight, either as excess baggage or as freight, or on a specially chartered flight.

The decision to be made on these three options – charter, excess baggage or freight – is further affected by the factors of time and cost.

8.3.3 Chartering

The fastest and probably the most expensive way of getting to a story is usually on an aircraft chartered especially for the assignment. It probably seems amazing that with the number of scheduled aircraft and the route capacity covered in the statistics mentioned earlier, that there are places in the world that cannot be quickly and easily reached by scheduled flights. However, there are locations that cannot be reached within the desired time-scale except by charter.

There is also a potential problem that although a flight may be available at the right time with the required number of seats for the newsgathering team, there is not the aircraft hold capacity to take an SNG flyaway, which can take a significant amount of baggage space. Then a decision has to be made as to whether to wait for the availability of a flight that meets all the required criteria, or seek an alternative in using a special charter.

Figure 8.8 Charter plane from Nairobi into Kigali being loaded with SNG flyaway system. *(Photo courtesy of Paul Szeless)*

But on the other hand, this may not be so expensive if the story is significant enough that other press organizations are willing to share the cost for space on the flight. This may, in fact, turn out to be the cheapest option for all parties if the aircraft is large enough to get everyone (reporters, producers, technicians) and their equipment on board – and that includes camera and editing equipment as well as the SNG uplink. Typically for a large party of press with a number of organizations involved, chartering an aircraft the size of, say, a Boeing 737 may be the most cost-effective way of travelling out to the story. This has happened on a number of major stories.

8.3.4 Excess baggage

For stories of lesser significance, the fastest method, after chartering, is to travel on a scheduled flight with the equipment carried as excess baggage. Bearing in mind that the typical weight of a SNG flyaway is in excess of a tonne and that excess baggage rates are typically 1% of the first class passenger fare per kg on long-haul flights, the bill could run into tens of thousands of dollars. This is for just a one-way trip, and will probably exceed the cost of the airline seats for the news team by a significant factor. However, many large newsgathering organizations have negotiated special bulk rate deals with a number of carriers, and this rate can drop significantly if such a deal is in place.

For example, typical excess baggage costs (at time of writing) for shipping an SNG system from New York to Tel Aviv are US$35 per kg, and for a journey from London to Delhi, US$40 per kg. Assuming a system shipping weight of 1500kg, this equates to US$52 500 for New York to Tel Aviv, and US$60 000 for London to Delhi – and team air fares are on top of this!

8.3.5 Freighting

The third option is to air-freight the equipment. This is possible where the operation can be pre-planned to cover a known event (such as an international conference or national elections). It also an option at the end of a major news story where the equipment can be brought back at a more leisurely pace. Although significant cost savings can be achieved by moving equipment by this method, the drawbacks are:

- The time taken for the whole process – this can be considerably longer than just the actual travel time.
- The inaccessibility of the equipment while it is in transit.

The process of freighting equipment is more involved because of the customs procedures at the point of departure and at the destination. It is essential to use a freight-handler, who will locate shipping agents and find and book the freight capacity on a flight which will meet as closely as possible the required date for arrival and, crucially, the date of clearance through customs at the destination, before passing the equipment to the embarkation shipping agent.

The equipment has to be delivered to a shipping agent with full documentation that lists each piece of equipment and includes the following information:

- A description of the item and its serial number.
- The country of origin/manufacture.
- The value.

The shipping agent at the point of embarkation will need to know the flight details and arrange for the equipment to be checked by customs. The shipment will be issued an air waybill with a unique number, which is an identifier that will be used to track the shipment through its journey. The shipper will arrange for a shipping agent to handle the equipment at the destination. Once the equipment has been cleared by customs, it is ready for loading onto the flight.

On arrival at the destination, the equipment similarly has to be handled by the destination shipping agent, who will arrange for clearance through customs. The time taken at each end to process the shipment can be between 2 and 4 days, and it should be remembered that in many countries there may be national holidays on which no customs activity for freight occurs. So it may take 7–10 days to air-freight equipment to a destination, even though it may only be on a flight of less than 24 hours, i.e. it takes considerably longer than the actual travel time. This is of course a significant factor to take into account, for this is ‘dead’ time for the equipment. Once it has embarked on this route, it is virtually irretrievable if it is suddenly required for another story.

The advantage of course is the costs, which are likely to be a fraction of excess baggage costs – typically less than 10–20%. Using the previous journey examples, typical freight costs for an SNG system from New York to Tel Aviv are US$5.30 per kg, and for a journey from London to Delhi, US$3.90. Assuming a shipping weight of 1500 kg (and with excess baggage costs in brackets for comparison), this equates to US$8000 for New York to Tel Aviv (US$52 500) and US$5850 for London to Delhi (US$60 000). This is just for a single journey.

Although the speed of processing at each end officially cannot be accelerated, a competent and skilled shipping agent will often have contacts to be able to get the equipment ‘cleared’ from customs and available for collection in faster than average time. Local knowledge and contacts are vital qualities for such a shipping agent.

8.3.6 Carriage of systems on aircraft

No matter how the equipment is going to be transported, it has to be packaged correctly for travel, and the issue of the packaging of flyaway systems was discussed in Chapter 3. The industry body that deals with international civil aviation is the International Air Transport Association (IATA). It establishes the regulations and standards for safe international air transport, and its membership is made up of all the international air carriers. One of the areas it regulates is how and what goods are carried, and SNG flyaway systems are treated no differently to any other type of cargo. The flyaway flight cases have to be within certain dimensions, as covered by the IATA rules on the dimensions of packaging cases, and of course have to be rugged enough to withstand the rigours of international air transport.

One of the items typically carried as part of an SNG flyaway system is a petrol generator for powering the system. Unfortunately, it is a difficult piece of equipment to transport by air, as it is classed as 'dangerous goods'.

8.3.7 Dangerous goods

IATA has regulations that cover the carriage of all hazardous goods[4]. This includes internal combustion engines fitted in machinery, under which category petrol generators are classified (a petrol generator is essentially a petrol engine driving a small electrical generator).

In the packing and transportation of generators of this type, the regulations require that the fuel tank and the fuel system be completely drained. In practice, most carriers require more than this, in that it is expected that there is no aroma of fuel vapour at all. This means that either a brand-new, unused generator has to be shipped, or that the fuel system has to be completely flushed and deodorized so that no smell of petrol can be detected. Failure to comply with this requirement will result in the generator not being loaded onto the aircraft. If the generator is packed into a packing case, then a special label as prescribed by IATA must be affixed to the crate showing the contents. Some carriers place further restrictions under their own rules.

In practice, many carriers are not keen to carry petrol-engined generators, even if all the above precautions have been taken. If there is the slightest smell of petrol vapour they will not permit them to be loaded onto the aircraft. It is possible to obtain generators that have no built-in fuel tank, and can be fuelled from jerry cans directly instead, but they are not as commonly available.

Diesel does not have the same degree of flammability as petrol, and so there are not the same restrictions. The drawback is that it is desirable that the generator is 'quietened' in its design, so that when it is running it cannot be heard in the background of a live report. The only diesel generators that are quietened are bigger than required (10 kW or larger) and very heavy (over 250 kg).

Liquefied petroleum gas (LPG) powered generators are just as difficult to transport by air as petrol-engined generators.

Of course, there is the option to try to hire or buy locally on arrival, but this leaves too much to chance. The availability of generators, particularly in a disaster zone or a hostile environment, is likely to be poor and much time can be wasted trying to source a generator while stories cannot be transmitted.

The generator is not the most obvious component in a flyaway system, yet it is so essential to most overseas SNG flyaway operations to provide flexibility. It is also the one item that has been left behind on the airport tarmac on more occasions than most newsgathering teams care to remember.

Figure 8.9 Transporting an uplink by road in Rwanda. *(Photo courtesy of Paul Szeless)*

8.3.8 Use of road transport

SNG systems can be transported both locally, regionally, nationally and internationally by road. For local stories, the vehicle-based SNG systems – usually a purpose-built SNG vehicle (SNV) – are, as has already been said, the obvious choice. Flyaway systems can be transported by road, and even if the majority of the journey has been by air, often have to be transported by road for the local connections at the beginning and end of the journey. There is little more to be said on this mode of transportation that has not already been covered in Chapter 3 or earlier in this chapter when discussing safety issues.

8.4 Operating in hostile environments

8.4.1 Background

A 'hostile' environment in terms of newsgathering can range from a civil riot to an international war. Operating SNG uplinks in these environments brings a new set of problems in terms of safety and logistics. Obviously, the greatest challenge is in operating in war zones – be it a civil war, such as in Bosnia, or an international conflict as in the Persian Gulf region.

These situations are the areas where the greatest problems occur with respect to both safety and logistics in deploying SNG systems, and these aspects are so

Figure 8.10 Uplink and sandbags in Baghdad. *(Photo courtesy of Paul Szeless)*

intertwined that we have to deal with them as one. It is also the area of newsgathering where satellite transmissions from the field have had the greatest impact in shaping news bulletins and so in turn our view of these significant events. The following discussion relates to the situation where it has been decided to send a newsgathering team that includes an SNG uplink to cover a story in a war zone, as this reveals the range of decisions that have to be made. Lesser stories, or smaller scale situations, will still require some difficult decisions, but neither the number nor to the degree that covering a war demands.

8.4.2 Decisions. . .decisions. . .

SNG systems are typically deployed into these areas with consideration of a larger set of factors than considered earlier. As outlined above, the logistical (e.g. personnel and equipment) and technical (e.g. availability of space segment) criteria must be met, and in addition all the decisions relating to the dangers and consequent protective measures need to be taken. The following issues are involved.

- How to get the personnel and the SNG uplink to the location, weighing up cost and time considerations.
- Is a pool arrangement valid?
- Assessing the level of risk to personnel and the availability of personal protection equipment if required.
- Insurance with respect to special risks.

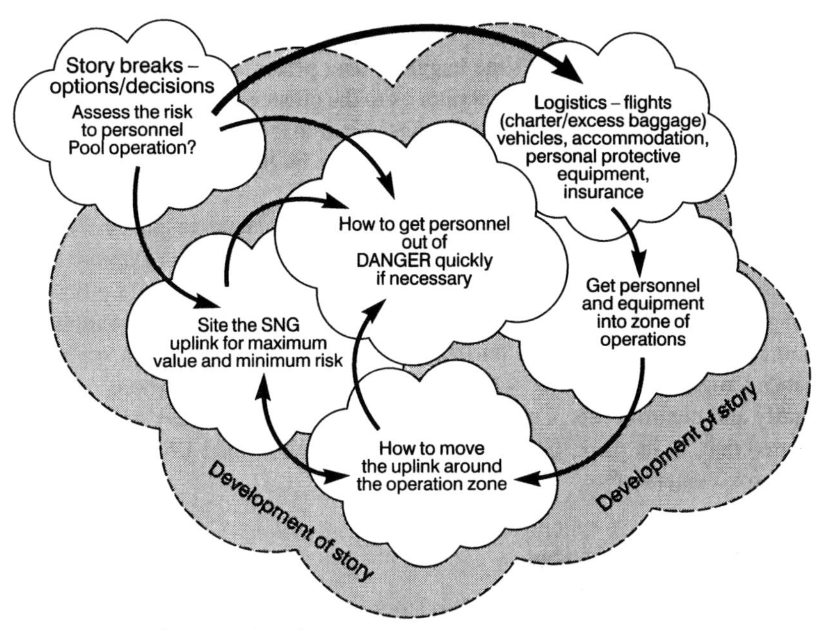

Figure 8.11 Deploying to hostile environments

- How to get the personnel and the SNG uplink into the zone of military operations.
- Where to site the SNG uplink for minimum risk and maximum usefulness.
- How to move the SNG uplink around if necessary.
- How to get the personnel (and hopefully the equipment) out to safety in a hurry.

As was said earlier, these are parallel processes, so Figure 8.11 shows how these interrelate in a summary overview.

8.4.3 *The immediate response*

For newsgathering in hazardous or difficult and remote locations, the first major problem faced is getting into the area with a means of reporting 'live' as quickly as possible. The first 24 hours for any newsgatherer in this situation can be the most difficult, while they attempt to 'ramp up' their own resources or hire in resources form the nearest available supplier in the region. The demand for coverage from even the most difficult and remote location is intense, particularly as there will be fierce competition from the other major newsgatherers to secure facilities for coverage. Some newsgathering organizations have prepared for this eventuality by signing up to arrangements with local providers in various regions around the world. For most it is very often a race as to who gets to book the nearest SNG uplink facilities, and this can be intense. In the pressure of the breaking of the story, no-one can afford to spend too much time haggling over prices, as speed of reaction is just as important to the independent operator as to the client newsgatherer. Standard 'rate cards' tend to go out of the window in these circumstances, and it becomes a matter of the highest bidder who can secure the facilities by being the fastest on the phone or the fax machine.

We then enter the next stage, where having somehow responded to provide coverage in the first 24 hours, the story develops and grows larger and more resources need to be organized and brought to the location – be they hired or the newsgatherer's own facilities. In the succeeding days, more and more in the way of resources tends to be poured into the story, and it can become a logistical and financial nightmare to track what equipment has been sent where, and then to identify and control costs. Covering events on major scale is very expensive: it was reported that, at its peak, CNN were spending an additional US$150 000 a day on the Kosovo situation[6].

8.4.4 Risk to personnel

As was said earlier in the chapter, 'no story is worth a life', and that has to be the overriding consideration when deploying any newsgathering team to cover a story in a hostile environment. Taken to the extreme, the story may not be covered, so it is a matter of the inherent risks being minimized as much as is possible. There are some situations where unfortunately the risks are too high, as it has been obvious that during the 1990s the media themselves became targets in some conflicts – for example in Chechnya, Kosovo and, most recently, in East Timor. This is a new phenomenon, as in conflicts in previous decades the presence of the media was generally tolerated. That can no longer be assumed, and is an added factor when trying to assess the risks a newsgathering team will face.

The typical newsgathering team – usually a minimum of reporter, producer, cameraman, video editor and sound technician, as well as the uplink operators – should all ideally be equipped to deal with such situations, and strategies worked out to deal with the various possible scenarios.

This may imply that the planning involved in covering a war or civil disturbance is formulaic, but that is not the case. Nor is there necessarily much time to debate and prepare as much as would be ideal. It is a matter of ensuring that certain steps are taken to ensure, as much as possible, the safety of the team. Circumstances permitting, this may include picking team members who have had previous experience of working in war zones, and ideally who have had relevant survival training. At the very least, the planning will include the engaging of a local 'fixer' – a local civilian contact who can keep abreast of the situation, possibly cultivating military, police and other contacts, and act as translator to the newsgathering team in the field. Any source that can be used to develop a body of intelligence about the situation, including reports from other press coverage, all adds to the quality of the planning process, which may also involve discussion with other 'friendly' newsgatherers involved in covering the same story to share hazard assessments.

The hazards faced by the team obviously vary from situation to situation. The reporter, producer, cameraman and sound technician will be working a lot of the time at the 'front line', while the video editor and uplink operators will probably be in a situation further away, at the 'base' position. Alternatively, the base position may also be liable to attack, and the whole team equally vulnerable.

Many newsgathering teams were trained to be able to defend themselves for NBC attacks (nuclear, biological and chemical) in the Persian Gulf conflicts, and travelled to such areas with appropriate personal protection equipment. Other training that some media personnel have undertaken includes battlefield first aid, how to deal with hostage situations (where they are the hostages), anti-ambush techniques and survival training. Flak jackets and helmets, as well as other personal protective equipment, are basic tools nowadays for the well-travelled newsgathering team, and the well-prepared newsgathering team has usually had some degree of paramilitary training in personal protection. There are a number of specialist companies that have sprung up in the 1990s, often run by ex-military special forces personnel, who have developed short, intensive training programmes for media personnel who are to be deployed into war zones or other types of hostile environment.

Other equipment that is commonly used includes hand-held GPS navigation aids and typically the compact Inmarsat Mini-M satphones. It should be noted that GPS units are position-reporting equipment, and in some conflicts a hostile military force

Figure 8.12 Operating an uplink under fire: Sarajevo TV station. *(Photo courtesy of Paul Szeless)*

may consider this as espionage equipment if a newsgathering team is caught with such equipment in their possession.

Each member of the team should accept the assignment knowing the risks, and ideally a full risk assessment will have been completed to protect personnel. As the overall objective is journalistic, the producer or the reporter will obviously lead the team, but every member of the team has their part to play, and they will be working together in an intense and frightening environment for days or weeks. The whole team is also under considerable pressure to 'make it' on time in terms of transmissions. Reports are often only prepared shortly before the transmission time, as there is always a feeling of having to work 'up to the wire' for the transmission to contain the very latest information and situation report.

8.4.5 Getting into the war zone

Having made the decision to try to cover the story, the next problem to be solved is how to get into the conflict zone. Normally, as soon as a country is approaching a war footing, scheduled airline flights are cancelled and civilian airports are closed. It may still be possible to get in by chartered aircraft, although chartering fees are likely to be increased due to the high risks of aircraft either being attacked or impounded on landing. It is also possible that the situation is so volatile that although there seemed no obstacle on landing, and having dropped off the newsgathering team, by the time the aircraft is ready to take off, the situation has deteriorated preventing departure.

In recent conflicts, newsgatherers have on occasion been able to 'hitch a ride' with aircraft involved in UN peacekeeping activities in the conflict. This can be difficult, as it may be a matter of waiting until an aircraft is available with enough payload to carry all the equipment.

As discussed in Chapter 1, particularly since the Gulf conflict in 1990, it has become generally accepted that the electronic media serve a role in publicizing the events in a conflict, and contribute to the overall political process – although whether positively or negatively is a matter of opinion and debate. All conflicts have some degree of news reporting from inside the conflict zone and some media will always find a way in. Nevertheless, it is unlikely that any newsgathering organization will wish to send an SNG flyaway uplink into a conflict zone where the team has to be able to move quickly and at short notice. Dragging over a tonne of equipment around flies in the face of this, but as discussed in Chapter 5, there are other compact SNG systems which can still enable newsgatherers to deliver timely news reports.

An interesting facet of the coverage of the conflict in Kosovo is that the deployment of SNG trucks into an area of major international conflict was seen for the first time (in addition to the use of flyaways). Due to the fact that the Balkans are part of Europe, with countries having developed infrastructures on all sides, a significant number of newsgatherers and independent operators were able to get trucks in either overland via Romania, Bulgaria, Hungary and Greece, or by ferry

Figure 8.13 Uplinks on the border the night before the UN advance from Macedonia into Kosovo. *(Photo courtesy of Simon Atkinson)*

across the Adriatic from Italy. Many of the trucks waited in Albania, Macedonia and Montenegro until the NATO forces were ready to move in, providing coverage of the story, and some were at the vanguard of the push into Kosovo with NATO forces as the Serbian army and militia left (see Figure 8.13). A few trucks were damaged, either by Kosovan Serbs (for revenge) or by Serbia on the grounds of no permission to operate having been given (mostly NATO nationality operators). Others were impounded by Macedonia on the way out as the story died down, on the grounds of customs violations. (Several newsgatherers also had flyaway systems damaged or lost in Serbia – either damaged inadvertently in the NATO attacks or again impounded by the authorities.)

This development in the deployment of SNG trucks was a sign of the rapid expansion of the number available in Europe: trucks came from the UK, the Netherlands, Germany, Austria, Turkey and Italy, among others. A few of the 'SNG trucks' were no more than vehicles with SNG flyaway systems hurriedly and temporarily rigged-up in the back, with the enterprising help of local carpenters! The conflict was also notable for the way the number of independent SNG hire companies took systems in (in addition to the major newsgathering organizations), with the development of a dynamic market for providing uplink facilities and open touting for business (some of it on the Internet[5]). This again had not been seen before in a major conflict.

However, it has to be realized that the availability of SNG trucks in the Balkans is unusual, and it is unlikely to be repeated. It was simply a reflection that a major conflict had occurred in Europe, which has in general very good road connections, and this is not a trend that is likely to be repeated unless a conflict occurs in another relatively developed area of the world.

8.4.6 Locating the uplink

Having taken an SNG uplink into a war zone, it needs to be in as safe a position as possible which still allows relative ease of access to the news of interest. It may be in a hotel, at a government facility or even in a private house rented for the operation. In some situations, all the newsgathering operations are gathered together in one place while in others they are scattered around. From the newsgatherer's point of view, the advantage of the media being together is that they are able to help each other out (as long as it does not interfere with editorial competitiveness), and it also facilitates the sharing of material where such arrangements exist. What determines where the media gather can vary from government dictat to wherever happens to be a natural congregating point.

For instance, during the recent Iraq crises from 1997 to 1999, the foreign media in Iraq were required by the government to be grouped together at the Ministry of Information in Baghdad, and they were all required to stay at the nearby al-Rashid Hotel. During the conflict in Bosnia, most of the media in Sarajevo were grouped at the local TV station, not because of any dictat, but because that was a natural congregation point. In the Gulf crisis of 1990, the main centre of media activity was in Dharhan in Saudi Arabia as that was where the Saudi government had allowed the UN headquarters to be established.

The uplink is often located on a rooftop or balcony of a building, and it is usually rigged so that in case of attack it can be remotely controlled from inside the building, enabling live transmissions to continue. INMARSAT satphones are also similarly rigged so that communications can be maintained at all times. The advantage of this arrangement has been seen in the Iraqi crises in particular, where live reports from Baghdad have been sustained during air attacks from American and British aircraft. Of course, the fact that western media are gathered in particular locations means that these locations are not targeted.

The decision as to where to site the SNG uplink for minimum risk and maximum usefulness is regularly reviewed during an operation (at times daily), to take account of changing military and editorial circumstances. Plans on how to move the SNG uplink around have to be devised, and we will come onto this next.

8.4.7 Moving the SNG uplink around

During the course of a conflict, it is sometimes necessary to be able to move the SNG uplink to other locations within the zone. This will inevitably be by road, and usually involve at least two sturdy four-wheel drive vehicles. The use of the word 'road' here is meant in its loosest sense, as wars tend to destroy road systems, and therefore four-wheel drive vehicles are an absolute necessity. It may also be prudent to use armour-plated vehicles if there is a serious risk of the team coming under fire,

Figure 8.14 Travelling light with an uplink in Macedonia. *(Photo courtesy of Simon Atkinson)*

and the vehicles can be brought in. This will obviously give a greater degree of protection to the reporter and camera crew as they move around, as well as offering a way of moving the SNG uplink if required. The degree of armour protection can vary from 'semi' to 'full', with the available payload on the vehicle decreasing (rapidly!) as the level of protection is increased, and this will also affect the power and speed of the vehicle. There may also be importation or exportation issues relating to (essentially) military equipment.

It is worth noting here that the mere possession of such vehicles can also make the newsgathering team subject to attack. This is either because they can be mistaken for legitimate targets or because, particularly in a civil war, one or both sides is likely to be short of arms and military equipment and desirous of any such attractive transport available for the taking. The fact that the vehicle is being operated by a news organization is of no consequence, as the following story illustrates.

During the Bosnian conflict, an independent news facilities company bought a full military specification armoured personnel carrier (APC), and they were contracted by one of the US networks for deployment to Bosnia, keen to gain a competitive edge. After having (at great expense) equipped it with a lot of equipment, including a flyaway SNG system, they showed it off with great pride to other newsgatherers at a publicity event in London. The team could thus move around in safety and venture into zones that other newsgatherers could not. Unfortunately, once deployed into Bosnia the vehicle broke down, was abandoned

Figure 8.15 Armoured vehicles at the UN HQ in Sarajevo ('Miss Piggy' is on the left). *(Photo courtesy of Martin Cheshire)*

by the team while they went to get help, and while they were gone the APC was appropriated by some local guerrillas. It was never returned to its rightful owners (who wince even now when it is mentioned).

Other newsgatherers have also used heavily armoured vehicles – the BBC famously had one called 'Miss Piggy', named after the *Muppet Show* character, as 'she' was rather large and cumbersome (Figure 8.15). These vehicles have typically been based on four-wheel drive Land Rovers and Toyotas, and have not therefore been quite as attractive to local connoisseurs of military vehicles as an APC.

In 1999 in Kososvo, the use of armoured vehicles became *de rigueur* but they were in short supply, and the prices of these vehicles were typically around $US30 000. With the increasing death toll of journalists and cameramen in these types of conflict, every organization now realizes the importance of these vehicles, and unlike in the early 1990s, those who make use of them are not now considered 'wimps'.

8.4.8 Evacuation

It would be irresponsible for any newsgathering management not to have carefully laid plans for evacuation of their team in a hostile environment for when the situation deteriorates to absolute danger – this would be part of the risk assessment. Judging the moment when the time is right to pull out is difficult to determine remotely, and detailed discussions with the team usually occur before the decision is taken. Because of the financial (and strategic) value of the SNG uplink, the means of escape usually includes measures to get the uplink out as well.

As with getting the team in, the quickest way of getting people out is generally primarily by air, or as a second choice by road. There will not be any scheduled airline traffic from a war zone, so if it is appropriate to use aircraft as a means of escape, the two usual methods are either by military (often UN) air transport or by having a chartered aircraft available near at hand. Even if such an arrangement is in place, it is common to have a secondary method planned of getting personnel and equipment out by road.

Alternatively, escape by road may be the more feasible option if access to a landing strip is not possible, and as we have seen, newsgathering teams normally have access to four-wheel drive vehicles in war zones. Evacuation plans by road will include a pre-planned route with alternatives, fuel reserves and emergency communications with INMARSAT satphones.

8.4.9 Pool arrangements

In some situations, the forming of a media pool is either by mutual consent of the newsgathering organizations, or it is a condition imposed on the media by an external authority.

A pool formed voluntarily is usually because of overwhelming safety considerations rather than cost (which is the more commonplace reason in non-hazardous situations), and tends to occur particularly once a situation has significantly worsened either because of military action or because of a direct threat of military action against the media.

The pool arrangement is likely to involve minimizing the number of personnel on the ground while still providing enough material for the differing news interests. It can also be formed and dissolved in different phases of the conflict, and constantly reviewed. The personnel most exposed are likely to be reporters and camera crews, and the focus will be on providing enough reporting crews to cover the different facets of the conflict. Nevertheless, the number of active SNG uplinks will be kept down to minimize the number of personnel in the area. This may involve either withdrawing or 'mothballing' (storing) uplinks for a period until the situation improves.

Where a pool is demanded by an external authority, for instance a national government on the grounds of security, the newsgathering organizations will be forced to comply or risk having all their press access withdrawn. Newsgatherers generally comply with these demands, though it has been known for individual 'maverick' operations to be undertaken to try to bypass such restrictions.

8.5 Conclusion

It can be seen that operating SNG uplinks has a number of factors to be considered that go beyond the merely technical and that have a direct impact on the safety of people – both operators and the public – in a number of different dimensions. The influence of time pressure on covering news events can push the significance of some of these factors into the background unless there is a commitment not to ignore them.

International travel with the typical volume of SNG flyaway equipment escalates the stakes, and deploying to a hostile environment pushes the decision-making processes to a very high level of pressure, both in terms of safety and costs. In the fiercely competitive news environment, everyone in the field wants to be first and 'live' – but above all, they need to be first, alive.

References

1. FCC (1996) Office of Engineering and Technology (OET) Bulletin 65, Evaluating compliance with FCC guidelines for human exposure to RF electromagnetic fields.
2. Radiocommunications Agency (1996) RA 172 Site Clearance Manual Code of Practice No.1, Annex B (Rev. 2).
3. Radiocommunications Agency (1996) RA 172 Site Clearance Manual Code of Practice No.1, Annex D (Rev. 2).
4. IATA (1999 and updated annually) Dangerous Goods Regulations.
5. www.tvnewsweb.com
6. www.tvnewsweb.com (*Newstalk*, 25 June 1999).

9

On the horizon

9.1 Finale

So we have arrived at the final chapter, and we'll spend some time looking at how the existing means and methods of gathering news by satellite (or otherwise) might develop in the future. Technical developments as well as changes in working practices will inevitably change the current situation. However, although these technical developments may promise change, some may turn out not to be superior to current techniques. Unfortunately for the reader, these developments are occurring so quickly that by the time this chapter is read some of them may not be so prominent.

The developments touched on in this chapter may seem at first glance as not necessarily having a great impact on 'traditional' SNG – but that would be complacent. Any technology that offers the opportunity of gathering news more cheaply, more effectively and more easily (although possibly at lower quality) is always worth considering. As it should now have become apparent, the requirement to deliver ever higher volumes of news material within a time-frame and location ever closer to the event is very pressing – even if it means material is delivered at a level of quality lower than would have been accepted ten or even five years ago.

9.2 GMPCS

According to independent estimates, the mobile satellite communications industry is expected to grow to between 4 and 8 million subscribers by 2002, and to generate revenues of between US$4 and US$13 billion per annum. Much of this will come from Global Mobile Personal Communication by Satellite (GMPCS), a largely new family of commercial satellite systems that will compete in the area of voice communications, and principally in traditional INMARSAT markets (effectively the original GMPCS system).

GMPCS promises global seamless mobile voice and data communications as well as broadband Internet-type services. The potential of these new services to improve both the telecommunications accessibility in under-served areas of the developing world, as well as their high-bandwidth capabilities which will support new types of multimedia applications, has generated a great deal of interest from investors, governments and certain niche markets. The public is as yet by and large oblivious to these developments.

The use of satphones generally is expanding rapidly in all existing niche market segments (including the media), and all of the new GMPCS systems are planned to come into service in the first five years of the new millennium. Iridium is the first of these systems and is already in service, though, as we shall see later, is suffering severe setbacks, which does not bode well in general for either Iridium or the other new GMPCS companies for the future.

Since the late 1980s, a number of large telecommunications corporations have been involved in this next generation of satellite communication services. Some of the biggest names in the telecommunications, satellite and space technology industries are deeply involved, with billions of US dollars being invested – so the stakes are very high. These include Alcatel (France), Boeing (US), General Electric (US), Hughes (US), Lockheed Martin (US), Loral (US), Matra Marconi Space (France), Motorola (US), Raytheon (US) and TRW (US). The dominance of the US may be noticed in this list of participants, and it is no coincidence that all the US companies involved have historically had significant commercial interests in the US military market. The technology being used for GMPCS is the same as that which might have been used for purely military applications if it had not been for the ending of the Cold War – a case of swords into plowshares.

The new services broadly fall into two categories: those offering low data rate voice, data and fax services (some with data rates high enough for slow Internet access); and those offering high bit-rates focused on Internet access. They are further subdivided into those that have satellite systems in low Earth orbit (LEO), medium Earth orbit (MEO) and geostationary Earth orbit (GEO).

The advantages and disadvantages of each type of orbit can be summarized as follows:

- LEO (below 5000 km) reduces delay (latency) to 20–25 ms, transmit powers are low, but requires more satellites for the widest possible coverage.
- MEO (10 000–20 000 km) has greater delay, needs greater transmit power, but needs fewer orbits for equivalent coverage to LEO systems.
- GEO, the conventional orbit for existing commercial communications satellites, offers the traditional attributes of a relatively small number of satellites providing very large zones of coverage, but with the problems of delay and large amounts of transmit power due to the distances involved.

Capital investment is still being sought on all the systems, and not all the technical details are finalized – some systems are still lobbying the ITU for spectrum allocation. The profusion of systems has been bewildering and it is hard to see how they can all survive commercially (assuming they even make it into the sky at all). Even while in development, there have been casualties; for example, the TRW Corporation started development of a system called Odyssey, but in 1997 merged the development with the INMARSAT-backed ICO system. Likewise, Motorola's efforts on their Celestri system (which was born out of their original M-Star system) merged with Teledesic in 1998. Financial issues bedevilled these schemes, and as we shall see later, these financial problems continue.

Low bit-rate (telephony) systems

For newsgathering, these are of interest for use in the same way as GSM or Inmarsat Mini-M services currently support activity, i.e. as a means of keeping in touch, and possibly for filing low audio quality reports for radio news. These services will be of use for keeping in touch with newsgathering teams particularly in remote and/or hostile locations. The current systems (mostly only 'paper' systems) are:

- Iridium – in service
- Globalstar – due in service October 1999
- Ellipso
- Constellation
- ICO
- Thuraya – a regional system

Many of these systems are intrinsically similar in function, varying only in the types of handsets and mobile terminals to be used and the configuration of the satellite constellations that carry the signals.

High bit-rate broadband (Internet) systems

The systems outlined below may offer high data rate Internet-type (broadband) access, suitable for newsgathering for high-quality radio contributions, or for television with either store and forward or video-conferencing, or bit-rates that are high enough to allow 'live' video. Many of the traditional C- and Ku-band satellite frequency allocations are already almost at capacity, particularly in North America and Europe. This spectrum scarcity has pushed satellite operators into the Ka-band frequencies. All bar one of the systems below are planned to operate in the Ka-band, which may limit their usefulness for newsgathering in the field (Skybridge is the exception, as it will be operating in the Ku band). The current systems – all of which essentially only exist on paper at present – are:

- Teledesic – one experimental satellite in orbit
- Skybridge
- Spaceway
- KaStar (US only initially)
- Astrolink

9.2.1 A brief history...

Because of the dominance of US companies in the initially proposed GMPCS systems, these systems had to secure approval from the FCC, who had to agree to the establishment of these new systems and the technical parameters in the US before they could be further developed. Iridium was the first system to apply to the FCC for permission to operate, and the FCC then invited applications from other systems companies to share the newly allocated band for the new Mobile-Satellite Service (MSS)[1]. An additional four proposals for non-geostationary mobile telephony systems were submitted in 1991: Loral Globalstar, TRW Odyssey, MCHI Ellipso and Constellation Communications. Collectively, these four non-geostationary satellite systems along with Iridium became known as the 'Big LEOs'.

The FCC sought spectrum allocation form the ITU for these new services, and at WRC-92, L-band spectrum around 1.6 GHz was internationally allocated for MSS for the uplinks on a primary basis in all three ITU regions. WARC-92 also allocated another part of the 1.6 GHz band to MSS on a secondary basis, and spectrum in S-band at around 2.5 GHz on a primary basis for downlinks from satellites to 'gateway' satellite ground stations.

All the systems must gain approval and access rights in each country where service is to be provided. For instance, Iridium aimed to achieve agreements with ninety priority countries to begin service and altogether Iridium is seeking access to some 200 countries through negotiation.

In achieving the allocation of the different parts of the MSS bands agreed for NGSO systems at WRC-92, the FCC conducted a conference in 1993 to discuss with the applicant systems the frequency-sharing plan and to make recommendations. The biggest debate was the method of transmission, between frequency division multiple access (FDMA) and code division multiple access (CDMA).

In FDMA, the spectrum is subdivided into smaller bands allocated to individual users. Of all the systems at the time, Iridium uniquely extended this multiple access scheme further by using time division multiple access (TDMA) within each FDMA sub-band. Each user is assigned two time slots – one for sending data and one for receiving data – within a repeating time-frame. During each time slot, the digital data signals are 'burst-transmitted' between the mobile handset and the satellite.

In CDMA, a code number identifies the signal from each user with all users sharing the same frequency spectrum. At the receiver, the desired signal for each user is extracted from the entire population of signals by replicating the same code number. The major advantages of CDMA are inherently greater capacity and higher spectral efficiency. Generally, the more closely the transmission is shared among users the more efficiently and flexibly the system can respond to demand.

The FCC also proposed that LEO and MEO satellite systems could only use the MSS spectrum, and that any GEO systems would have to use spectrum already allocated in the Ku- or Ka-bands. Agreement was finally reached on all the technical aspects of the initially proposed services in 1995. Since then, there have been a number of additional systems proposed and actively planned.

9.2.2 Voice systems

Iridium

Motorola originally conceived the Iridium system in 1987 (although Iridium's other investors now include Raytheon and Lockheed), and planned to have seventy-seven active satellites within it (thus its naming after the element iridium, which has seventy-seven electrons). Iridium was later redesigned to need fewer satellites so that the Iridium constellation now consists of sixty-six satellites in LEO at 780 km. The satellites are allocated into six orbital planes around the equator, with eleven satellites in each plane. Each satellite has an orbital period of 100 minutes, and so is only in view to any user on the ground for around 9 minutes, and each 'cell' spotbeam from the satellite passes over the user in 1 minute. As each satellite passes over the user, there is a sophisticated 'handing-off' procedure from one cell to the next, and from one satellite to the next. Each satellite has a capacity of 1100 communication channels, and the complexity of the system in achieving this is quite remarkable.

As Iridium is, at the time of writing, the only system in service, we will spend a moment examining how it works. In principle, the other systems are broadly similar, and thus by looking at Iridium one can gain an understanding of how these systems work in general. The voice signal is digitally encoded in the users handset and at the satellite gateway ground station. The method of digitizing the voice is by use of voice coding based on psycho-acoustics, which produces a set of parameters that emulate speech patterns, vowel sounds and acoustic level. The resulting data rate of 2.4 kbps is thus capable of transmitting clear, intelligible speech, comparable to the performance of existing GSM cellphones and good quality connections on Inmarsat Mini-M, but not quite the quality of traditional telephones. The signal strength is not sufficient to penetrate buildings, and so users have to stand near windows or go outside to place a call. Handing over from cell to cell within the field of view of an orbiting satellite is imperceptible, though hand-over from satellite to satellite every 9 minutes may occasionally be detectable by a quarter-second gap in transmission.

The Iridium satellite system is based on the principle of a large degree of 'intelligence' being in the satellite rather than the gateway satellite ground station. Hence, a call can be routed through the satellite constellation without 'touching the ground' by use of inter-satellite links (operating in the Ka-band at 23 GHz). There are twelve gateway satellite ground stations around the world, and these are connected to the local telephone networks in their country of location.

The principle of operation is as follows. A user dials the destination telephone number with the Iridium handset using an international 13-digit number, as one would do normally using a standard telephone, and presses the 'Send' button. The system identifies the user's position and authenticates the handset at the nearest gateway, and once the user is validated, the call is sent to the satellite. The call is routed through the Iridium constellation of satellites and drops to the gateway closest to the destination. There it is completed over standard telephone network circuits. The process is reversed for a call from a fixed location to a handset, so after the call is placed the system identifies the recipient's location and the handset rings, no matter where the user is in the world.

Iridium handsets are made by Motorola and Kyocera (a leading Japanese manufacturer of cellphones). Handsets permit both satellite access and terrestrial cellular roaming capability within the same unit, and handsets include a 'subscriber identity module' (SIM) card, as with GSM cellphones. Interchanging a 'cellular cassette' for each standard covers the various cellular standards. The Iridium system also offers satellite paging, which by 2000 will also offer two-way paging (i.e. 'message read' acknowledgements).

Initial comments on Iridium service indicated that the satellite telephone service was extremely variable in quality, whilst the pagers have proved very effective. For newsgathering purposes, the first major news event on which Iridium phones were used was the conflict in the Balkans in 1999, closely followed by the earthquake in Turkey. Users in the Balkans reported poor performance, though there was conjecture that there was some jamming of the signals from the UN forces. At the time of writing, the second generation of satphone handsets were just being introduced, promising improved service quality.

Globalstar

Loral and Qualcomm founded Globalstar, which nominally came into service in autumn 1999. The system is based on a constellation of forty-eight satellites in a LEO at 1400 km, designed to provide voice, data, fax and messaging services worldwide. By November 1999, all forty-eight satellites had been launched (despite a disastrous loss of twelve satellites in one launch in Russia in 1998) and the service was demonstrated in December 1999, with Globalstar handsets due on sale in early 2000. Globalstar is based on what is known as a 'bent-pipe' system, where a single satellite must 'see' both the handset user and the nearest 'gateway' satellite ground station simultaneously, rather than using inter-satellite links as with Iridium. Consequently, many more gateway satellite ground stations are needed, and connection from the anticipated sixty satellite gateway stations around the globe will be to the local telephone service in virtually every populated area of the world. Globalstar advocates that their system places the complexity on the ground rather than the satellite, and therefore offers greater flexibility in building and upgrading the system.

As with most of the other operators, Globalstar is chasing after the same markets such as the mineral exploration industry, disaster relief, media and the 'cellular extension' market – users of cellphones who find they are often out of the coverage zone. Globalstar has always had a stated intention to undercut Iridium's prices, although whether this is still feasible after the dramatic cut in the cost of handsets and air-time rates by Iridium in July 1999 remains to be seen. Two types of handset are planned: triple-mode (GSM/US analogue cellular [AMPS]/Globalstar satellite) and dual-mode (GSM/Globalstar satellite).

Ellipso

The Ellipso system is one of a number of GMPCS systems that plans to provide geographically selective coverage, by dividing its coverage into two zones each served primarily by its own constellation of satellites. Ellipso is a MEO-based system, utilizing ten satellites in two inclined elliptical orbital planes (called the 'Borealis' plane) providing Northern Hemisphere service, and seven satellites in the Equatorial region and Southern Hemisphere (the 'Concordia' plane). Its system is designed to concentrate on populated areas of the world, and it uses a variation of an elliptical orbit principle termed a 'Molnya' orbit. The Molnya orbit is named after the series of Russian satellites originally used to distribute television signals to the high latitude polar regions of the former USSR.

The Earth's distribution of land and population by latitude serves as the basis for the design of Ellipso. Ellipso maintains that, due to the distribution of the global landmasses, the Northern Hemisphere contains many times the area of landmass north of latitude 40°N than the Southern Hemisphere has south of 40°S. For example, virtually all of Europe is north of 40°N, and almost one half of the US and all of Canada lie north of 40°N.

In contrast, the Southern Hemisphere contains much less landmass at high latitudes than the Northern Hemisphere. For example, continental Australia reaches south only to 39°S, and continental South America to 52°S. Moreover the amount of land south of 50°S is very small in global terms, comprising only the southern tip of Argentina and Chile and Antarctica. All areas south of 50°S are very sparsely populated, and therefore virtually all of the Earth's populated landmasses lie north of 50°S.

In view of this asymmetry of populated landmasses, any satellite system offering coverage of populated landmasses must provide extensive coverage to northern latitudes. To do so requires inclined orbits for all low and medium Earth orbit satellite systems. But if circular orbits are used, the inclined orbits mean that equal coverage is also given to the far southern latitudes, which Ellipso therefore concludes is largely wasted.

The Ellipso system is intended to match its capacity more closely to the populated landmasses than would be possible using constellations of satellites in circular orbits. It does so by using two complementary constellations of satellites, and this therefore selectively covers a significant proportion of the world's population able to afford to indulge in wireless communication.

The primary thrust of Ellipso's marketing has been on call costs, where it aims to offer an air-time rate at a wholesale price of US$0.35 to service providers. Ellipso is owned by MCHI, with other investors such as Lockheed Martin, Harris, Arianespace, Israel Aircraft Industries and Boeing.

Ellipso has recently proposed an enhancement called Ellipso 2G (second generation). Compatible with the current proposed Ellipso satellite system, it will have much greater capacity with data rates increased to 64 kbps from Ellipso's current proposal of 9.6 kbps. The Ellipso 2G (second generation) constellation will consist of twenty-six non-geostationary satellites arranged in five orbital planes. However, this proposal is still at a very early stage and may well be overtaken by events.

Constellation

The Constellation system is another selective coverage system, and will consist of a single orbital plane of eleven operational satellites (with one spare) in circular orbit around the Earth's equator, in a LEO of 2000 km. The system will provide service to all areas around the world located between 23°N and 23°S latitude. This is a band which broadly covers South America, Africa (excluding South Africa), India, South East Asia, Indonesia and central and northern Australia. As with the other systems, there will be a number of gateway satellite ground stations, connecting to the telephone network of the country of their location. Investors in the company include Orbital Sciences, Bell Atlantic and Raytheon.

ICO

The formation of ICO in 1995 was the culmination of Project 21, an INMARSAT-led initiative launched in 1991. The primary objective of Project 21 was the implementation of global satellite service to pocket-sized satphones by the end of the decade, originally dubbed 'Inmarsat-P'.

ICO is planned as a constellation of ten satellites in a MEO of 10 000 km. There will be two 45° inclined orbital planes of five satellites each, with operation of the full system beginning in 2001. The satellites will communicate with the twelve gateway satellite ground stations located around the world and linked by broadband fibre-optic routes. The orbital pattern of the ICO constellation is designed for significant coverage overlap, ensuring that two or more satellites will always be in view of a user and a satellite ground station for more than 80% of the time in most latitudes. Each satellite will cover more than 25% of the Earth's surface at a given time. INMARSAT is the largest investor in ICO, with other investors including Hughes and TRW.

To date, ICO has signed service agreements that will give them access to more than 180 countries, and ICO's own forecasts have put total demand for satellite services at between 30 and 40 million subscribers by 2005.

Thuraya

Thuraya is an Arab-based consortium which is based on initially a single geostationary satellite placed at 44°E. The satellite is due to be launched in May 2000, and the system will start full commercial operation from September 2000. A second satellite will be launched later and placed at 28.5°E. The mobile communication services offered by Thuraya will support the use of hand-held, vehicle-based and fixed terminals, with dual-mode handsets (GSM and satellite) integrated with both existing terrestrial and the new satellite services.

Thuraya will be offering voice, short messaging, low-speed data and fax services as well as a Global Positioning System (GPS), providing coverage to ninety-nine countries spanning Europe, North and Central Africa, the Middle East, Central Asia and the Indian subcontinent – nearly 40% of the world's population. The system is estimated to cost US$1.1 billion to implement.

9.2.3 Broadband Internet systems

Teledesic

Essentially the 'Internet-in-the-sky', the Teledesic system is a high-capacity broadband network that offers the global coverage and low latency of a LEO constellation of satellites with the flexibility of the Internet with fibre-like service quality – an impressive claim. In the initial constellation, Teledesic will consist of 288 operational satellites, divided into twelve orbital planes, each with twenty-four satellites. The configuration of the Teledesic constellation assures that any satellite is always at least 40° above the horizon within its entire service area.

American communications entrepreneur Craig McCaw and Microsoft's Bill Gates originally formed the company in 1994, later joined by Boeing, Motorola (abandoning its US$13 billion Celestri project) and Saudi Arabian royal family interests. It aims to bring affordable access to interactive broadband communication to all areas of the Earth, including those areas that cannot be served economically by any other means.

Covering nearly 100% of the Earth's population and 95% of the landmasses, Teledesic is designed to support millions of simultaneous users. Teledesic terminals will support a wide range of data rates and interface with the full range of standard network protocols, including IP (Internet protocol), ISDN and ATM (asynchronous transfer mode). Although optimized for service to fixed-site terminals, Teledesic aims to serve transportable and mobile terminals, such as those for maritime and aviation applications. This also opens up possibilities for terminals for use in the field for newsgathering. Most users will have two-way connections that provide up to 64 Mbps on the downlink and up to 2 Mbps on the uplink. Broadband terminals will offer 64 Mbps of two-way capacity.

Each satellite is a node in the network and has inter-satellite communication links with other satellites in the same and adjacent orbital planes. This interconnection arrangement is designed to form a robust mesh, or 'geodesic' network, that is tolerant to faults and local congestion.

The lowest frequency band with sufficient spectrum to meet Teledesic's broadband service, quality and capacity objectives is the Ka-band. The terminal-satellite communication links operate within the portion of the Ka-frequency band that has been identified internationally for non-geostationary (NGSO) fixed-satellite service. Downlinks operate in the 19 GHz band and uplinks operate in the 29 GHz band. Communication links at these frequencies are degraded by rain and blocked by obstacles in the line-of-sight. To avoid obstacles and limit the portion of the path exposed to rain requires that the satellite serving a terminal be at a high elevation angle above the horizon.

Latency is a critical parameter of communication service quality, particularly for interactive communication and for many standard data protocols. To be compatible with the latency requirements of protocols developed for the terrestrial broadband infrastructure, Teledesic satellites operate at a low altitude, under 1400 km. The combination of a high access angle and low Earth orbit results in a relatively small satellite footprint and requires a large number of satellites to serve the entire globe. Using this design, Teledesic aims to achieve availability of 99.9% or greater.

Skybridge

SkyBridge is a system of eighty LEO satellites designed to provide global broadband services, including Internet access and high-speed data communications. SkyBridge aims to provide downlink data rates of up to 20 Mbps and 2 Mbps on the uplink. Alcatel (France) is the principal partner of SkyBridge, and other partners in SkyBridge include Loral (US), Toshiba (Japan), Mitsubishi (Japan) and Sharp (Japan). The investment required is estimated to exceed US$4 billion, and service is aimed at beginning in 2001.

Spaceway

Spaceway is a hybrid GEO/LEO system to provide high-speed communications for broadband applications. Started by Hughes with a US$1.4 billion investment, Spaceway will consist of eight geostationary birds to serve most of the world by the end of 2003. There is also a LEO system planned to cover Europe, the Middle East, Africa, Latin America and Asia. Spaceway will initially begin operations over the US, but Hughes is looking for other partners to fund the full system. Like the other services, Spaceway will provide bandwidth-on-demand and the system will also integrate with existing land-based systems.

KaSTAR

KaSTAR plans to operate a global Ka-band satellite system that will provide low cost, on-demand, interactive broadband services – another 'Internet-in-the-sky' project. KaSTAR will deliver bandwidth to end-user service providers initially throughout North, Central and South America and subsequently around the world, augmenting terrestrial and broadcast satellite networks. KaSTAR's target end-customers are residential and small office users who want broadband communications services at affordable prices.

The two satellites will provide a large number of small spotbeams (versus a single CONUS beam in conventional satellites covering North America) with a large degree of frequency reuse providing greater capacity from the satellite. KaSTAR's first phase is for the launch and deployment of two satellites at 109°W and 73°W. These two satellites will reach all of North, Central and South America as well as (eventually) linking to Europe.

Astrolink

Astrolink is a broadband satellite system based on a constellation of five Ka-band geostationary satellites, estimated to cost US$3.6 billion. The system is a partnership between Lockheed Martin, Telecom Italia and TRW, with the first satellite scheduled to launch in 2002 to cover the Americas and Europe, followed by the launch of three additional satellites at six-month intervals to extend the network worldwide. It is planned that if market demand increases, up to nine satellites will eventually be placed in five orbital positions.

9.2.4 The use of Ka-band for SNG

While newsgatherers will continue for the present to use geostationary satellites for Ku-band SNG uplinks, the development of the Ka-band for broadband Internet applications, as already discussed, has prompted the question amongst the technically inclined as to whether there is any advantage in using this type of capacity for newsgathering. The pure physics tend to suggest not. It may offer the potential of 60 cm or smaller antennas, but there could be an issue of how rugged the error correction is because of the propagation limitations of this band. The Ka-band is more susceptible to the effects of rain fade and this would necessitate an increase in uplink power to compensate. There seems no advantage (and a number of disadvantages) in a move to the use of SNG in the Ka-band. However, since none of the Ka-band systems outlined in this chapter are yet in service, we will have to wait to see if any of the service offerings are suitable for gathering news in the field.

9.2.5 Orbcomm

Finally, a brief look at one of the more interesting of the so-called 'little LEO' systems which offer data-only services on a 'store and forward' basis. Although not a voice or broadband data service, 'little LEO' services (such as ORBCOMM) are of interest to newsgatherers because they offer a 'keep-in-touch' service that can be so vital in hostile environments. This type of service is similar to the Inmarsat-D+ service (see Chapter 5). The ORBCOMM constellation currently has thirty-five LEO satellites offering near global coverage. The system is capable of sending and receiving two-way alphanumeric packets, similar to two-way paging or e-mail. The satellites are effectively orbiting 'packet-routers' configured to receive small data packets from 'subscriber communicators' – hand-held devices for personal messaging – as well as fixed and mobile units for remote monitoring and tracking applications. ORBCOMM units are available integrated with a GPS receiver (Magellan GSC100 hand-held communicator) and can therefore offer the life-saving capability of sending position information. The messages are then relayed to a ground earth station and delivered onto the Internet.

9.3 The development of GMPCS

The GMPCS systems are primarily aimed at mass markets rather than niche markets such as newsgatherers (who have been one of the groups of users traditionally targeted by INMARSAT). Around the world, people living in remote communities have little means of communication beyond their villages, and so one of the markets is therefore the 'village phone market'. Of the world's population, there are an estimated 50% (3 billion out of an estimated total population of 6 billion) who have no access to a telephone at all. Of those populations that do have access, 80% of telephones are available to 20% of the population in urban areas, while 80% of the population in villages and small towns are served by only 20% of the telephones.

Interestingly (but perhaps not surprisingly), the majority of potential telephone traffic in the rural communities is only to the next village or town ('over-the-hill' traffic), as these communities are still relatively close-knit. Globalstar has always maintained that one of their primary targets is this village phone market, and have been developing satellite phone booths. Iridium originally targeted the much vaunted (but largely mythical) business traveller market, which is probably nowhere near as large or profitable as many of the GMPCS operators anticipate, due to the rapid spread of GSM cellphone systems around the world. Iridium failed to generate anywhere near the anticipated volume of business, with just 20 000 subscribers within the first nine months of operation, instead of the 500 000 required to keep the network solvent. Subsequently, Iridium switched to supporting the village phone market and now propose selling portable solar-powered satellite phone booths in several countries[2].

Marketing is plainly an issue for all the GMPCS operators, but so far, apart from the obvious challenges of technology and designing systems, the development of GMPCS has revolved around three key issues: spectrum allocation, financing and licensing.

9.3.1 Spectrum allocation

In the early 1990s, when the various systems were still in their infancy, the long struggle to obtain spectrum for the new services began – and continues to this day. GMPCS systems were key agenda item at the ITU WRC-95 and WRC-97, and GMPCS systems received their first substantial spectrum allocations at WRC-95. The majority of the systems are US-based, and their petitioning for spectrum with the ITU was regarded with great suspicion by many ITU members, particularly those from the Third World. Their opposition centred around two issues: the 'bypass' effect and national sovereignty.

Firstly, the threat from the so-called 'bypass' effect is the perception that the new systems would remove substantial revenue from national PTTs and PTOs by GMPCS users, who would completely bypass national Public Switched Telephone Network (PSTN) systems (as with Iridium by use of inter-satellite links). GMPCS can be a drawback in developing countries that are trying to modernize their terrestrial infrastructures, and who are therefore counting on revenues from existing telecommunications traffic to achieve their goals. Calls initiated on a GMPCS system will generally need, at some point in their journey to their destination, to interconnect with the fixed terrestrial network or with other cellular networks. However, it is also possible in some systems for users to make end-to-end calls which travel directly over the GMPCS satellites without touching the ground, thereby depriving poorer countries of essential income from telecommunications.

Secondly, there is the issue of sovereignty and control of information by governments, already covered with reference to SNG in Chapter 6. The issues are the same, and the resolution is just as difficult.

For instance, during the lengthy civil war in Sri Lanka in early 1999 the Sri Lankan government approved the use of Iridium satphones in Sri Lanka so long as there was no service offered in guerrilla-held areas. The defence ministry opposed the extension of the use of Iridium satphones (or those of any other system) to the zone where Tamil Tiger guerrillas were fighting government forces. The Tamil Tigers were already known to operate INMARSAT satphones to communicate with their offices abroad. Iridium agreed that Sri Lanka's embattled north and eastern territories would be excluded from their system because of national security considerations. However, despite similar requests from the government to INMARSAT, the technology used by INMARSAT could not restrict coverage to a particular region.

It is often extremely difficult to meet this type of request, on technical grounds, but this is symptomatic of the problems faced by all the GMPCS and broadband systems when dealing with many governments. The WTO GMPCS MoU (see Chapter 5 and Appendix D) is intended to facilitate the free movement of GMPCS equipment, but it will still be at least several years before this can be achieved on a global scale.

9.3.2 Financing

Of all the issues facing the establishment of the GMPCS systems, financing has proved to be the major problem – though this had been conjectured from the beginning as their biggest issue. By mid-1998, there were an estimated sixty systems on the drawing board, requiring over $US130 billion of investment. Finding this level of investment has been a major problem for every system from the beginning, and even if every system raised the money required, it is difficult to see with the size of the markets and the potential revenues how more than a handful can survive commercially – and there has already been some merging of efforts.

The future of at least the first generation of LEO systems (Iridium and Globalstar) is the subject of debate. Some industry sources suggest that high capital and service costs will cause their demise (for example, all sixty-six of Iridium's satellites will reach the end of their lives and need replacing in 2003), whereas ICO and the regional GEO systems will emerge as the winners. LEO satellites can typically only 'see' between 1 and 3% of the Earth at a time, whereas ICO's MEO satellites can see up to 10% of the Earth, and are therefore more attractive to potential users. Thus ICO is thought to have a potentially better architecture by using fewer satellites and locating them in higher orbits where they remain visible longer during each pass.

The projected end-user air-time charges for ICO were one-quarter of Iridium's pricing, and therefore some industry pundits felt that Iridium and Globalstar would be driven out of business. During the early summer in 1999, in an effort to encourage subscribers (and to counter the fall in its stock price from US$72 per share in May 1998 to less than US$13 in May 1999), Iridium slashed its end-user costs. The prices of satellite telephone handsets dropped from US$3000 to under $US1000, and pagers from US$600 to under US$350. The air-time charges were also cut, with international calls now costing about US$3 per minute to the end-user and national calls below US$2.50 per minute, and calls between Iridium handsets about US$1.50 per minute[2]. With these pricing moves Iridium hoped that they could recover lost ground, but as the share price fell steadily to US$3 in August 1999, Iridium filed for 'Chapter 11' bankruptcy in the US[3]. However, this is not as drastic or fatal as it might initially sound.

Under the US Bankruptcy Code, a case filed under Chapter 11 is frequently referred to as a 'reorganization' bankruptcy. Effectively, a business ceases trading on the stock exchanges and goes into a holding pattern while it attempts to reschedule its debt, the rationale being that the value of a business as an ongoing concern is greater than it would be if its assets were sold. If the business can extend or reduce its debts or drastically lower its operating costs, it often can be returned to a viable state. Generally, it is more economically efficient to reorganize than to liquidate, because doing so preserves jobs and assets.

Then, to the surprise of many, two weeks after Iridium filed, ICO also filed for Chapter 11 bankruptcy after failing to achieve the level of investment required, partly blaming their problems on the Iridium bankruptcy filing – now dubbed the 'Iridium effect'. ICO was seeking a further investment of US$1 billion (after two months earlier failing to raise a further US$500 million[2]), as it needed more than US$1 billion to launch its twelve satellites by late 2000. These funds would be in addition to the US$3 billion ICO has already raised from investors. So to date, Iridium has cost some US$5.2 billion and ICO has run to US$5 billion, while Globalstar is estimated to have cost US$3.8 billion before it has even come into service. (In late 1999, ICO was bought out by a group of investors led by Craig McCaw, Chairman of Teledesic.)

These financial problems were not restricted to the voice-service GMPCS systems. Due to the scale of investment required and the difficulties in obtaining this in pace with the technical development of systems, Teledesic had obtained only US$1.5 billion by mid-1999, while estimates put the cost of the Teledesic system anywhere between US$9 billion and US$13 billion[4]. Hence the future of both Iridium and ICO are seriously in question at the time of writing, but Globalstar has announced it will commence a limited public service in the first quarter of 2000. This uncertainty is creating a good deal of caution on the part of consumers, and even many of those in the niche markets – including newsgathering – are holding off on any expensive purchases of equipment until the scene becomes less confused and the clear winner(s) emerge.

9.3.3 Regulation and licensing

The issue of licensing has proved to be a headache for GMPCS operators, though perhaps to a lesser degree than the issues of financing. There is no mechanism for the global licensing of such a service, for although the ITU can achieve agreement on general spectrum allocations, GMPCS operators have to obtain operating licences for each country through the respective national administrations. It has been claimed that licensing agreements between each of the ITU member countries would add up to more that 16 000 separate agreements. The problems of licensing have been rated as being even greater than designing, launching and operating a large, moving constellation of satellites. Since each country manages its own radio frequency spectrum allocations, each administration also has to approve spectrum assignments for the new service. This may mean reassignment of existing services out of the appropriate band to make way for the GMPCS service, and may have widespread implications for existing systems in a given country.

A related problem is that of type-approvals for the handsets and terminal equipment. At present, each country licences operation of telecommunications equipment for its own telecommunications environment and countries each have their own safety standards, so operators need to modify their handsets to suit the regulations of a number of different markets.

The work being done on GMPCS under the aegis of the WTO and the ITU offers some encouragement to the future of the satellite industry. Administrations, satellite operators, service providers and manufacturers agreed a Memorandum of Understanding (MoU) in 1997 which should facilitate the trans-border use of satellite terminals and the provision of service around the world; by mid-1998, forty nations had signed the MoU and it was implemented by the ITU[5]. The GMPCS MoU therefore offers a mechanism for achieving mutual benefits to both the satellite industry and individual countries.

9.4 Cellphones and third generation wireless services

Through the 1990s, mobile telephony grew faster than anybody could have predicted with the introduction of the digital Global System for Mobile Communications (GSM) in 1992, and it is now the world's most successful wireless communications standard. It has more than 100 million subscribers in 120 countries, and it attracts more than five million new users every month. It is predicted that by 2005 the figure will exceed one billion. GSM cellphones are cheap to buy and cheap to operate, and the reliability of services is very high.

For newsgathering, the GSM cellphone has become as much a tool of the trade as the tape recorder, the camera, the notepad and the laptop PC. It enjoys usage across continents, from sophisticated metropolitan areas in the developed world to less-developed areas of central Eastern Europe and Asia, where users have leapt from the traditional landline telephone to the digital GSM network in one bound. In 1999, GSM cellphone calls could be made from the middle of a war zone, from Belgrade, Skopje, and Pristina, whereas before the reliance on telephone communication in such a location would have been via INMARSAT satphones alone. Journalists carry cellphones everywhere and expect them to work everywhere, which, despite the different variations on the GSM system across the globe, is becoming easier to achieve with dual- and tri-band cellphones.

The obvious advantage of GSM over any satellite-based system is that GSM cellphones can be used from inside buildings, as opposed to GMPCS systems that require hand-held terminals to have direct line of sight to a satellite. A user trying to call using GMPCS may not even be able to obtain a connection if standing at a window if the window has multiple glazing. Conversely, there are large areas of the world that will never have GSM service due to very low population densities, which make it uneconomic for any GSM system to be installed. There is therefore a place for both systems, but it is questionable whether the number of GMPCS systems would have been planned if the rapid rate of take-up of GSM had been envisaged, but by then it was too late.

Current GSM systems currently provide a data rate of just 9600 bps, barely adequate for anything more than basic e-mail. However, the telecommunications service providers around the globe see extending and increasing the use of data to as many people as possible as the next strategic priority, and to do that, data rates must rise. Thus increasing the data rate will bring the GSM networks into the realm of providing newsgatherers with a more sophisticated tool for them to use beyond purely voice calls.

A new mobile system for worldwide use is being developed to enhance and supersede current systems, referred to as the 'third generation mobile system' (3G), the first generation being analogue and the second digital. First generation cellular systems have one frequency channel assigned to one user for as long as they need it. Second generation systems use TDMA, and third generation systems will use CDMA. 3G will be an enhanced digital system that will provide universal personal communications to anyone, anywhere, enabling wireless Internet access, video-conferencing and other broadband applications – including newsgathering. In fact, it is a direct challenge to the objective of GMPCS, which was conceived just before GSM was born and before the explosive growth in GSM in the late 1990s could even have been contemplated.

In Europe, the 3G system being developed under the regulatory framework of ETSI is called the Universal Mobile Telecommunication System (UMTS). Similarly, the ITU is formulating IMT-2000 (International Mobile Telecommunications) in parallel, which is to be a group of systems that will allow users to roam worldwide with the same handset, with UMTS as part of this group. UMTS is expected to support up to 144 or 384 kbps for mobile use, with fixed applications supporting up to 2 Mbps in either a symmetrical or asymmetrical configuration (identical or different bandwidths in each direction according to service requirements). UMTS with a full 2 Mbps service will use frequencies in the 2 GHz band, but will inter-operate with the older systems as well as being available via satellite. Japan plans to launch its UMTS network in the year 2000, and the UK plans to have UMTS working alongside and enhancing GSM networks by the year 2002. By 2005, the UK expects its first fully working UMTS network to operate compatibly with its 'legacy' systems, shortly followed by the rest of Europe.

Meanwhile, IMT-2000 will ensure that 3G systems are globally compatible and provide uniform communications. It is envisaged that IMT-2000 will include satellite services for global access. All interested parties are being encouraged to work towards convergence of technologies that otherwise might compete against each other – again, a contradiction of the way that GMPCS has developed, with no interoperability or co-operation between systems.

An important aspect of UMTS is the sequential upgrade path from current GSM systems, and the incremental increase in data capacity is of particular interest. The first significant development is called high-speed circuit switched data (HSCSD) and is being introduced in the UK in 2000. This works by taking the time slots GSM normally allocates – one per subscriber – and amalgamating them so one subscriber can use up to four slots, providing 56 kbps aggregated rate.

Following HSCSD, the next step forward will be the general packet radio services (GPRS). This will provide data rates up to 115 kbps, and is expected by the end of 2000 in the UK. As it is packet-based, it allows permanent connection and will compete with ISDN in mobile applications. Major telecommunications manufacturers such as Motorola, Ericsson, and Nokia see GPRS as increasing user demand for high-speed wireless data services and GPRS is expected to be in service by the end of 2001.

After GPRS, a further development will be enhanced data rates for GSM evolution (Edge). This changes the modulation method for GSM data transfer but leaves the rest of the system unchanged. This will increase the data rates to 384 kbps and will be the first true 3G service. All the enhancements will be workable in any of the existing GSM bands across the world – 900 MHz, 1800 MHz and 1900 MHz – and will follow the existing system of roaming between cells.

There is still some way to go before the full impact of these new services can be gauged and we can tell just how much use they will be to newsgathering. In particular, as with GMPCS systems, one of the areas where they may have the most significant effect is in the inter-connectivity with electronic newsroom computer systems. The dependence on newsroom computer systems is such that increasingly they need to be extended into the field. Plainly, these systems will also have application in the transfer of video and audio data files.

9.5 The future of the transmission medium

The term 'SNG' implies by its very name that satellites are the transmission medium, but this may not necessarily be true in the future. As we reach the end of the 1990s, a multiplicity of telecommunication routings are available – terrestrial microwave, 'local-loop' through local telephone companies, intra- and trans-continental fibre – and some of these challenge the traditional satellite territory of the last forty years.

Since 1987, the undersea cable industry has produced and installed an amount of fibre-optic cable nearly equal to all of the undersea copper cable installed by 1958. In 1995, it was estimated that 129 countries would have direct undersea links to the global fibre-optic network by 1998[6]. Countries not already projected to be connected by fibre-optic undersea cables either have a population of less than 10 million and relatively low volumes of outgoing international traffic, are landlocked, or are considered to be politically sensitive by nations with companies capable of providing the connectivity. This dramatic growth in the use of fibre-optic cables inevitably challenges many of the traditional trans-continental satellite routes, and the technology that enables this is ATM.

9.5.1 ATM

Asynchronous transfer mode (ATM) transmission is revolutionizing communications in general. ATM is a new technology that allows bandwidth to be allocated on demand with a switching technique, achieving faster transmission rates. The ATM protocol has a significant advantage in that it is not tied to any single transmission medium, and so can be used with all existing types of transmission routes, including satellite networks.

In 1988 the ITU defined ATM as the means for future broadband ISDN (B-ISDN) services, which includes MPEG-2 transmissions. The basic idea behind ATM is the division of the digital signal into small packets (and these can be digital video and audio signals), where data is broken up and transmitted in small, fixed sized data 'cells'. The packets can then be routed through a network and, at each switching node in the network, the ATM 'header' part of the packet identifies a virtual path or virtual circuit that the cell contains data for, enabling the switch to forward the cell to the next location. A virtual path is set up through the involved switches when two endpoints wish to communicate. This makes ATM 'switches' and routings very fast, allowing transmission rates of up to 155 Mbps. Furthermore, because ATM can provide bandwidth on demand, effective dynamic management of bandwidth can be achieved. Thus simple applications such as video-conferencing get a small allocation of bandwidth while full broadcast-quality video is allocated a larger bandwidth, maximizing usage of the bandwidth, unlike other schemes that have fixed bandwidth allocation.

Although ATM is aimed at the full range of existing networks, whether satellite or terrestrial, the spread of ATM has gone hand in hand with the development of fibre-optic networks. The evolution of the 'wired world' has received a significant boost with the advent of fibre-optic transmission, and the result has been that many international telecommunication companies have developed integrated ATM networks, where part of the network is in the sky (satellite) and part under the ground (fibre). This means that entry points into the ATM network – commonly termed POPs (points of presence) – are literally 'popping' up in a large number of places. It is therefore feasible that some aspects of newsgathering currently served by satellite could be replaced by the use of ATM, as connectivity is available particularly in metropolitan areas.

There is concern among some broadcasters and newsgatherers regarding the prioritizing of traffic on an ATM network. Newsgatherers have become arrogant in their assumption and demand that they should get bandwidth when they want it, and the speed of delivery is taken for granted. A question that many are asking is whether ATM can offer to a niche group such as newsgatherers the seamless non pre-emptible capacity they have come to expect. Some newsgatherers have said that ATM cannot be all things to all users, suggesting 'not all bits are equal' (naturally, in their view, their bits are more important than others). However, it is possible to buy ATM service at different levels, by assigning data streams with QoS (quality of service) 'flags', so that provided a newsgatherer is prepared to pay, there is no reason why they should not get the same level of service as via traditional 'real' satellite routes. QoS defines either the limit of the range of delay that is acceptable or the maximum bit-error rate that can be tolerated. The drawback is that provision of a 'high priority' connection only works over a network where everything connected to it supports QoS. To achieve end-to-end service that supports QoS, all applications on the network must support QoS – and currently that is often not the case.

There is a distrust amongst some newsgatherers of digital networks that create 'virtual paths' of connectivity, as opposed to the supposed 'real paths' via satellite. However, ATM fibre-optic routes are no less reliable than satellite routes currently used by newsgatherers; in fact, they are probably far more dependable, as weather or galactic influences do not affect them. Added to this, sections of contribution and distribution between the broadcast centres and the ground earth stations are already partly or fully served by ATM fibre-optic services, and the fear of ATM seems even less justifiable.

It has been suggested that ATM could largely replace SNG in the long term. While this is certainly true of coverage by satellite links of known and planned events in areas of the world with a developed infrastructure, there are problems when trying to extrapolate its potential to replace satellite newsgathering. The first hurdle is that many news stories currently covered by SNG often occur in places that barely have a telephone system, let alone an infrastructure supporting ATM. The second problem is that many news stories occur at very short notice, and there can never be – at least within the foreseeable future – enough POPs to be able to serve a modest number of requirements in even the metropolitan areas of the developed world.

PanAmSat has initiated a service for customers, particularly newsgatherers, offering store and forward delivery of material – not because of a lack of bandwidth, but as a cost-saving measure so that material can be sent at off-peak times when the PanAmSat system is not as busy. The concept is that material is transmitted via PanAmSat capacity to their facility and recorded on a server, and overnight the material is then played out via PanAmSat capacity again to the final destination. For non-urgent material, this may be a very attractive option, and makes time an elastic commodity, as power and bandwidth have been traditionally with satellite capacity.

9.5.2 Internet

So the use of ATM data networks for the transfer of news programme material is already in use by many newsgathering organizations between fixed points – and that now includes the Internet. With the fall in cost of video and audio editing software for PCs, many newsgathering organizations are already sending edited reports from the field as compressed files, either as e-mail attachments or using FTP (file transfer protocol) via the Internet. This is particularly attractive from those parts of the world where traditional delivery of reports in quality via satellite is difficult or prohibitively expensive. There is an added benefit in that these reports can be sent in a very secure manner, so that the problems of censorship are not encountered, as they would be if the reports were sent by more open means. However, data networks have problems dealing with live working because of path predictability and delay.

As we all know by now for live broadcasting, real-time fixed and low latency are an absolute must. In simple terms, Internet protocol (IP) works by sending packets of data across the Internet and LAN in a series of hops in a discontinuous stream, which are assembled at the destination. As with ATM, data is sliced into variable-sized packets with addresses on the front for the destination. The packets are sent through a series of nodes (switches, routers etc.) on a hop-by-hop basis. At each node, every packet is analysed to decide which hop it should take next. This is decided on parameters completely unrelated to any importance that may be attached to this packet as part of the whole stream it is part of (a complete anathema to newsgatherers, who regard their packets as more important that anyone else's). This process of packet-switching is similar to ATM but is different to the traditional broadcast process of 'circuit-switching' (which of course everyone is comfortable with).

With IP, as each packet arrives at a node, it is treated as a fresh event, as the intermediate nodes do not keep track of what connections are actually flowing through them. This lack of information 'state' means that a special priority cannot be remembered and applied to a specific data packet flow. Unlike ATM, which can assign a QoS flag to each packet with a unique identifier, IP has no such way of assigning priority. Hence, all packets are equally important and treated equally – and that is the problem for newsgatherers where material is absolutely time-critical in 'live' working.

On the one hand with dedicated satellite paths, we have channels that are specifically switched as required, that have high bandwidth with no variation during transmission, and consistent delay behaviour. They are dedicated and terminate in fixed locations. On the other hand, we have a technology that struggles to approach the bandwidths required for even compressed digital video, that aggregates bandwidth together on a statistical basis and cannot guarantee bandwidth or latency performance – but is available on desktops in every developed country. Therefore, it can provide a particular way of delivering news material from remote locations. There are a number of bespoke applications designed to transfer video and audio data files in high quality, and although currently these are not widely used, the pervasive nature of the Internet will increase their significance for newsgathering.

The balance between the two technologies is difficult to weigh up. ATM is expensive and complex, with a relatively small range of products, but ATM does create a 'virtual circuit' whereas IP doesn't. An ATM virtual circuit has a fixed delay (as with SNG), which can be coped with in a 'live' environment, whereas with IP it is continuously variable, which makes it almost impossible to cope with in production terms. But IP is a technology that is driving ever lower costs, with a burgeoning array of applications that can be used for newsgathering – certainly in terms of easy file transfer. In an attempt to address the issue of prioritizing traffic, in the IP world there is something akin to QoS being developed called 'class of service' – effectively a poor man's QoS. Data that requires low delay is placed in a queue which is served more frequently. However, although this might improve the use of IP for newsgathering for file transfer, it is still unlikely to bring IP to the fore for use with 'live' and direct-to-air transmissions.

Nevertheless, the disadvantage of both ATM and IP is access to POPs, and although for non real-time transfer of material as files it is a low-cost solution, it cannot be seen that it will replace the relative speed and immediacy of SNG (admittedly at a much higher cost). Some newsgatherers have said that data networks cannot be all things to all users, as 'not all bits are equal', and in their view their 'bits' are more important than others. This view is likely to prevail for the moment.

9.5.3 Lower the bit-rate. . .

The issue of the acceptability of lower bit-rates is a constantly recurring question, which has been often discussed in the latter part of the 1990s. However, the extensive use of video and audio processing in studio and transmission centres, including transition through computer-based server architectures, means that even for news it may not be possible to tolerate bit-rates significantly lower than those that are accepted now. In Chapter 4, reference was made to the issue of concatenation of digital signals through multiple processes, and although MPEG-2 algorithms are constantly being improved and refined, using very low bit-rates at the start of the chain is not going to make matters any easier – quite the contrary.

It has already been seen that the original 'standard' use of 4:2:0 MPEG-2 signals for SNG has been replaced by the higher quality 4:2:2 processes in order that quality can be maintained. This has forced the use of 4:2:2 coding even for news feeds because of the fragility of 4:2:0 signals through video server manipulation. Experience has already shown how multiple encode/decode/re-encode processes have affected output, and as more broadcasts are transmitted digitally to audiences – either via satellite DTH or digital terrestrial television (DTT) services – any defects in the chain are more obvious even to the untrained eye and ear. In a sense, the use of servers may determine that the quality of the contribution process be kept as high as possible, at least until server technology matures and improves.

Current strategies in designing news production systems focus on the use of automation and hence reducing the staffing levels previously required to produce news bulletins. All the costs of the contribution process (staff and resources) are also subject to this downward pressure. Even though lower bit-rates mean less bandwidth and power, and hence fewer dollars, the drive to reduce costs to achieve 'more for less' may not be as productive if lower quality becomes too objectionable to the viewer and listener.

9.6 Challenges to traditional working practices

The way in which SNG has traditionally been used, and the whole mechanics of the newsgathering process, is being challenged in a bid to reduce costs. This has resulted in a number of existing practices being challenged and alternative (cheaper) processes being suggested or implemented. In this section the principal issues are briefly examined.

9.6.1 De-skilling the SNG process

One method to reduce cost and increase productivity that has emerged through the 1990s is 'multi-skilling' or 'multi-tasking'. Much of newsgathering management has seen that this may offer the potential to reduce the most expensive part of the newsgathering process – employment of people. The concept is that equipment (and this includes everything in the acquisition process from the camera and audio recorder, to editing equipment and the SNG uplink) becomes easier to operate as it becomes more 'intelligent' and 'intuitive'. Therefore the number of people in the field can be reduced, and in particularly the traditional craft-skilled technicians – cameramen, sound recordists, uplink operators – can be reduced in number or eliminated, thereby driving down costs.

Crudely, the theory is that one person, who is trained primarily as a journalist, can arrive at the scene of a news event, find out the facts of the story, present and shoot a report using a digital camcorder (and edit the report on the ground if it is not being sent back 'live'), and then transmit the story back using a very compact satellite link. If the story can be covered in such a sequential way, and the time-scale allows, then this is possible under these circumstances – perhaps.

Unfortunately, the nature of most news stories is that such a process would fail at or near the very beginning. In addition, on a big story many journalists often have to produce 'output' for a number of outlets almost simultaneously (this applies to those working for large networks or as freelance 'stringers'), and the pressures to simply keep up with the editorial demands makes meeting the technical demands incompatible.

Ironically, the logical culmination of this concept was seen in the late 1980s on US television. There was an entertainment show called *Saturday Night Live* on NBC, and it featured some comedy sketches of spoof news reports called *Weekend Update* from the field. These featured the comedian Al Franken as 'the one man mobile uplink', playing a reporter with a small antenna fixed to his helmet, a satellite uplink transmitter on his back and a camera mounted on his Steadicam™ harness. Al Franken was supposed to be able to report items as well as operate all the equipment, and for those involved in the newsgathering industry, it presents an eerie fantasy of what many television news editors then, now, and in the future would like to see.

So the question has to be posed: are journalists able and willing to operate a lot of equipment in the field themselves as well as doing their own 'real job'? There are some journalists in the field – typically freelance 'stringers' – who are able to offer this type of service within certain limitations. Certainly, a number of the global newsgatherers have succeeded in making this work, particularly with journalists working solely in radio where the technological and operating demands of the equipment are lower. However, both corporate culture and trade union opposition have prevented some organizations from moving very far down this path to date.

9.6.2 'Send the flyaway!' – or not

Many of the global and regional newsgatherers are finding that as markets develop, there are an increasing number of local SNG services available for hire. As described in Chapter 8, the problems of shifting SNG uplink flyaway systems around the world are often significant, both in terms of the amount of work involved and in the costs. The question has been asked as to whether the concept of shipping the 'thirty box/1200 kg' flyaway around the globe is going to become extinct as more SNG service providers across the globe means there is less need to ship systems around the world. In particular, are broadcasters – whose principal objective can be seen as news dissemination rather than newsgathering – going to continue to provide their own facilities or look to 'outsource' from local service providers to a greater degree in the future? We have described the 'comfort factor' that many organizations feel by having their own people and facilities, which supersedes considerations of cost on very large stories. It is also arguable as to whether it is appropriate for broadcasters in particular to continue to try and 'play all the positions on the field', or concentrate on their core strengths of journalism and editorial expertise, leaving the provision of the technical facilities to independent service providers.

As we noted in the last chapter, the development of the SNG uplink hire market in a number of regions of the world is coming on apace, so that there is often a viable alternative to newsgatherers sending their own SNG uplink systems. It is certainly the case that in some countries the number of local SNG service providers is so great that it is questionable for any organization to want to take in their own uplink equipment. A classic example is the US, where the SNG uplink hire market is of such size and is so competitive that it makes it highly uneconomic for newsgathering organizations from outside North America to go to the effort and expense of shipping their own systems there. Added to this are the hurdles in obtaining a temporary uplink licence from the FCC (though a friendly US network might be able to help).

However, the local SNG market cannot always be relied upon even if there are a number of operators offering service. In the hours after the disastrous earthquake in Turkey in August 1999, it was impossible for foreign newsgatherers to contact the local providers by voice or by fax because of the major disruption to telecommunications, which had put the telephone service out of action. Even e-mail was used to try to contact the operators but proved futile. Foreign broadcasters (and news agencies), in the absence of being able to guarantee local hire, were therefore reliant on being able to ship their own systems out to Turkey to cover this major story.

For the time being, it seems likely that those broadcasters who already have their own flyaway systems will continue to operate them, while smaller broadcasters will perhaps not need to invest in systems as the availability of both truck-based and flyaway SNG uplinks for hire increases and costs are driven down.

9.6.3 Do we send a truck or a laptop?

The development of the new laptop store and forward video systems described in Chapter 5, combined in particular with the arrival of the INMARSAT-M4 satphones, has now made it practical for a journalist to travel with two attaché-sized cases; one containing the laptop store and forward unit, and the other the M4 satphone. Combine this with the development of relatively high-quality video-conferencing units (just one more attaché-sized case unit), and the journalist is truly self-sufficient. The rhetorical question in the title is somewhat tongue-in-cheek, but does indicate that within a very short time, many stories may well be covered by this compact type of equipment – particularly if it offers considerable cost savings. There is of course the issue of quality, but as we have already discussed, nothing is ruled out on the grounds of quality if it gets the breaking 'live' update back.

So what SNG platforms are we likely to see in the future? In so far as the use of trucks, of particular interest on both sides of the Atlantic is the direction in which the construction of SNG vehicles is going. The development of news-gathering using SNVs appears to be polarizing into a choice between a small vehicle, in the range of the 3500–4500 kg (10 000 lbs) GVW, or a larger vehicle that has the capability of handling more complex news productions, based on vehicles approaching 12 000 kg (26 000 lbs). Some voices in the industry are questioning whether anything is required in the middle – customer requirements seem to gravitate towards one extreme or the other.

Other tasks now drive the size of the vehicle more than the uplink capability itself, such as handling multi-camera operations, field editing facilities, and serving multiple outlets simultaneously through multi-path working. For the SNV constructors, this is proving to be a challenge as almost invariably every customer wants the smallest vehicle in physical size while fitting in the maximum amount of equipment.

There also seems to be a need to allow for scaling of facilities, with systems needing to be modular in approach and design. This is so that the vehicle can be fitted with enough equipment for the immediate news response requirement, yet be able to be easily upgraded to cope with a higher level of production activity by just 'sliding in' the extra equipment. Many have noticed that production demands often require an SNG vehicle or flyaway that effectively becomes an integrated location production facility, coping with the scaling of a news 'story' into a news 'event'.

Particularly in the US, which has come late to DSNG, there are questions as to whether small DSNG vehicles will replace traditional terrestrial microwave ENG vehicles. Using direct (or near) line-of-sight communication, terrestrial microwave links have been the traditional means of covering domestic news stories for many local stations in the US for twenty years (and in Europe for almost as long). There is an extensive existing ENG infrastructure of receiver sites in every major town and city, as well as transmitters fitted to every ENG vehicle.

Coincidentally, due principally to the spectrum demands for 'feeder' links of the Globalstar satellites, and the development of IMT-2000 and UMTS, the existing ENG infrastructure will probably have to be replaced by around 2002. In the US, the existing ENG terrestrial 2 GHz microwave frequency allocation is to disappear, to be replaced by allocation in another part of the spectrum, and in addition the new individual channels are likely to be reduced in size. This will force new investment, probably in digital equipment to achieve better spectral efficiency. Therefore, there is a question as to whether the use of terrestrial microwave should be abandoned in the face of the high cost of this re-investment if the cost of digital capacity on US domsats falls in price.

In Europe, where DSNG caught on much earlier than in the US, there has already been a move away from terrestrial ENG. Europe too faces the same pressure to vacate the traditional 2 GHz ENG band, which is widely used around the world.

9.7 Outlook

So what is the future of satellite newsgathering? We have touched on a number of technological developments and changes in working practices that will directly impact the process of newsgathering and therefore ultimately what is seen on the television screen or heard on the car radio. The speed of development is so fast that by the time this book appears on the bookstall, many of the projected changes will have either occurred or fallen by the wayside.

As new projects steadily roll out, stock prices are falling, and a spate of failed launches in 1998 and 1999, plus tight money markets, means there is in reality a brake being applied to the development of the satellite communications industry. Forecasts are optimistic, with some predicting that the global market for broadband communications will reach US$200 billion by 2005, while those GMPCS voice systems that are in their infancy are struggling. It is estimated that a total of almost 1500 satellites will be launched on between 850 and 900 launch vehicles between 2000 and 2009, of which almost 900 will be commercial satellites worth about US$56 billion. The cost of the launches alone needed to put these commercial satellites into orbit is estimated at US$25 billion[4].

The development of GMPCS promises to be the greatest change that will be seen over the next three to five years, but progress to date has been disappointing. Ironically, if the development of UMTS comes to fruition, it will effectively provide (at potentially much lower costs) exactly the same services that GMPCS is aiming to do. The GMPCS systems face a serious challenge to their business in many countries from this GSM technology, which further undermines the already shaky commercial foundation of GMPCS. For newsgathering, the development of GMPCS and GSM services just increases the range of tools that are available, and the deeper the penetration into the consumer market, the lower the cost to everyone.

For the journalists in the field, there is an ever-increasing demand to 'go live', to be in the midst of the story rather than just reporting from the edges, and for those who work for the major networks, to produce the maximum amount of output for every dollar spent. Unfortunately, in many cases journalists are becoming simply 'content providers', and the investment in satellite technology is there to feed this drive. More facilities are being demanded in the field on the major stories; while at the other end of the scale, domestic camera equipment is being thrust into the hands of journalists who are individually expected to provide what a team of three or four people would have produced a decade ago.

Even in the middle of a war zone, there is the desire to give the audience the feeling of 'being there', and this requires the use of wireless cameras, short-hop microwave and laser links, and ever-more mobile SNG uplinks, to combine instant mobility with guaranteed 'live' transmissions. Just being at the nearest hotel reporting the story is no longer enough.

So it seems that for the foreseeable future, satellites will continue to play a vital role in delivering news to audiences from remote locations. Hopefully the subjects covered in this book will have given you some insight into the very wide range of issues that face the satellite newsgatherer, whether it is for coverage of local, national, international or global news.

References

1. Nelson, R.A., Phillips Business Information Inc. (1998) Iridium: from concept to reality, *Via Satellite*.
2. Phillips Business Information Inc. (1999) *Satellite Today*, **2** (118); www.satellitetoday.com.
3. Phillips Business Information Inc. (1999) *Satellite Today*, **2** (156); www.satellitetoday.com.
4. Phillips Business Information Inc. (1999) *Satellite Today*, **2** (131); www.satellitetoday.com.
5. ITU (1998) Ninth Plenary Meeting, Resolution 1116, Implementation of the Gmpcs MoU arrangements.
6. ITU (1995) *ITU Newsletter*, No. 9.

Appendix A
Frequency bands of operation

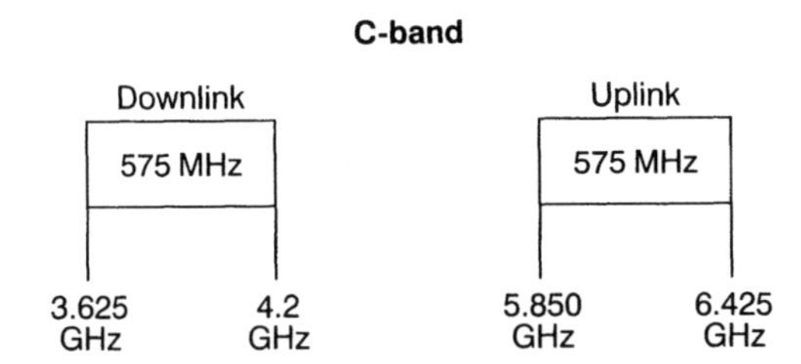

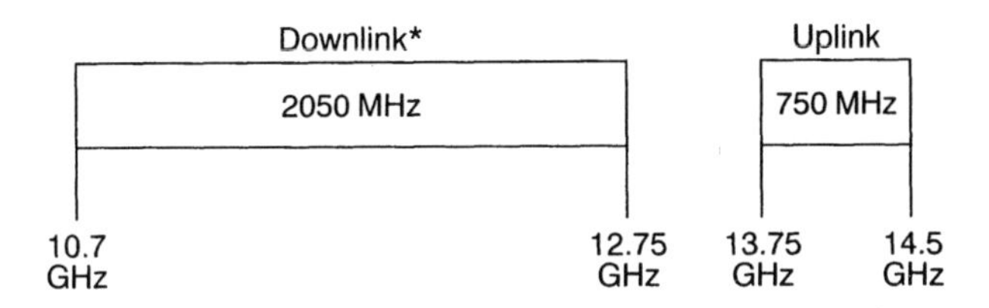

* Includes bands used for DTH services as well as mobile uplinks

C-band	3.6–6.5 GHz
Ku-band	10.7–18 GHz
Ka-band	18–300 GHz
L-band	1.0–2.0 GHz
X-band	7.25–8.4 GHz

There are a number of different definitions for frequency bands, depending on whether for commercial, military or space use – the above are those broadly accepted for commercial satellite usage.

Appendix B
Sample link budgets

Scenario: A major news event has occurred at Buckingham Palace in London, and News-TV are going to deploy one of their SNG uplinks to provide live coverage. The uplink is working onto a EUTELSAT satellite, and the transmission is going to be downlinked at a number of locations, including a teleport in London.

We need to calculate link budgets for both an analogue and a digital transmission, as well as calculating positional information for setting up the uplink.

The link budgets shown are of the simplified type – fully detailed link budgets run to many tens of pages where every possible parameter is calculated and included.

(Note: many of the trigonometric calculations have to be performed with values in radians rather than degrees.)

Uplink position and pointing

Firstly, we need to calculate certain parameters related to the position of the uplink relative to the satellite being used.

Longitude of uplink: Latitude of uplink:

$LO_{es} = 0° \, 8' \, 18''$ $\qquad LA_{es} = 51° \, 29' \, 56''$

Converting from degrees and minutes to decimal form:

$LO_{es} = 0.14°$ $\qquad LA_{es} = 51.5°$

Satellite being used is EUTELSAT 2 Flight 4 at 10°E, satellite longitude (LO_{sat}) = 10°.

We need to calculate the difference in longitude between the uplink and the satellite (uplink at 0.14° West compared to satellite at 10° East – we're only interested in the size of angular difference at this stage and need not worry about the sign).

$$(LO_{es} - LO_{sat}) = LO_{diff} = 10.14°$$

Let E = an angle that describes the relative longitudinal difference to the uplink latitude.

$$\cos E = (\cos LO_{diff} \cos LA_{es})$$

$$= (\cos 10.14° \cos 51.5°)$$

$$= (0.9844)\ (0.6225)$$

$$= 0.6128$$

$$E = \arccos (0.6128)$$

$$= \mathbf{52.2°}$$

This is of no direct significance – it is an intermediate calculation that we require for computing the angle of elevation (EL_{es}) of the uplink, and slant path length (S) from the uplink to the satellite.

Azimuth angle to satellite (AZ_{es})

The equation varies according to the part of the hemispheric quadrant in which the uplink is located compared to the satellite:

- If uplink is West and North of satellite position over the Equator, true azimuth = 180 – Az
- If uplink is East and North of satellite position over the Equator, true azimuth = 180 + Az
- If uplink is West and South of satellite position over the Equator, true azimuth = 360 – Az
- If uplink is East and South of satellite position over the Equator, true azimuth = Az

This uplink is West and North of the EUTELSAT 2 satellite, so true azimuth = 180 – Az:

$$AZ_{es} = 180 - \arctan\left(\frac{\tan LO_{diff}}{\sin LA_{es}}\right)$$

$$= 180 - \arctan\left(\frac{\tan 10.14}{\sin 51.5}\right)$$

$$= 180 - \arctan\left(\frac{0.1788}{0.7826}\right)$$

$$= 180 - \arctan(0.2285)$$

$$= 180 - 12.87$$

$$= \mathbf{167.13°}$$

This is the bearing in degrees ETN (East True North), i.e. measured clockwise from True North looking down on the uplink. Note that magnetic variation between True to Magnetic North at this part of the UK is +4°, and this must be allowed for and the compass reading taken accordingly.

Elevation angle to satellite

$$EL_{es} = \arctan\left(\frac{\cos E - 0.1513}{\sin E}\right)$$

$$= \arctan\left(\frac{0.6128 - 0.1513}{\sin 52.2}\right)$$

$$= \arctan\left(\frac{0.4616}{0.79}\right)$$

$$= \arctan(0.5842)$$

$$= \mathbf{30.3°}$$

Polarization skew

Having got the azimuth and elevation figures for the antenna, we need also to calculate the skew angle θ:

$$\theta = 90° - \arctan\left(\frac{\tan LA_{es}}{\sin LO_{diff}}\right)$$

$$= 90° - \arctan\left(\frac{\tan 51.5°}{\sin 10.14°}\right)$$

$$= 90° - \arctan\left(\frac{1.257}{0.176}\right)$$

$$= 90° - \arctan (7.14)$$

$$= 90° - 82°$$

$$= \mathbf{+8°}$$

The antenna has to be adjusted by +8° in skew; however, this is a theoretical calculation which has to be adjusted in operation as it does not take into account how the satellite has been 'parked' in orbit. It is adjusted in the field by examining signals received on the uplink from the satellite on a spectrum analyser.

Slant range

Slant range path length from uplink to satellite (S) is given by:

$$S = 35758\sqrt{1 + 0.4199(1 - \cos E)}$$

$$= 35758\sqrt{1 + 0.4199(1 - 0.6128)}$$

$$= 35758\sqrt{1.163}$$

$$= 35758(1.078)$$

$$= \mathbf{38\,555\,km}$$

Analogue link budget

The uplink is to transmit a single analogue 625-line TV carrier to the transponder; the channel bandwidth of the transponder is 36 MHz.

The uplink frequency (f) is 14.25 GHz $= 14.25 \times 10^9$ Hz

Speed of light (c) $= 2.99793 \times 10^8$ m/s

Wavelength (λ) $= \dfrac{c}{f}$

$$= \frac{2.99793 \times 10^8}{14.25 \times 10^9}$$

$$= \mathbf{0.02104\,m}$$

Antenna diameter (D) $= 1.5$ m

Antenna efficiency (η) $= 0.6$

Antenna gain

$$G^{ANT} = 20\ \log_{10}\left(\sqrt{\eta}\ \frac{\pi D}{\lambda}\right)$$

$$= 20\ \log_{10}\left(\sqrt{0.6}\ \frac{\pi 1.5}{0.02104}\right)$$

$$= 20\ \log_{10}\ (173.5)$$

$$= \mathbf{44.8\ dBi}$$

HPA power

$$\text{HPA maximum output } (P^{W}) = 400\,\text{W}$$

$$= 26\,\text{dBW}$$

$$\text{Waveguide losses } (W^{loss}) = 0.75\,\text{dB}$$

Uplink EIRP

The maximum EIRP capability of the uplink is given by:

$$P^{eirp} = P^{HPA} + G^{ANT} - W^{loss}$$

$$= 26 + 44.8 - 0.75\ dBW$$

$$= \mathbf{70\,dBW}$$

Free space attenuation (FSA)

$$FSA = 20\ log_{10}\left(\frac{4\pi S}{\lambda}\right)$$

$$= 20\ \log_{10}\left(\frac{4\pi 38584 \times 10^{3}}{0.02104}\right)$$

$$= 20 \log_{10} \left(\frac{4.8486 \times 10^8}{2.104 \times 10^{-2}} \right)$$

$$= 20 \log_{10} (2.304 \times 10^{10})$$

$$= \mathbf{207.25\,dB}$$

Uplink power to the satellite

London lies on the +6 dB/K G/T satellite receive antenna's gain advantage contour, and EUTELSAT has specified that an IPFD of –82 dBW/m^2 is required on the –0.5 dB/K uplink reference contour to achieve saturation of the transponder. (This can vary from transponder to transponder and this information needs to be sought from the satellite operator.)

$$\text{Saturated IPFD} = -82 - 6.5\,\text{dBW/m}^2$$

$$= \mathbf{-88.5\,dBW/m^2}$$

In addition, an IBO of 6 dB is specified for operation of the transponder (resulting in an OBO of 2 dB).

$$\text{Required IPFD} = -88.5 - 6$$

$$= \mathbf{-94.5\,dBW/m^2}$$

Spreading loss (L_{spread}) = +10 $\log_{10}$ ($4\pi S^2$) dB(m^2), where S (slant range) is in *metres* (rather than in kilometres as we have shown):

$$L_{spread} = +10 \log_{10} [4\pi\, (38555 \times 10^3)^2]$$

$$= \mathbf{+162.72\,dB(m^2)}$$

We also need to consider some loss attributable to 'pointing' error of the antenna and the slight movement of the satellite within its station-kept box. Most of the pointing loss is not because you cannot point a dish accurately at a satellite – it is because the satellite is not always in the same place. So we will estimate L_{point} as being 0.5 dB. The atmosphere will introduce some loss (L_{atmos}), which we will estimate as 0.6 dB. (As this is a simplified link budget, we will consider all other losses as negligible.)

Uplink EIRP required ($EIRP^{up}$) = IPFD + spreading loss + atmospheric loss + pointing loss

$$EIRP^{up} = \text{IPFD} + L_{spread} + L_{atmos} + L_{point}$$

$$= -94.5 + 162.72 + 0.6 + 0.5$$

$$= \mathbf{+\,69.32\,dBW}$$

This is 0.68 dB less than the maximum EIRP capability of the uplink, thus leaving a very small amount of power in hand. This does not mean that if the uplink is unable to achieve the +69.32 dBW figure (and therefore fail to reach the desired satellite IPFD) that it could not be used – just that the operating margins are likely to be less than ideal. In practice, this means that the margin for rain fade or any other unpredicted atmospheric attenuation would be reduced.

Carrier-to-noise ratio

We can now calculate the C/N of the uplink and the downlink. We need to introduce a constant called Boltzmann's constant (k), which is equal to 1.38×10^{-23} J/K (joules per kelvin).

Expressed in dB:

$$k = +10 \log_{10} (1.38 \times 10^{-23})$$

$$= \mathbf{-228.6\,dB\ (J/K)}$$

This is related to noise temperature, and for the purposes of this link budget is needed for the C/N calculation. (For a full explanation of its significance, refer to a communications engineering textbook.)

We also need to calculate the noise power in the bandwidth of the uplink signal:

$$B^{power} = +\ 10 \log_{10} \mathrm{B}\ (\mathrm{Hz})$$

where B is the full bandwidth occupancy of the transponder channel, which in this case is 36 MHz (36×10^6 Hz):

$$B^{power} = +10 \log_{10} (36 \times 10^6)\ (\mathrm{Hz})$$

$$= \mathbf{75.56\,dB\ (Hz)}$$

We can sum all the uplink losses, including free space attenuation (FSA), into one figure:

$$L_{up\text{-}total} = \mathrm{FSA} + L_{point} + L_{atmos}$$

$$= 207.25 + 0.5 + 0.6$$
$$= \mathbf{208.35\,dB}$$

So the uplink C/N is:

$$\left(\frac{C}{N}\right)_{up} = EIRP^{up} - L_{up-total} + \left(\frac{G}{T}\right)_{sat-up} - k - B^{power}\ (\mathrm{dB})$$

$$= 69.35 - 208.35 + 6 - (-228.6) - 75.56$$

$$= \mathbf{20.04\,dB}$$

The downlink we are going to use is also in London, just 6.4 km away from the uplink, so the same azimuth, elevation and slant range figures can be used as the difference in distance is so small. On the downlink, we will assume for convenience that atmospheric losses are the same – in practice, these losses would be smaller due to the lower frequency of the downlink. The downlink antenna is 9 m in diameter, with a G/T of +36 dB/K, and London lies within the 47.5 dBW EIRP saturated receive contour of the widebeam from EUTELSAT 2 Flight 4. However, the pointing loss at the downlink antenna is likely to be smaller, because of the use of an automated tracking system that is essential for larger antennas. So let us assume the pointing loss is now 0.3 dB, and as the downlink frequency is 11.5 GHz, the FSA has now altered to 205.4 dB, and thus:

$$L_{down\text{-}tot} = FSA + L_{down\text{-}point} + L_{atmos}$$

$$= 205.4 + 0.3 + 0.6$$

$$= \mathbf{206.3\,dB}$$

Previously, we have said that the IBO of 6 dB results in an OBO of 2 dB; therefore this has to be subtracted from the saturated EIRP of the satellite:

$$\left(\frac{C}{N}\right)_{down} = EIRP_{sat} - OBO - L_{down-tot} + \left(\frac{G}{T}\right)_{downlink} - k - B^{power}\ \text{(dB)}$$

$$= 47.5 - 2 - 206.3 + 36 - (-228.6) - 75.56$$

$$= \mathbf{28.24\,dB}$$

Looking at the uplink and downlink C/N ratios, it can be seen that the worse figure is on the uplink, and therefore this link is ‘uplink limited’ – no matter what improvements are made on the downlink, the overall quality is mainly unaffected.

The overall C/N **as expressed in power ratios** for the complete link path is:

$$\left(\frac{C}{N}\right)_{overall} = \frac{1}{\dfrac{1}{\left(\frac{C}{N}\right)_{up}} + \dfrac{1}{\left(\frac{C}{N}\right)_{down}}}$$

In dB, the expression looks like:

$$\left(\frac{C}{N}\right)_{overall} = -10 \log_{10}\left(10^{-\frac{\left(\frac{C}{N}\right)_{up}}{10}} + 10^{-\frac{\left(\frac{C}{N}\right)_{down}}{10}}\right) \text{ (dB)}$$

So substituting the figures we have derived:

$$\left(\frac{C}{N}\right)_{overall} = -10 \log_{10}\left(10^{-\frac{20.04}{10}} + 10^{-\frac{28.24}{10}}\right)$$

$$= -10 \log_{10} (10^{-2.004} + 10^{-2.824})$$

$$= \mathbf{19.42\ dB}$$

Finally, we want some measure of the quality of the video signal (generally speaking, if the video quality is satisfactory, so the audio will also be satisfactory).

FM theory gives an FM improvement factor (FMI) which adds to the overall C/N:

$$FMI = 10 \log_{10}\left[\frac{3}{2}\frac{(f_d)^2}{(f_m)^3} \times B\right] \text{dB}$$

where: f_d is the peak-to-peak deviation of modulation; f_m is the bandwidth of the video signal; and B is the channel bandwidth.

Hence, substituting with f_d of 25 MHz/V, f_m is 5 MHz (normally the video would be assumed to be 5.5 MHz, but the ITU-R recommendation is to use 5 MHz) and B is 36 MHz:

$$FMI = 10 \log_{10}\left[\frac{3}{2}\frac{(25 \times 10^6)^2}{(5 \times 10^6)^3} \times (36 \times 10^6)\right] \text{dB}$$

$$= \mathbf{24.3\ dB}$$

Add to this 2 dB improvement gained by using video pre- and de-emphasis, and also with PAL-I (the analogue TV standard being used), an 11.2 dB improvement when using a weighting network to measure the unified (luminance-weighted) S/N ratio.

$$\text{Unified S/N} = \left(\frac{C}{N}\right)_{overall} + FMI + 2 + 11.2 \text{ (dB)}$$

$$= 19.42 + 24.3 + 2 + 11.2$$

$$= \mathbf{56.92\ dB}$$

This is a perfectly adequate quality link, acceptable for both 'news' and 'non-news' events (anything over 52 dB luminance-weighted received at the MCR is considered 'good quality'). Note that this is only true if the effects of interference can be ignored. The satellite operator should be able to provide guidance on what allowance should be made for this: usually (not always) the frequency plan for analogue services is managed so that the overall carrier-to-noise-plus-interference is no more than 1 dB worse than the carrier-to-noise figure alone.

We haven't as yet mentioned the impact of rain on the link budget. The reason to plan for such high S/N ratios is to have enough margin to cope with rain-fading. This link has about 5 dB difference between the calculated S/N and the desired target of 52 dB which means that it should meet the desired performance reliably, as 5 dB lost to rain does not happen very often (but it *can* happen). In many real-world cases (particularly SNG), you have to accept less margin and carry a greater risk of failure.

Digital link budget

Now we are going to calculate a digital link budget from the same location. This is very much a theoretical calculation, as to aid understanding, we are not going to take into account the change in IBO which is necessary with multiple digital carriers in the same transponder. Note that unless otherwise stated, all parameters are as for the analogue link budget.

We are going to uplink an 8.448 Mbps signal. For the digital link budget we need to know how much of this is carrying useful information, the so-called information rate (IR). In this case, the 8.448 Mbps includes Reed-Solomon error-protection code which limits the usable information carried to 188/204ths of the total, so that the IR = 7.785 Mbps.

The overall FEC rate in a system with concatenated coding such as the Reed-Solomon code mentioned above, followed by the convolutional coding used in the satellite modem, is the product of the two coding rates. In our example, we have an RS code of (204,188) followed by 3/4 convolutional coding. Therefore the overall FEC rate is:

$$FEC = \frac{3}{4} \times \frac{188}{204} = 0.6912$$

The symbol rate (SR) is:

$$SR = \frac{0.5 \times IR}{FEC}$$

$$= \frac{0.5 \times (7.785 \times 10^6)}{0.6912}$$

$$= \mathbf{5.632\ Mbps}$$

We shall assume that the noise bandwidth (NBW) of the digital signal is a bandwidth equal to the symbol rate, i.e. 5.632 MHz.

The IPFD at the satellite to cause saturation from its +6 dB/K contour is given by:

$$\text{Saturated IPFD} = -82 - 6.5\,\text{dBW/m}^2$$

$$= \mathbf{-88.5\,dBW/m^2}$$

However, an IBO of 12 dB is required to achieve a more linear operation of the transponder, which is needed for digital carriers to prevent interference between the multiple carriers that would normally share a transponder. This results in an OBO of 7 dB.

$$\text{Required IPFD} = -88.5 - 12$$

$$= \mathbf{-100.5\,dBW/m^2}$$

Uplink EIRP required ($EIRP^{up}$) = IPFD + spreading loss + atmospheric loss + pointing loss

$$EIRP^{up} = \text{IPFD} + L_{spread} + L_{atmos} + L_{point}$$

$$= -100.5 + 162.72 + 0.6 + 0.5$$

$$= \mathbf{+\,63.32\,dBW}$$

We also need to calculate the noise power in the bandwidth of the uplink signal, which is:

$$B^{power} = 10\log_{10}\left(0.5\,\frac{IR}{FEC}\right)(\text{dB}) = 10\log_{10} NBW\ (\text{dB})$$

$$= 10\log_{10}\left(0.5\,\frac{7.785\times10^6}{0.69}\right)$$

$$= \mathbf{67.5\,dB}$$

Remembering uplink losses haven't changed, the uplink C/N is:

$$\left(\frac{C}{N}\right)_{up} = EIRP^{up} - L_{up-total} + \left(\frac{G}{T}\right)_{sat-up} - k - B^{power}\ (\text{dB})$$

$$= 63.32 - 208.35 + 6 - (-228.6) - 67.51$$

$$= \mathbf{22.07\ dB}$$

Previously, we have said that the IBO of 12 dB results in an OBO of 7 dB; therefore this OBO has to be subtracted from the saturated EIRP of the satellite:

$$\left(\frac{C}{N}\right)_{down} = EIRP_{sat} - OBO - L_{down-tot} + \left(\frac{G}{T}\right)_{downlink} - k - B^{power}\ (\text{dB})$$

$$= 47.5 - 7 - 206.3 + 36 - (-228.6) - 67.5$$

$$= \mathbf{31.3\,dB}$$

$$\left(\frac{C}{N}\right)_{overall} = -10\log_{10}\left(10^{-\frac{\left(\frac{C}{N}\right)_{up}}{10}} + 10^{-\frac{\left(\frac{C}{N}\right)_{down}}{10}}\right)\ (\text{dB})$$

So, substituting the figures, we have derived:

$$\left(\frac{C}{N}\right)_{overall} = -10\log_{10}\left(10^{-\frac{22.07}{10}} + 10^{-\frac{31.3}{10}}\right)$$

$$= -10\log_{10}(10^{-2.207} + 10^{-3.13})$$

$$= \mathbf{21.58\ dB}$$

Overall E_b/N_0 is:

$$\left(\frac{E_b}{N_0}\right)_{overall} = \left(\frac{C}{N}\right)_{\text{overall}} + 10\log_{10} NBW - 10\log_{10} IR$$

$$= 21.58 + 10\log_{10}(5.632\times10^6) - 10\log_{10}(7.785\times10^6)$$

$$= \mathbf{20.17\ dB}$$

This figure takes no account of any interference, either from other carriers within the transponder, or cross-polar interference (XPI), which can have a very much more drastic effect on a digital link than an analogue link. It should be noted that this could reduce this figure by as much as 10 dB. In the real world, a digital link budget has to be carried out by the satellite operator, who can take into account all the potential interference parameters on the satellite.

Note that the very strong concatenated error-protection codes used for this link exhibit a very rapid failure at an E_b/N_0 value of about 3–5 dB depending on equipment performance. During a rain-fade, the link will typically change from perfect reception to complete failure over a very narrow range of value of E_b/N_0. This differs from an analogue link where, even during severe rain-fades, a watchable if far from perfect picture can still be received.

Hopefully the link budgets shown here have given you some idea of the influence of the various basic parameters that are accounted for in such budgets. The calculations are very tedious, and some satellite operators make software tools available from their website (e.g INTELSAT's LST application) to carry out these calculations. There are a number of commercially available satellite link budget programs as well.

Appendix C
Digital video

ITU-R BT.601

	25 Hz	30 Hz
Frame rate	25	30
Field rate	50	60
Line scanning rate	15 625	15 750
Lines per frame	625	525
Luminance samples per line	864	858
Chrominance samples per line	432	429

Luminance sampling frequency = line scan rate × luminance samples per line

For 625/50 standard TV signals:
= 15 625 000 × 864
= 13.5 MHz

For 525/60 standard TV signals:
= 15 750 000 × 858
= 13.5 MHz

Chrominance sampling frequency = line scan rate × chrominance samples

For 625/50 standard TV signals:
= 15 625 000 × 432
= 6.75 MHz

For 525/60 standard TV signals:
= 15 750 000 × 429
= 6.75 MHz

Therefore, for both standards, the total (composite) sampling frequency
= 13.5 + 6.75 + 6.75 = 27 MHz.

Data rate is number of samples × sampling frequency:

8 bits = 8 × 27 MHz = 216 Mbps
10 bits = 10 × 27 MHz = 270 Mbps

Active video

Additionally, in both 525- and 625-line pictures, not all the lines contain 'active' picture information. Due to the use of interlace in TV, in 525/60 systems there are 39 lines per frame of primarily synchronization information for the television display (49 lines per frame in 625/50). These can be replaced in a digital system with a simpler timing signal. The number of active picture lines is therefore 486 lines in 525/60 and 576 in 625/50 systems, and for both systems, 720 samples/line for luminance and 360 samples/line for chrominance.

With 4:2:2, there are 8 luminance bits, and 8 bits for each chrominance sample which occurs at every other luminance sampling point in the horizontal plane, i.e. half the luminance sampling rate (but the same number of samples in the vertical plane); therefore the uncompressed bit-rates are:

$$\text{Bit-rate} = Y^{\text{sample}} + C_b^{\text{sample}} + C_r^{\text{sample}}$$

For 525/60:

= (720 pixels × 486 pixels × 8 bits × 30 fps) + (720 pixels × 244 pixels × 8 bits × 30 fps) + (720 pixels × 244 pixels × 8 bits × 30 fps)

= **168 Mbps**

For 625/50:

= (720 pixels × 576 pixels × 8 bits × 25 fps) + (720 pixels × 288 pixels × 8 bits × 25 fps) + (720 pixels × 288 pixels × 8 bits × 25 fps)

= **166 Mbps**

For 4:2:0, there are 8 luminance bits, and 8 bits for each chrominance sample which occurs at every other luminance sampling point in the horizontal plane and in between each luminance sample in the vertical plane; therefore the uncompressed bit-rates are:

For 525/60:

= (720 pixels × 486 pixels × 8 bits × 30 fps) + (360 pixels × 243 pixels × 8 bits × 30 fps) + (360 pixels × 243 pixels × 8 bits × 30 fps)

= **126 Mbps**

For 625/50:

= (720 pixels × 576 pixels × 8 bits × 25 fps) + (360 pixels × 288 pixels × 8 bits × 25 fps) + (360 pixels × 288 pixels × 8 bits × 25 fps)

= **124 Mbps**

Appendix D
INMARSAT certificate for transportation of equipment for SNG

INMARSAT has worked hard to ease the restrictions of trans-border use of some types of equipment for use on its system, and in 1997 the ITU ratified Recommendation ITU-R SNG.1152, which encouraged national administrations to allow the passage across their borders and the use of INMARSAT satphones for satellite newsgathering for radio. INMARSAT now issues this certificate in seven languages for carriage with INMARSAT satphones to ease the problems of importation into a country for use by newsgatherers.

Nevertheless, anyone considering taking an INMARSAT terminal to a foreign country is advised to contact the telecommunications licensing authority in the country they plan to visit, to get up-to-date information on any conditions attached to use of the terminal in the country visited. Failure to secure the appropriate approvals is likely at the very least to result in confiscation of the equipment at the point of entry to the country or even, in some countries, arrest on the grounds of suspected espionage.

Inmarsat Equipment Nominated for Satellite News Gathering (SNG)

MES ID Number ____________________

ITU Recommendation ITU-R SNG.1152 considers that the nature of Satellite News Gathering (SNG) requires that mobile earth stations (MESs) are used for short periods of time and for specific news gathering purposes and hence must be activated in an expedient manner.

The recommendation also recognises that:
- some equipment and services for SNG purposes are already used in Mobile Satellite Service (MSS) bands, e.g. Inmarsat Mobile Earth Stations (MES), under special license from administrations, and
- the requirement for expeditious permission to operate the equipment.

The Inmarsat MES ID Number indicated above fully meets Inmarsat Type Approval Procedures and those General Characteristics outlined in the Annex of ITU Recommendation ITU-R SNG.1152. The MES has been nominated by its owner for SNG in coordination with the undersigned Routing Organisation.

Рекомендация МСЭ МСЭ-Р SNG.1152 считает, что характер сбора новостей с помощью спутниковой связи (СНС) требует, чтобы подвижные земные станции (ПЗС) использовались в течение коротких промежутков времени и для определенных целей сбора новостей, а следовательно, должны включаться на обслуживание практически целесообразным способом.

Рекомендация также признает, что
- некоторое оборудование, а также некоторые службы в целях СНС уже используются в диапазонах подвижной спутниковой службы (ПСС), например, подвижные земные станции (ПЗС) системы связи "Инмарсат", по специальной лицензии от администраций и
- в соответствии с требованиями ускоренного получения разрешения на эксплуатацию оборудования.

Идентификационный номер ПЗС системы связи "Инмарсат" (указанный выше) полностью отвечает процедурам одобрения типа организации "Инмарсат" и общим характеристикам, изложенным в приложении к Рекомендации МСЭ МСЭ-Р SNG.1152. ПЗС была представлена владельцем для СНС в координации с нижеподписавшейся направляющей организацией.

La Recomendación UIT-R SNG.1152 estima que la naturaleza misma del Periodismo Electrónico por Satélite (SNG) exige que las estaciones terrenas móviles sean utilizadas durante cortos periodos de tiempo y para fines específicos de transmisión de noticias y que, por tanto, deben ser activadas de una manera apropiada.

Del mismo modo, dicha recomendación reconoce
- que se están ya utilizando algunos equipos y servicios en las bandas del servicio móvil por satélite (por ejemplo: las estaciones terrenas móviles de Inmarsat) para el SNG bajo licencia especial de las diferentes administraciones, y
- que también es necesario obtener permiso con rapidez para poner en marcha el equipo.

La ETM de Inmarsat, con Número de Identificación indicado más arriba, satisface rigurosamente los Procedimientos de Homologación de Inmarsat, así como todas las Características Generales descritas en el Anexo de la Recomendación UIT-R SNG.1152. El propietario de dicha ETM la ha propuesto para suministrar los servicios SNG, en coordinación con la Organización de Encaminamiento abajo firmante.

إن توصيات الإتحاد العالمي للإتصالات (ITU-R SNG. 1152) إذ تضع في إعتبارها أن طبيعة تجميع الأخبار عن طريق الأقمار الإصطناعية (SNG) تتطلب إستخدام المحطات المتنقلة (MESs) لفترات قصيرة من الزمن ولهذه الإهداف بشكل خاص، ولذلك يجب تجهيز هذه المحطات للخدمة في أسرع وقت ممكن.

كما وتأخذ هذه التوصيات بعين الإعتبار كل من الآتي:
أن بعض الأجهزة والخدمات المخصصة للإستخدام لهدف تجميع الأخبار عبر الأقمار الإصطناعية تستخدم حالياً ضمن نطاقات التردد المستخدمة لخدمات الإتصالات المتنقلة عبر الأقمار الإصطناعية بموجب رخصة خاصة تمنحها بعض الدوائر الرسمية، مثالاً على ذلك المحطات المتنقلة العاملة عبر أنظمة إنمارسات للإتصالات، وأنه لمن المطلوب الإسراع في عملية ترخيص إستخدام هذه المحطات.

تنسجم المحطة المتنقلة (ذات الرقم المبين اعلاه) مع المتطلبات الفنية والخطوات التي توصيها إنمارسات والمتعلقة بالموافقة النوعية لهذه المحطات ومع الخواص الفنية العامة المبينة في ملحق توصيات الإتحاد العالمي للإتصالات (ITU-R SNG. 1152). حيث وقع الإختيار على هذه المحطة المتنقلة من قبل مالكها للإستخدام لهدف تجميع الأخبار عبر الأقمار الإصطناعية بالتعاون مع منظمة ربط الإتصال (Routing Organisation) الموقعة أدناه.

Die ITU Empfehlung ITU-R SNG.1152 berücksichtigt, daß die Berichterstattung der Nachrichtenagenturen über Satelliten (Satellite News Gathering, SNG) es erfordert, Mobile Erde-Funkstellen (MES) für kurze Zeit und zu bestimmten Zwecken der Berichterstattung zu benutzen, und daher müssen diese in einer angemessenen Weise in Betrieb gesetzt werden.

Die Empfehlung nimmt auch zur Kenntnis
- daß einige Ausrüstungen und Dienste für SNG-Zwecke bereits mit spezieller Genehmigung der jeweiligen Verwaltungen in den Frequenzbereichen für Mobilfunkdienste über Satelliten betrieben werden, z.B. Mobile Inmarsat Erde-Funkstellen (MES),
- und von dem Erfordernis für eine schnelle Erteilung der Genehmigung zum Betrieb der Geräte.

Die Mobile Inmarsat Erde-Funkstelle (MES) mit der Registriernummer wie oben angegeben, erfüllt alle Anforderungen der Inmarsat Typmusterprüfung und weist die allgemeinen Merkmale auf, welche im Anhang zu der ITU Empfehlung ITU-R SNG.1152 zusammengefaßt sind. Die MES ist durch ihren Eigentümer in Absprache mit der unterzeichnenden Routing Organisation für SNG-Zwecke nominiert worden.

ITU 的 ITU-R SNG.1152 建议考虑到卫星新闻搜集（SNG）的性质要求移动终端（MESs）实际使用时间短，并用于突发的新闻搜集目的，因此必须能快捷地采取行动。

此建议也承认:
- 一些用于SNG 目的的设备和业务已经使用移动卫星业务频段（MSS），例如使用特殊执照管理的 Inmarsat 移动终端（MES）；
- 对迅速批准使用设备的要求。

Inmarsat MES ID 号码（同上）完全满足Inmarsat 类型批准程序和ITU 的ITU-R SNG.1152 建议的附件中所述的一般特性，此MES 已被它的拥有者在与以下签字的经办机构协调后而指定为SNG 的用途。

La Recommandation UIT-R SNG.1152 de l'UIT considère que par sa nature même, le reportage d'actualités par satellite (SNG) fait intervenir des stations terriennes mobiles (STM) qui sont utilisées pendant de courtes périodes, à des fins de reportages d'actualités spécifiques, et qui doivent donc être mises en service rapidement.

Cettre recommandation reconnaît également :
- que certains équipements et services de SNG sont déjà utilisés dans les bandes du service mobile par satellite (SMS), par exemple les stations terriennes mobiles Inmarsat, sous licences spéciales accordées par certaines administrations, et
- la nécessité d'obtenir promptement l'autorisation d'exploiter l'équipement.

La STM Inmarsat, dont le numéro d'identification est indiqué ci-dessus, satisfait totalement aux procédures d'homologation d'Inmarsat et présente les caractéristiques générales résumées à l'annexe de la Recommandation UIT-R SNG.1152 de l'UIT. La STM a été désignée par son propriétaire pour des activités de SNG, en collaboration avec l'organisme d'acheminement soussigné.

Date ____________________ **Signature** ____________________

Routing Organisation ____________________

Appendix E

ITU emission codes for the classification of transmissions

The ITU has formulated an internationally recognized code for electromagnetic emissions, which are frequently used on all types of applications for SNG transmissions, including when applying for a telecommunications licence or permission to operate. Transmissions are designated according to their bandwidth and signal classification.

The basic characteristics are as follows.

- First symbol: type of modulation of the main carrier.
- Second symbol: nature of signal(s) modulating the main carrier.
- Third symbol: type of information to be transmitted.

The following two characters are optional.

- Fourth symbol: details of signal(s).
- Fifth symbol: nature of multiplexing.

Necessary bandwidth

The full designation of a transmission and the necessary bandwidth, indicated in four characters, is added just before the classification symbols. The letter occupies the position of the decimal point and represents the unit of bandwidth.

- Between 0.001 and 999 Hz shall be expressed in Hz (letter H): 0.1 Hz = H100; 400 Hz = 400H.
- Between 1.00 and 999 kHz shall be expressed in kHz (letter K): 6 kHz = 06K0; 12.5 kHz = 12K5.
- Between 1.00 and 999 MHz shall be expressed in MHz (letter M): 8 MHz = 08M0; 30 MHz = 30M0.
- Between 1.00 and 999 GHz shall be expressed in GHz (letter G): 5.95 GHz = 5G95; 14.25 GHz = 14G0.

Format of code

Number	Number	Letter	Number	Modulation	Signal	Information	Details	Multiplexing
[	BANDWIDTH		]					

Examples

Analogue SNG transmission: 30M0F3WWN

- 30M0 describes the signal occupying a 30 MHz bandwidth.
- F describes frequency modulation.
- 3 describes that it is a single channel containing analogue information.
- W describes that the signal has a combination of video and sound components.
- W describes that the detail of the signal is a combination of several elements.
- N defines no multiplexing.

Digital SNG transmission with comms carrier: 08M0G7WWN

- 08M0 describes the signal occupying an 8 MHz bandwidth.
- G describes phase modulation (QPSK).
- 7 describes that the signal carries two or more channels containing digital information (this signal has a main programme carrier and a comms carrier).
- W describes that the signal has a combination of video and sound components.
- W describes that the signal is complex, with a combination of several elements.
- N defines no multiplexing.

First symbol: modulation of the main carrier

N Unmodulated carrier
A Amplitude-modulated – double sideband
H Amplitude-modulated – single sideband, full carrier
R Amplitude-modulated – single sideband, reduced or variable level carrier
J Amplitude-modulated – single sideband, suppressed carrier
B Amplitude-modulated – independent sidebands
C Amplitude-modulated – vestigial sideband
F Frequency modulation
G Phase modulation
D Transmission in which the main carrier is amplitude- and angle-modulated either simultaneously or in a pre-established sequence
P Sequence of unmodulated pulses
K Pulse amplitude modulation
L Pulse modulated in width/duration

M Pulse modulated in position/phase
Q Pulse modulation in which the carrier is angle-modulated during the period of the pulse
V Pulse modulation which is a combination of the foregoing or is produced by other means
W Pulse modulation – combinations of amplitude, angle and pulse modulation
X Cases not otherwise covered

Second symbol: nature of signal(s) modulating the main carrier

0 No modulating signal
1 A single channel containing quantized or digital information without the use of a modulating sub-carrier
2 A single channel containing quantized or digital information with the use of a modulating sub-carrier
3 A single channel containing analogue information
7 Two or more channels containing quantized or digital information
8 Two or more channels containing analogue information
9 Composite system with one or more channels containing quantized or digital information, together with one or more channels containing analogue information
X Cases not otherwise covered

Third symbol: type of information to be transmitted

N No information transmitted
A Telegraphy – for aural reception
B Telegraphy – for automatic reception
C Facsimile
D Data transmission, telemetry, remote command
E Telephony (including sound broadcasting)
F Television (video)
W Combination of the above
X Cases not otherwise covered

Fourth symbol: details of signal(s)

A Two-condition code with elements of differing numbers and/or durations
B Two-condition code with elements of the same number and duration without error-correction
C Two-condition code with elements of the same number and duration with error-correction
D Four-condition code in which each condition represents a signal element (or one or more bits)

E Multi-condition code in which each condition represents a signal element (of one or more bits)
F Multi-condition code in which each condition or combination of conditions represents a character
G Sound of broadcasting quality (monophonic)
H Sound of broadcasting quality (stereophonic or quadraphonic)
J Sound of commercial quality
K Sound of commercial quality with the use of frequency inversion or band-splitting
L Sound of commercial quality with separate frequency-modulated signals to control the level of demodulated signal
M Monochrome
N Colour
W Combination of the above
X Cases not otherwise covered

Fifth symbol: nature of multiplexing

N None
C Code-division multiplex
F Frequency-division multiplex
T Time-division multiplex
W Combination of frequency-division multiplex and time-division multiplex
X Other types of multiplexing

For further information, see *ITU-R Radio Regulations*, Appendix S1 – Classification of emissions and necessary bandwidths.

Appendix F

Sample uplink registration forms for INTELSAT and EUTELSAT

In the following sample applications, the station News-TV in London is applying to INTELSAT for registration for its new compact SNV, and to EUTELSAT for registration of a new SNG flyaway system. (News-TV is a mythical company.)

The SNV has a 1 m circular antenna fed by a single 350 W HPA – as a transportable, it is classified as a Standard G system by INTELSAT (antenna size under 4.5 m).

The flyaway is a phase-combined system with 2 × 350 W HPAs feeding a 1.5 m diamond-shape antenna – this is classified as a Standard L system by EUTELSAT.

Both applications have been made on behalf of News-TV by British Telecom, the UK Signatory for both INTELSAT and EUTELSAT. In the UK, these applications could have been made via other entities, but News-TV has chosen to use BT.

It can be seen that the type of information required in each of these application processes is similar, and other satellite system operators typically ask for this or similar information.

The INTELSAT form is available from INTELSAT from its website (www.intelsat.int) in either Microsoft™ Word or Adobe Acrobat™ file format. It can also be supplied from INTELSAT by mail or fax on request.

The EUTELSAT form is available from the home country EUTELSAT Signatory or an authorized customer.

SSOG 200
Earth Station Registration

Instructions and Forms for Earth Station Registration

06 October 1998, Rev 3

Companion documents SSOG 210, 220 and IESS 207, 208, 410, and 601 may be obtained through Conference Affairs at the address listed on the back cover.

Planning Documents

101 SSOG References
102 Glossary of Terms
103 Operational Management
200 Earth Station Registration
210 Earth Station Verification
220 Type Approval
600 Transmission Plan

Line-up Procedures

306 TV/FM
307 TDMA
308 QPSK/IDR
309 IBS
310 TCM/IDR
311 DAMA
317 TDMA/DDI
403 Digital ESC
419 HDR
501 DCME

CONTENTS SUMMARY

FORMS

HELPFUL INFORMATION

Submit to: INTELSAT Sales Support
3400 International Drive NW
Washington, D.C 20008-3098
FAX +1 202 944 7005
TELEX 39-2707

Earth Station Registration Form (page 1 of 4)

To be submitted by the Signatory, DATE, or Authorized Direct Access Customer for the country in which the earth station is located.

Circle One ▶ (Signatory) DATE Direct Access Customer

Today's Date DD/Month/YYY
Sig./DATE/DAC Name
Country Registered
Sent by (individual)
Reference
Telephone
FAX
EMAIL

Details filled in here by Signatory, DATE, or DAC

Earth Station Name
Antenna Number
Country Located

Earth Station Operator NEWS-TV, London, UK
Telephone +44 207 123 4000
FAX +44 207 123 4001
EMAIL sng@news-tv.co.uk

24 Hour Remote Contact MCR, NEWS-TV, London, UK
Telephone +44 207 123 4100
FAX +44 207 123 4101
EMAIL mcr@news-tv.co.uk

If previously approved, provide the name of approval authority and designator code below. Earth Stations not previously approved by INTELSAT will need to provide certified test results.

Approval Authority
Designator Code

Contact Information
Owner NEWS-TV, London, UK
Telephone +44 207 123 4000
FAX +44 207 123 4001
EMAIL sng@news-tv.co.uk

Antenna Type circle all which apply
C Band — Circular / (Linear) / Switchable — (Ku Band) — Dual Band (C and Ku Band)

Circle all which apply:
C Band: A B D1 D2 E1 F2 F3 G H2 H3 H4
Ku Band: C E1 E2 E[illegible] (G) [illegible]2 K3

For identically configured earth stations, include this last page of this form.

Earth Station Registration Form (page 2 of 4)

Type Approved circle one (Yes) No

If Type Approved, complete items 1-4 in addition to the remainder of this form. Depending on the approval authority, complete item 1, or both items 2 and 3.

1. INTELSAT Type Approval # KU IA049AAA
2. Authority Name INTELSAT
3. Other Type Approval # None
4. Type Approved as (circle one) Antenna Model Antenna System (Earth Station)

Code	Approval Level	Tests Required
IANNNXOO	Antenna Model	G/T + Stability
IANNNXXO	Antenna System	Stability
IANNNXXX	(Earth Station)	No Tests

Attach or FAX a copy of the manufacturer's type approval certificate or shipping document for this earth station.

Antenna Information

Antenna Manufacturer Continental Microwave Ltd
Antenna Model # DST140
Feed Manufacturer Continental Microwave Ltd
Feed Model # DST140

Antenna Manufacturer's Specifications (circle one)

C Band	500	575	800
Ku Band	250	500	750

Provide Usable Frequency Ranges

HPA	13750	MHz	to	14500	MHz
LNA/LNB/LNC	10950	MHz	to	12750	MHz
Number to Feed Ports	2	Transmit		2	Receive

Antenna Shape

Circular	1.0	meters		
Rectangular		by		meters
Elliptical		by		meters
Diamond		by		meters

Antenna Feed Type

Circle which applies: (Offset) Center Fed Cassegrain Gregorian Other ______

Tracking System

Circle all which apply:

Autotrack	Manual
Monopulse	Handcrank
Step	(Drive Motors)
Step w/memory	Fixed
Program Only	Other

If other tracking system, describe:

Design Parameters

Transmit			Receive		
Transmit Axial Ratio	35 dB		LNA/LNB/LNC Noise Temperature		°K
Transmit Antenna Gain	42	dBi	Antenna Noise Temperature		°K
HPA Maximum Rated Power	350	W	Receive Antenna Gain	40.7	dBi
Maximum e.i.r.p.	67.5	dBW			

G/T=18.5 dB/K at 10 °Elevation Angle

Earth Station Registration Form (page 3 of 4)

To be submitted by the Signatory, DATE, or Authorized Direct Access Customer for the country in which the earth station is located.

Earth Station Geographical Information

Nearest Town TRANSPORTABLE

Altitude Above Sea Level

Transportable (yes/no) YES If transportable, no longitude or latitude coordinates are required.

Latitude (circle one) North/South degrees minutes seconds

Longitude (circle one) East/West degrees minutes seconds

*Usable geostationary arc °E to °E

*For example, 42° E to 165° E. or Satellite 310 ° E to 60 ° E.

Earth Station Type (Circle all which apply)

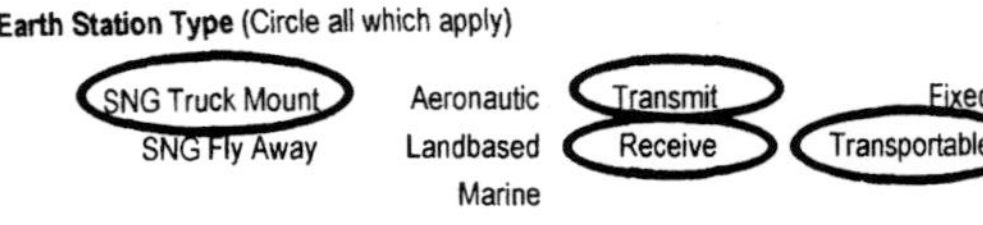

SNG Truck Mount Aeronautic Transmit Fixed

SNG Fly Away Landbased Receive Transportable

Marine

Earth Station Operation

Manned Full-time ✓

Manned Part-time ☐ If manned part-time, earth station must have remote control capability.

Remotely Controlled ☐

Service Information

Planned Operating Satellite Location °E

Circle all services that apply.

IDR	TDMA	Lease
VISTA	CFDM	Domestic
FDM	SCPC	International
DAMA	IBS	TCM/IDR
TV/Digital	Cable Restoration	TV/FM

Please attach any additional comments.

Requesting INTELSAT services for Verification Testing? Verification is required for all earth stations except Standard Gx antennas and some Type Approved antennas. See Section 2, *Special Earth Station Approvals*, for more information. Indicate below if you plan to use INTELSAT facilities for testing. Circle one YES NO

If "yes", complete "Request to Use INTELSAT Facilities for Verification Testing" form.

If "no" please note which facility you plan to use under below:

Planned Facility

Please list all other orbital or frequency constraints:

TYPE-APPROVED SYSTEM

Application Completed and Submitted by

Name A.SMITH

Title CHIEF ENGINEER

If this had been submitted to INTELSAT previously as a planned antenna, please indicate the planned antenna code.

Planned Antenna Code N/A

Identically Configured Earth Stations
Earth Station Application (VSATs & Type-approved) page 4 of 4

Identical earth stations are those of the same manufacture, model number and aperture size configured with the same RF equipment.

	Antenna Information				Location Information		
	Serial Numbers				Latitude	Longitude	Altitude Above
Name and Number	Antenna	Feed	LNA	City and Town	Deg:Min:Sec (N)orth/(S)outh	Deg:Min:Sec (E)ast/(W)est	Sea Level (Meters)
G00945G	102	0102	1251	TRANSPORTABLE			

Please note total of identically configured earth stations here 1

Please note total of attached pages here. 4

Today's Date (DD/Month/YYYY) 15 July 1999

APPLICATION FOR APPROVAL TO ACCESS THE EUTELSAT SPACE SEGMENT

To : Head of EUTELSAT Systems Operations Division

Applicant : British Telecom/Signatory Affairs
14 Farringdon St, LONDON EC4A 4DX

Date : 27 June 1999
Ref. :

1. GENERAL

1.1 Earth Station Name : SNG-07

- [ ] Mobile
- [ ] Fixed
- [X] Transportable

1.2 Standard :

T2	V1	S1	S2	S3	(L)	M	I1	I2	I3

- [X] Transmit
- [ ] Receive

1.3 EUTELSAT Type Approved [1] (if applies), Certificate N° :

> This particular type of antenna is actually type-approved, (EA-A001) but for the purposes of this form we have regarded it as unapproved.

1.4 Previous station code (if registered before with EUTELSAT) : N/A

1.5 Location Latitude : Deg. : .... Min. : .... Sec. : .... [ ] N
Longitude : Deg. : .... Min. : .... Sec. : .... [ ] E [ ] W

TRANSPORTABLE

Altitude : .... meter
Nearest Town : ....
Country : ....

1.6 Earth Station Address : **TRANSPORTABLE**

P.O. Box : .... Postal Code : ....
Town : .... Country : ....
Telephone : +.. Facsimile : + ...
Telex : +.. E - Mail : ....

1.7 If not manned 24h/day, state single point of contact : A. Smith, Chief Engineer
+44 207 123 4000

1.8 Operator Name : NEWS-TV Ltd.
Address : NEWS-TV House, Garland Street
P.O. Box : Postal Code : WC1 2XX
Town : LONDON Country : UK
Telephone : +44 207 123 4000 Facsimile : +44 207 123 4001
Telex : +.. E - Mail : sng@news-tv.co.uk

2. SERVICE

2.1 Planned Service Type (according to EUTELSAT Tariff Manual) : SNG TV

2.2 Planned Commencement and Period of Service : 1ST JULY 1999

[1] For type approved E/S only pages 1 and 4 need to be completed.

A1, may 1995

3. ANTENNA DATA

3.1 Manufacturer of main reflector : Continental Microwave Ltd

3.1.1 Model (if appl.) : SNG 140T

3.1.2 F/D : 0.7

3.2 Manufacturer of feed: Continental Microwave Ltd

3.3 Source of Performance Data

- [X] Manufacturer
- [X] Applicant measurements
- [ ] ESVA
- [ ] Other, please state ....

3.4 Main Reflector

- [ ] Circular — Diameter : .... m
- [X] Non. Circ. — Hor. Axis : 1.5 m; Ver. Axis : 1.6 m

3.5 Tracking

- [ ] Monopulse
- [ ] Monopulse (with memory)
- [ ] Program
- [ ] Step
- [ ] Step (with memory)
- [X] Manual
- [ ] Other (describe)
- [ ] None

3.5.a Slew Speed
AZ : ....°/s
EL. : ....°/s

3.6 Type

- [ ] Front fed
- [ ] Cassegrain
- [ ] Gregorian
- [X] Offset Front fed
- [ ] Offset Cassegrain
- [ ] Offset Gregorian
- [ ] Other

- [ ] Pointing accuracy .... °
- [ ] Depointing loss .... dB

3.7 Steerability :
- [ ] Limited over AZ West : ....° EL Min : ....°
- [X] Full AZ East : ....° EL Max : ....°

3.8 Feed
- [X] Tx Nr. Ports: 2
- [ ] Rx Nr. Ports: ....

3.9 Frequency Bands

		Frequency Bands	Gain	Tsys	G/T	EL
Rx	[X]	10.70-10.95	..............dBi at .... GHz	.........K	.... dB/K at	...........°
Rx	[X]	10.95-11.20	..............dBi at .... GHz	.........K	.... dB/K at	...........°
Rx	[X]	11.20-11.70	44.2 dBi at 12.5 GHz	.........K	.... dB/K at	...........°
Rx	[X]	11.70-12.50	..............dBi at .... GHz	.........K	.... dB/K at	...........°
Rx	[ ]	12.50-12.75	..............dBi at .... GHz	.........K	.... dB/K at	...........°
Tx	[ ]	12.75-13.00	..............dBi at .... GHz			
Tx	[ ]	13.00-13.25	..............dBi at .... GHz			
Tx	[ ]	13.75-14.00	..............dBi at .... GHz			
Tx	[X]	14.00-14.50	45.5 dBi at 14.5 GHz			
Tx	[ ]	17.30-18.10	..............dBi at .... GHz			

ANT Noise [K]	LNA Noise [K]	Feed Loss [dB]	
....	....	....	at .... GHz
....	....	....	at .... GHz
....	....	....	at .... GHz

3.10 Transmit Sidelobe Pattern

3 dB Beamwidth : 0.85 °
CCIR mask[1] 29-25logθ
Attach patterns @ 14, 14.25, 14.5 GHz YES

A1, may 1995 2

3.11 Transmit Polarisation Isolation	On axis	40	dB
	Worst case value[2]	35	dB
	@ Angle	0.2°	
	Attach patterns @ 14, 14.25, 14.5 GHz	YES	
3.12 Receive Sidelobe Pattern	3 dB Beamwidth :	....	°
	CCIR mask[1]	....	
3.13 Receive Polarisation Isolation	On axis	....	dB
	Worst case value[2]	....	dB
	@ Angle	....	°

4. TRANSMIT EQUIPMENT

4.1 HPA's

[X] TWTA	Number of units 2 Rating (Watt) 350	[X] Phase combined
[ ] Klystron	Number of units .... Rating (Watt)	[ ] Phase combined
[ ] SSA	Number of units .... Rating (Watt)	[ ] Phase combined

4.2 EIRP (in the direction of the satellite) : Maximum capability : 73.5 dBW
Overall RMS stability : +0.5 dB
-0.5 dB

4.3 If Up-Path Power Control (UPPC) available, give maximum range : +............................... dB

UPPC mechanism: [ ] Receive beacon level
[ ] Other

4.4 Up-Converters

	Number	Filters 3 dB Bandwidth	
12.75-13.00 GHz	....	....	MHz
13.00-13.25 GHz	....	....	MHz
13.75-14.00 GHz	....	....	MHz
14.00-14.50 GHz	1	*see below	MHz
17.30-18.10 GHz	....	....	MHz

*switchable to in accordance with EUTELSAT EESS-400

4.5 Local Oscillator

[X] Synthesizer, Step Size : 100 kHz
[ ] Crystal
Stability: ±......... kHz

5. RECEIVE EQUIPMENT **RECEIVER FOR LINE-UP AND MONITORING ONLY**

5.1 Down-Converters

	Number	Filters 3 dB Bandwidth	
10.70-10.95 GHz	....	....	MHz
10.95-11.20 GHz	....	....	MHz
11.20-11.70 GHz	....	....	MHz
11.70-12.50 GHz	....	....	MHz
12.50-12.75 GHz	....	....	MHz

5.2 Local Oscillator

[ ] Synthesizer, Step Size : ...... kHz
[ ] Crystal

---

1 S1-S2-S3-T1-T2 :32-25 log (θ) dBi for 1°<θ<48°, -10 dBi for θ>48°
L-M : 29-25 log (θ) dBi for 2.5°<θ<7.0°, 8dBi for 7.0°<θ<9.2°, 32-25 log (θ) dBi for θ>9.2°

2 Peak gain of the antenna sidelobe characteristics in the cross polarisation plane.

6. EARTH STATION VERIFICATION ASSISTANCE

6.1 Test requested : [ ] Yes [X] No

6.2 Requested Period :
Earliest start : .... /........./19.....
To be finished before :. ..../........./19

6.3 Tests to be conducted :

- [ ] Earth Station EIRP
- [ ] Transmit XPD
- [ ] Earth Station G/T
- [ ] Tx Sidelobe Pattern
- [ ] Rx Sidelobe Pattern
- [ ] Other (describe on separate page)

7. RECOMMENDED LINE-UP TESTS (according to service)

7.1 Test requested : [ ] Yes [X] No

7.2 Requested Period :
Earliest start : .... /........./19.....
To be finished before :. ..../........./19

8. DATA TO BE TREATED CONFIDENTIALLY

[ ] Yes [X] No

9. AGREEMENTS AND CERTIFICATION

9.1 Agreements

The applicant agrees with respect to the earth station of : SNG-07
for which he has submitted this application to be responsible to EUTELSAT for compliance with the requirements of the document of approval as specified by EUTELSAT.

Place : LONDON Signature

Date : 27TH June 1999

9.2 Certificate of Duly Authorised Telecommunications Entity (DATE) to Operating Agreement
(only applicable if the Applicant is not a Signatory to the EUTELSAT Operating Agreement)

This is to certify that the Government of : ....

has designated : ....

to act as the Duly Authorised Telecommunications Entity for the subject earth station

Place : .... Signature

Date : ....

N.B.

1. In the case that the Signatory (or DATE) applying is the entity to which an allotment is made, it must comply with all conditions and criteria contained in the approval document for the operation of the station.
2. If the earth station is operated under space segment capacity allocated by EUTELSAT to an entity other than the applicant, the latter shall inform this entity of the conditions contained in the document of approval related to this earth station, if and when required.

Appendix G
Sample INTELSAT SSOG 600 form (space segment booking request)

In the following sample applications, the UK Signatory – British Telecom in London – is applying to INTELSAT for approval of transmission plans, which are space segment booking requests, for a digital and an analogue transmission, with two different uplink and downlink stations.

These forms are available from INTELSAT from its website (www.intelsat.int) in either MICROSOFT™ Word or Adobe Acrobat™ file format. They can also be supplied from INTELSAT by mail or fax on request.

Satellite Systems Operations Guide **INTELSAT**

SSOG 600
Transmission Plan Approval

06 October 1998, Rev 3

This document is designed for use by INTELSAT Authorized Customers. To obtain information on becoming an authorized customer, please contact Sales and Marketing at the address listed on the back cover. Companion documents include: SSOG 200, *Earth Station Registration;* SSOG 210, *Antenna Verification; and IESS 410 and 601.*

Planning Documents

101 SSOG References
102 Glossary of Terms
103 Operational Management
200 Earth Station Registration
210 Earth Station Verification
220 Type Approval
600 Transmission Plan

Line-up Procedures

306 TV/FM
307 TDMA
308 QPSK/IDR
309 IBS
310 TCM/IDR
311 DAMA
317 TDMA/DDI
403 Digital ESC
419 HDR
501 DCME

CONTENTS SUMMARY

FORMS

HELPFUL INFORMATION

INTELSAT

Transmission Plan Form (Digital)

Facsimile: 1 202 944 7005 INTELSAT Sales Support/TOCC

3400 International Drive, NW Washington, DC 20008-3098 USA

To be completed by **Signatory, DATE, or Authorized Direct Access Customer** circle one

Today's Date Day 11 Month 10 Year 99 Satellite Location 342 °E SVO-L Number

Service Start Date Day 12 Month 10 Year 99 Transponder Tp 63/63 - INTELSAT 705

Company Name: British Telecom Uplink Beam Spot 2 Downlink Beam Spot 2

Country United Kingdom Bandwidth (MHz) 9.0

Your Name J.Jones Telephone +44 207 1111 FAX +44 207 1112 Email jjones@bt.vbs.com

Description of Carriers, Proposed Frequency Assignments (If Available) and Desired Performance (Digital Service Description)

Transmit Earth Station[1,2]	Receive Earth Station[2]	Information Rate(kbps)	Overhead (kbps)	Modulation Type[3]	FEC Rate[4]	R-S Code[5]	IF Noise Bandwidth (kHz)	Minimum[6] Frequency Step (kHz)	Transmit[6] Frequency (MHz)	Receive[6] Frequency (MHz)	Clear Sky Performance[7] (dB) C/N	Eb/No
G00955G 65dBW	UK-45	8.448	None	QPSK/ MPEG2	3/4	204,188	5632	22.5	14-14.25GHz			10

1. Indicate the maximum e.i.r.p. of the transmit earth station if this may impact your service.
2. Indicate the INTELSAT earth station code or enter the site name and complete the Gx application on page 3 of this form.
3. e.g. QPSK, BPSK, etc., and Digital TV (MPEG2), SCPC, MCPC, TDMA, etc.
4. Forward Error Correction (e.g. 1/2, 2/3, 3/4, 7/8...)
5. Reed-Solomon Code [e.g.(219, 201, 9)]
6. Please indicate the minimum frequency step, or both the transmit and receive frequency.
7. The desired clear sky Eb/No (data rate) or C/N value should include the up-link and down-link system margins for: tracking error, rain, inclination, adjacent satellite interference, terrestrial microwave interference,and HPA intermodulation.

INTELSAT

Transmission Plan Form (Analog)

Facsimile: 1 202 944 7005 INTELSAT Sales Support/TOCC
3400 International Drive, NW Washington, DC 20008-3098 USA
To be completed by **Signatory, DATE, or Authorized Direct Access Customer** circle one

Today's Date Day 20 Month 10 Year 99
Service Start Date Day 20 Month 10 Year 99
Company Name: British Telecom
Country United Kingdom

Satellite Location °E 332.5 SVO-L Number
Transponder Tp 72/72 - INTELSAT 605
Uplink Beam E Spot Downlink Beam E Spot
Bandwidth (MHz) 72

Your Name J.Jones
Telephone +44 207 1111

FAX +44 207 1112
Email Address jjones@bt.vbs.com

Description of Carriers, Proposed Frequency Assignments (If Available) and Desired Performance (Analog Service Description)

Transmit Earth Station[1,2]	Receive Earth Station[2]	CXR	Occupied Bandwidth MHz	Modulation Type[3]	EDF[4]	Transmit[5] Frequency MHz optional	Receive[5] Frequency MHz optional	Clear Sky Performance[6] (dB) C/N	S/N
G00950G 70dBW	UK-42	40	27-30	TV/FM Panda Audio	4MHz	14.0 - 14.25GHz		12	

1. Indicate the maximum e.i.r.p. of the transmit earth station if this may impact your service.
2. Indicate the INTELSAT earth station code or enter the site name and complete page 3 of this form.
3. e.g. TV/FM, FDM/FM, SCPC/CFM, etc.
4. Energy Dispersal Frequency in MHz.
5. Please indicate the minimum frequency step, if different from 22.5 kHz
6. The desired clear sky Eb/No (data rate) or C/N value should include the up-link and down-link system margins for: tracking error, rain, inclination, adjacent satellite interference, terrestrial microwave interference, and HPA intermodulation.

Appendix H

A typical service order for booking SNG space segment

SATELLITE BOOKING FORM
(Occasional Capacity)

DATE: 8th October 1999 **FROM:** News-TV, LONDON

CONTACT: S. Cole **TEL:** +44 207 123 4004 **FAX:** +44 207 123 4005

TO: PanAmSat Global Scheduling Center

TEL: +1 404 244 2340 **FAX:** +1 404 244 2335

REQUEST ☑ **AMENDMENT** ☐ **CANCELLATION** ☐

OUR BOOKING REFERENCE: SNG 4789/99

A. (Date) 15th October 1999

B. (Times) Start: 2200 **End:** 2215 **(UTC)**

C. (Uplink Location): Pristina, Kosovo **(Uplink Registration):** UKI-3003

D. (Band): Ku **(Max. EIRP):** 65 **(dBW)**

E. (Downlink Location): London, UK **(Downlink Registration):** G-111

(G/T): 36 **(dB/K)**

F. (Transmission Type): Unilateral ☑ **Multilateral** ☐

☐ **Analogue**

625/50 PAL ☐ **525/60 NTSC** ☐ **625/50 SECAM** ☐

Sound Subcarrier(s): **(MHz)** **Pre-emphasis:**

☑ **Digital**

Symbol Rate: 5.632 **(MSps)** **Occupied B/W:** 7.209 **(MHz)**

FEC Rate: 3/4

G. (Satellite/Txp): PAS-4 Txp 6K Slot E **Pol:** V

H. (Comments):

Please send invoices to: Accounts Dept, News-TV, PO Box 123, LONDON, UK

Appendix I

Sample information typically required for a request for permission to operate in a foreign country

The amount of information required by a country can vary widely, and the list given below is likely to be the most that might have to be provided. Because of the sometimes sensitive nature of the issuing of permissions, if in doubt, it is probably best to provide as much information as possible. This should then reassure any authorizing entity that the applicant is not trying to subvert the government or in any way interfere in the politics of the country.

General details

- The reason for the importation of the uplink.
- The dates when the equipment will enter the country and depart, and the ports of entry and departure.
- Where the uplink will be operate: the town, city or region.
- The types of transmissions; for example, to cover transmissions of news material for a named event.
- *Named contacts in the country during operation: this is optional information.*

Technical information

- The satellite system registration number of the uplink.
- The satellite(s) to be used for transmissions.
- The ITU emission code.
- A description of the types of transmission: analogue or digital.

- The serial number of the uplink: as an uplink can be made up of a number of components, it is best to quote the serial numbers of the principal components, including the antenna and HPA.
- The maximum EIRP of the uplink.
- The international technical standard to which the uplink complies, e.g. ITU-R S.465 (formerly CCIR Rec. 465).

Additional information

Sometimes the authorizing entity in the country of operation may require a letter from the satellite system operator confirming that the uplink will be permitted access to the space segment.

Request the contact and address to which any fees have to be paid. These may have to be paid in advance of the authorization being given to temporarily import the uplink, or at the time of entry into the country.

Appendix J

Formulae for calculating non-ionizing radiation levels

Non-ionizing radiation is measured in a number of ways, but in practice for microwave equipment, the safety limits are often expressed as power density levels, in W/m^2 or mW/cm^2 as these can be easily measured.

The near-field region is the region in general proximity to an antenna in which the electric (E) and magnetic (H) fields do not have a substantially plane-wave character, but vary considerably from point to point. In other words, as the signals are not necessarily in phase at any single point, the sum of the signals does not add up in-phase.

The far-field region is that region of the field of an antenna where the angular field distribution is essentially independent of the distance from the antenna. In this region (the free space region), the field has a predominantly plane-wave character, i.e. a locally uniform distribution of E and H field strengths in planes transverse to the direction of propagation. The signals are essentially in-phase and therefore add up at any particular point.

Calculations can be made to predict power density levels around antennas. These equations have to a certain extent been developed empirically and are generally accurate in the far-field of an antenna but will over-predict power density in the near-field, where they could be used for making a 'worst-case' prediction. This is because the power density decreases inversely with the square of the distance ($1/D^2$).

$$S_{ff} = \frac{PG}{4\pi R^2} \tag{1}$$

where:

S_{ff} = far-field power density (W/m)
P = power input to the antenna (W)
G = antenna power gain (relative to an isotropic radiator)
R = distance from the surface of the antenna (m)

It should be noted that the antenna power gain G in Equation (1) is the numeric gain. Therefore, when the power gain is expressed in logarithmic terms (dB), the power figure in dB has to be converted:

$$G = 10^{dB/10} \tag{2}$$

For example, a typical SNG flyaway antenna, 1.9 m in diameter, has a logarithmic power gain of 45 dB, which is equal to a numeric gain of 31 623. If the antenna gain is not known, it can be calculated from the following equation (using actual or estimated value for aperture efficiency):

$$G = \frac{4\pi\eta A}{\lambda^2} \tag{3}$$

where:
η = aperture efficiency (typically 0.55–0.75)
G = antenna power gain (relative to an isotropic radiator)
λ = wavelength
A = antenna area

$$A = \pi r^2 \tag{4}$$

where r = radius of antenna.

In the near-field zone of the antenna, the power density reaches a maximum before it begins to decrease with distance. It can be calculated thus:

$$R_{nf} = \frac{D^2}{4\lambda} \tag{5}$$

where:
R_{nf} = extent of near-field
D = antenna diameter (cm)
λ = wavelength (cm)

The maximum power density of the near-field antenna on-axis zone is:

$$S_{nf} = \frac{16\eta P}{\pi D^2} \tag{6}$$

where:
S_{nf} = maximum near-field power density (W/m)
η = aperture efficiency (typically 0.55–0.75)
P = power input to the antenna (W)
D = antenna diameter (m)

In the far-field region, power is distributed in a series of maxima and minima as a function of the off-axis angle (defined by the antenna axis, the centre of the antenna and the specific point of interest). For constant phase, or uniform illumination over the aperture, the main beam will be the location of the greatest of these maxima. The on-axis power densities calculated from the above formulae represent the maximum exposure levels that the system can produce. Off-axis power densities will be considerably less.

For off-axis calculations in the near-field it can be assumed that, if the point of interest is at least one 'antenna diameter' distance from the centre of the main beam, the power density at that point would be at least a factor of 100 (20 dB) less than the value calculated for the equivalent distance in the main beam. For practical estimation of RF fields in the off-axis vicinity of aperture antennas, use of the antenna radiation pattern envelope can be useful.

Use of the gain obtained from the antenna radiation pattern envelope in simple far-field calculations will generally be adequate for estimating RF field levels in the surrounding environment, since the apparent aperture of the antenna is typically very small compared to its frontal area.

Appendix K
Useful contacts

The following is a directory of useful contacts. It is by no means an exhaustive list, but is intended as offering points from which the reader can further explore. The author infers no particular endorsement of any of the companies listed, and cannot be held in any way responsible for the quality of the services or products provided. All the companies and organizations listed have varying degrees of involvement in the SNG market. The information was correct at time of writing.

Satellite system operators

The following is a list of satellite system operators who cater for the SNG market. The author acknowledges the use of The Satellite Encyclopaedia (www.tbs-satellite.com) in compiling the list of contacts in this section.

ASIASAT
Asia Satellite Telecommunications Company Ltd, 23–24/F,East Exchange Tower, 38–40 Leighton Road, Causeway Bay, Hong Kong
Tel: +85 2 805 6666 **Fax:** +85 2 504 3875
Web site: www.asiasat.com.hk

ANIK
Telesat Canada, 1601 Telesat Court, Gloucester K1B 5P4, Ontario, Canada
Tel: +1 613 748 0123 **Fax:** +1 613 748 8712 **Email:** info@telesat.ca
Website: www.telesat.ca

APSTAR
Asia Pacific Telecommunication Satellites Company Ltd, Room 3111–2, 31st floor, One Pacific Plaza, 88 Queens way, Hong Kong
Tel: +86 2 526 2281 **Fax:** +86 2 522 0419
Website: www.apstar.com

ARABSAT
Arabsat, PO Box 1038, Riyadh 11431, Saudi Arabia
Tel: +966 1 4820000 **Fax:** +966 1 4887999 **Email:** arabsat@arab.net
Website: www.arabsat.com

EUTELSAT
Eutelsat S.A., 70 rue Balard, F-75502 Paris Cedex 15, France
Tel: +33 1 5398 4747 **Fax:** +33 1 5398 3700
Website: www.eutelsat.org

GE SPACENET
GE American Communications, Inc. 4 Research Way, Princeton, NJ 08540–6684, USA
Tel: +1 609 987 4000 **Fax:** +1 609 987 4517
Email: americom.info@capital.ge.com
Website: www.ge.com/capital/americom/

Hispasat
Hispasat S.A., Apartado PO Box 95.000, 28080 Madrid, Spain
Tel: +34 1 372 9000 **Fax:** +34 1 307 7705
Website: www.hispasat.es

INMARSAT
Inmarsat Ltd, 99 City Road, London EC1Y 1AX, UK
Tel: +44 207 728 1000 **Fax:** +44 207 728 1044
Website: www.inmarsat.org

INTELSAT
Intelsat Ltd, 3400 International Drive, Washington, DC, 20008–3098, USA
Tel: +1 202 944 6800 **Fax:** +1 202 944 7898 **Website:** www.intelsat.com

Loral Skynet
500 Hills Drive, PO Box 7018 Bedminster, NJ 07921, USA
Tel: +1908 470 2300 **Fax:** +1
Website: www.loralskynet.com

JSAT
Japan Satellite Systems Inc. (Kabushiki-gaisha Nihon Sateraito Sisutemuzu Toranomon)
17 Mori Bldg. 5th Floor, 1–26–5 Toranomon, Minato-ku, Tokyo 105, Japan
Tel: +81 3 5511 7778 **Fax:** +81 3 3597 0601
Website: www.jsat.net

OPTUS
Cable & Wireless Optus Systems Pty Ltd, 101 Miller Street, North Sydney, NSW 2060, Australia
Tel: +61 2 342 7800 **Fax:** +61 2 342 7100
Website: www.optus.net.au

PanAmSat
Panamsat Corporation, 20 Westport Road, Wilton, CT 06897, USA
Tel: +1 203 210 8000 **Fax:** +1 203 210 8001
Website: www.panamsat.com

Superbird
Space Communications Corp., 2–8, Higashi-shinagawa 2-chrome, Shinagawa-ku, Tokyo 140, Japan
Tel: +81 3 5462–1370 **Fax:** +81 3 5462 1391
Website: www.superbird.co.jp

Telenor
Telenor ASA, PO Box 6701, St. Olavs pl., Oslo 0130, Norway
Tel: +47 22 77 86 60 **Fax:** +47 22 77 79 80
Website: www.telenor.com

Thaicom
Shin Satellite PLC, 41/103 Rattanathibet Road, Nonthaburi 11000, Thailand
Tel: +662 591 0736 **Fax:** +662 591 0714
Website: www.thaicom.net

TURKSAT
Turk Telekom, Aydinlikevler, Ankara 06103, Turkey
Tel: +90 312 555 1000 **Fax**: +90 312 313 1919
Website: www.telekom.gov.tr

Standards bodies

CETECOM
CETECOM GmbH, Untertürkheimer Strasse 6 – 10, D-66117 Saarbruecken, Germany
Tel: +49 681 598 8435 **Fax:** +49 681 598 9075 **Email:** info@ict.cetecom.de
Website: www.cetecom.de

ETSI
European Technical Standards Institute, 650 route des Lucioles Sophia Antipolis 06921, France
Tel.: +33 4 92 94 42 00 **Fax:** +33 4 93 65 47 16
Website: www.etsi.org

FCC
Federal Communications Commission, 445 12th St. SW, Washington DC 20554, USA
Tel.: +1 202 418 2555 **Fax:** +1 202 418 0232 **Email:** fccinfo@fcc.gov
Website: www.fcc.gov

US Government Regulations (including FCC documents)
Website: www.frwebgate.access.gpo.gov

IEEE
Institute Of Electrical & Electronic Engineers, Inc., 445 Hoes Lane, Piscataway, NJ 08855–1331, USA
Tel: +1 732 981 0060 **Fax:** +1 732 981 9667
Website: www.ieee.org

ITU
International Telecommunications Union, Place des Nations, Geneva 20, CH-1211, Switzerland
Tel: +41 22 730 51 11 **Fax:** +41 22 733 7256
Email (Radiocommunication Sector): brmail@itu.int
Website: www.itu.int

SMPTE
Society of Motion Picture and Television Engineers, 595 West Hartsdale Avenue, White Plains, NY 10607, USA
Tel: +1 914 761 1100 **Fax:** +1 914 761 3115 **Email:** smpte@smpte.org
Website: www.smpte.org

Safety: Organizations

ICNIRP
International Commission for Non-Ionising Radiation Protection Committee; Scientific Secretary – Dipl.-Ing. R. Matthes
Bundesamt für Strahlenschutz, Institut für Strahlenhygiene, Ingolstadter Landstrasse 1, D-85764 Oberschleissheim, Germany
Tel: +49 89 31 60 32 37 **Fax:** +49 89 31 60 32 89 **Email:** rmatthes@icnirp.de

NRPB
National Radiological Protection Board, Chilton, Didcot, Oxon, OX11 0RQ, UK
Tel: +44 1235 831600 **Fax:** +44 1235 833891 **Email:** nrpb@nrpb.org.uk
Website: www.nrpb.org.uk

OSHA
US Department of Labor, Occupational Safety and Health Administration (OSHA), 200 Constitution Avenue NW, Washington, D.C. 20210, USA
Website: www.osha.gov

Safety: Equipment

Radiation Protection Meters

Narda
L-3 Communications Corp., 435 Moreland Road, Hauppauge, NY 11788, USA
Tel: +1 516 231 1390 **Fax:** +1 516 231 1711
Website: www.nardamicrowave.com/east/

Raham
General Microwave, 425 Smith Street, Farmingdale, NY 11735–1156, USA
Tel: +1 516 631 630 2020 **Fax:** +1 631 630 2066
Website: www.generalmicrowave.com

SNG Antenna Manufacturers

Advent Communications Ltd
Preston Hill House, Nashleigh Hill, Chesham, Bucks HP5 3HE, UK
Tel: + 44 1494 774400 **Fax:** + 44 1494 791127
Email: sales@advent-comm.co.uk
Website: www.advent-comm.co.uk

Andrew
Andrew Corporation, 10500 W. 153rd Street, Orland Park, IL 60462, USA
Tel: +1 708 349–3300 **Fax:** +1708 349–5444
Website: www.andrew.com

AMP-C3C
AMP-C3C, 8 avenue de l'Atlantique, Zone de Activité de Courtaboeuf, BP 143, 91944 Les Ulis Cedex A, France
Tel: +33 1 69 07 80 93 **Fax:** +33 1 64 46 65 21
Website: www.amp-c3c.com

AvL
AvL Technologies, PO Box 8247, Asheville, NC 28814, USA
Tel: +1 828 250 9950 **Fax:** +1 828 250 9938
Website: www.avltech.com

CML
Continental Microwave Ltd., 1 Crawley Green Road, Luton, Beds LU1 3LB, UK
Tel: +44 1582 424 233 **Fax:** +44 1582 455 273
Website: www.continental-microwave.co.uk

Comtech
Comtech Antenna Systems Inc., 3100 Communications Road, St. Cloud, Florida 34769, USA
Tel: +1 407 892 6111 Fax +1 407 892 0994
Website: www.comtechsystems.com

ERA
ERA Technology Ltd, Cleeve Road, Leatherhead, Surrey KT22 7SA, UK
Email: info@era.co.uk
Website: www.era.co.uk

IGP
Naarderpoort 2–10, 1411 MA Naarden, The Netherlands
Tel: +31 35 699 0333 **Fax:** +31 35 699 0345 **Email:** info@igp.net
Website: www.igp.net

Vertex
Vertex Communications Corporation, 2600 North Longview Street, Kilgore, TX 75662–6842, USA
Tel: +1 903 984 0555 **Fax:** +1 903 984 1826

SNG Systems Manufacturers/SNG System Integrators

Advent Communications Ltd
Preston Hill House, Nashleigh Hill, Chesham, Bucks HP5 3HE, UK
Tel: + 44 1494 774400 **Fax:** + 44 1494 791127
Email: sales@advent-comm.co.uk
Website: www.advent-comm.co.uk

AMP-C3C
AMP-C3C, 8 avenue de l'Atlantique, Zone de Activité de Courtaboeuf, BP 143, 91944 Les Ulis Cedex A, France
Tel: +33 1 69 07 80 93 **Fax:** +33 1 64 46 65 21
Website: www.amp-c3c.com

BAF
BAF Communications Corp., 316 Northstar Court, Sanford, FL 32771, USA
Tel: +1 407 324 8250 **Fax:** +1 407 328 0513 **Email:** info@bafsat.com
Website: www.bafcom.com

Bickford Broadcast Vehicles
4313 Walney Road, Chantilly, VA 20151, USA
Tel: +1 703 818 8666 **Fax:** +1 703 818 9090
Website: www.bickfordbroadcast.com

CiSET
CiSET International SA, Rue Saint Jacques 350, 5500 Dinant, Belgium
Tel: +32 82 213 377 **Fax:** +32 82 22 43 16 **Email:** marketing@ciset-int.com
Website: www.ciset-int.com

CML
Continental Microwave Ltd., 1 Crawley Green Road, Luton, Beds LU1 3LB, UK
Tel: +44 1582 424 233 **Fax:** +44 1582 455 273
Website: www.continental-microwave.co.uk

CommSystems
CommSystems Inc.,6440 Lusk Blvd., D-100, San Diego, CA 92121, USA
Tel: +1 858 824 0056 **Fax:** +1 858 824 0057 E-mail: info@comm-systems.com
Website: www.comm-systems.com

Dawcom
Dawcom Ltd., Unit 2, Tything Park, Arden Forest Industrial Estate, Alcester, Warwicks B49 6ES,UK
Tel: +44 1789 765850 **Fax:** +44 1789 765855 **Email:** sales@dawcom.com
Website: www.dawcom.com

Eclips
Eclips nv., Mercelislaan 8, B-2275 Lille, Belgium
Tel: +32 3 3117795 **Fax:** +32 3 3091053 E-mail: info@eclips.be
Website: www.eclips.be

E-N-G
E-N-G Mobile Systems, 2245 Via De Mercados, Concord, CA 94520, USA
Tel: +1 925 798 4060 **Fax:** +1 925 798 0152 **Email:** engsales@e-n-g.com
Website:www.e-n-g.com

Extel
Extel Engineering, Paraná 275 3° Of.6 (1017), Buenos Aires, Argentina
Tel: +54 11 4371 2371 **Fax:** +54 11 4371 3454 **Email:** extel@interlink.com.ar
Website: www.extel.com

Frontline
Frontline Communications Corp., 12770 44th Street North, Clearwater, FL 33762, USA
Tel: +1 727 573 0400 **Fax:** +1 727 571 3295 **Email:** corp@frontlinecomm.com
Website: www.frontlinecomm.com

Harris
Harris Corporation, 1025 West NASA Boulevard, Melbourne, FL 32919–001, USA
Tel: +1 321 727 9207
Website: www.broadcast.harris.com

Gerling
Gerling & Associates Inc., 138 Stelzer Court, Sunbury OH 43074, USA
Tel: +1 740 965 2888 **Fax:** +1 740 965 2898 **Email:** gerling@qn.net
Website: www.gerlinggroup.com

IDB Systems
IDB Systems, 3236 Skylane Drive, Carrollton, TX 75006, USA
Tel: +1 972 407 7700 **Fax:** +1 972 407 7787 **Email:** info@idbsystems.com
Website: www.worldcom.com/idbsystems

L3 Communications – Satellite networks
125 Kennedy Drive, Hauppauge, NY 11788. USA
Tel: +1 631 272 5600 **Fax:** +1 631 272 5500 **Email:** mktg@lnr.com
Website: www.l-3com.com/satellitenetworks

Megahertz
Megahertz Broadcast Systems Ltd., College Business Park, Coldhams Lane, Cambridge CB1 3HD, UK
Tel: +44 1223 414101 **Fax:** +44 1223 414102 **Email:** sales@megahertz.co.uk
Website: www.megahertz.co.uk

NEC
NEC Corporation, 7–1, Shiba 5-chome Minato-ku, Tokyo 108–8001, Japan
Tel: +81 3 3454 1111 **Fax:** +81 3 3798 1510 / 1511 / 1512
Website: www.nec.com *(or in Japanese)* www.nec.co.jp

Nera
Nera Networks AS, Kokstadveien 23, PO Box 7090, 5020 Bergen, Norway
Tel: +47 55 225 100 **Fax:** +47 55 225 299
Website: www.nera.no

ND SatCom
NDSatCom GmbH, An der B31, D-88039 Friedrichshafen, Germany
Tel.: +49 7545 9390 **Fax** : +49 7545 939 8780 **Email:** sales@ndsatcom.de
Website: www.ndsatcom.com

Protec
Protec AV, Am Ockenheimer Graben 39, 55411 Bingen, Postfach 1528, 55405 Bingen, Germany
Tel. +49 6721 1818 0 **Fax:** +49 6721 1818 20
Website: www.protecav.de

Shook
Skook Electronic USA, Inc., 18975 Marbach Lane, Bldg.200, San Antonio, TX 78266, USA
Tel: +1 210 651 5700 **Fax:** +1 210 651 5220
Website: www.shook-usa.com

Sonera
Sonera Ltd, Sonera Event, PO Box 576, 00051 Sonera, Finland
Tel: +358 2040 2143 **Fax:** +358 2040 2011

SWE-DISH
Swe-Dish Satellite Systems AB , Hälsingegatan 40, P.O Box 6495, S-113 82 Stockholm, Sweden
Tel: +46 8 5879 5000 **Fax:** +46 8 5879 5005 **Email:** sales@ swe-dish.com
Website: www.swe-dish.com

TEC
Television Engineering Corporation, 101 Industrial Drive, Sullivan, MO 63080, USA
Tel: +1 573.860.4700 **Fax:** +1 573.860.4600
Website: www.tvengineering.com

Tecsat
Tectelcom Téc. em Telecomunicações Ltda., Av. Tecsat, 401 Jd. Pôr do Sol, São José dos Campos SP CEP, 12240–420, Brazil
Tel.: + 55 12 331–0199 **Fax:** +55 12 331–8122 **Email:** tecsat@tecsat.com.br
Website: www.tecsat.com.br

Thomson
Thomson MultiMedia, 17 Rue du Petit Albi, B.P. 8244, 95801 Cergy-Pontoise, Cedex, France
Tel: + 33 1 34 20 70 00 **Fax:** + 33 1 34 20 70 47
Website: www.thomsonbroadcast.com

Vertex
VertexRSI , 2600 North Longview Street, Kilgore, TX 75662–6842
Tel: +1 903 984 0555 **Fax:** +1 903 984 1826
Website: www.rsicom.com

Wolf Coach
Wolf Coach Inc., 7 B Street, Auburn, MA 01501, USA
Tel: +1 508 791 1950 **Fax:** +1 508 799 2384 **Email:** sales@wolfcoach.com
Website: www.wolfcoach.com

MPEG-2 Equipment Manufacturers

BarcoNet
BarcoNet, Luipaard Straat 12, B-8500 Kortrjk, Belgium
Tel: +32 56 445 000 **Fax:** +32 56 445 010
Website: www.barconet.com

Harmonic
Harmonic Inc, 549 Baltic Way, Sunnyvale, CA 94089, USA
Tel: +1.408 542 2500
Website: www.harmonic.com

Tandberg Television*
Tandberg Television Ltd., Strategic Park, Comines Way, Hedge End, Southampton, Hants S030 4DA, UK
Tel: +44 2380 484000 **Fax:** +44 2380 484330
Website: www.tandbergtv.com
** formerly the hardware manufacturing division of NDS Ltd.*

Scientific Atlanta
Scientific-Atlanta, Inc., PO Box 6850 4344 Shackleford Road Norcross, GA 30091–6850, USA
Tel: +1 770.903.2530 **Fax:** +1 770.903.2591
Website: www.sciatl.com

Scopus Network Technologies Ltd.
Scopus Network Technologies Ltd, 5 Hatzoref St., PO Box 267, 58102 Holon, Israel
Tel: +972 3 557 6200 **Fax:** +972 3 557 6249 **Email:** info@scopus.co.il
Website: www.scopus.co.il

Thomson
Thomson Broadcast Systems, 17 Rue du Petit Albi, B.P. 8244, 95801 Cergy-Pontoise, Cedex, France
Tel: + 33 1 34 20 70 00 **Fax:** + 33 1 34 20 70 47 **Email:** marketing@thomsonbroadcast.com
Website: www.thomsonbroadcast.com

Tiernan
Tiernan Communications Inc. 6340 Sequence Drive, San Diego, California, CA 92121, USA
Tel: +1 858 458 1800 **Fax:** +1 858 657 5403
Website: www.tiernan.com

Wegener
Wegener Communications, 11350 Technology Circle, Duluth, GA 30097, USA
Tel: +1 770 814 4000 **Fax:** +1 770 623 0698 **Email:** info@wegener.com
Website: www.wegener.com

Modulator/Modem/Multiplexer Manufacturers

Advent Communications Ltd
Preston Hill House, Nashleigh Hill, Chesham, Bucks HP5 3HE, UK
Tel: + 44 1494 774400 **Fax:** + 44 1494 791127
Email: sales@advent-comm.co.uk
Website: www.advent-comm.co.uk

CML
Continental Microwave Ltd., 1 Crawley Green Road, Luton, Beds LU1 3LB, UK
Tel: +44 1582 424 233 **Fax:** +44 1582 455 273
Website: www.continental-microwave.co.uk

Comtech
Comtech Communications Corp., 4666 South Ash Avenue, Tempe, Arizona 85282, USA
Tel: +1 480 831 7501 **Fax:** +1 480 831 7563 **Email:** sales@comtechcom.com

Comtech EF Data
2114 West 7th St Tempe, AZ 85281, USA
Tel: +1 480 333 22000 **Fax:** +1 480 333 2161
Website: www.efdata.com

Newtec
Newtec Cy, Laarstraat 5, B-9100 Sint-Niklaas, Belgium
Tel: +32 3 780 65 00 **Fax:** +32 3 780 65 49
Website: www.newtec.be

Paradise
Paradise Datacom Ltd., 1 Wheaton Road, Witham, Essex, CM8 3TD, UK
Tel: +44 1376 515636 **Fax:** +44 1376 533764 **Email:** sales@paradise.co.uk
Website: www.paradise.co.uk

Radyne ComStream
Radyne ComStream Corp., 6340 Sequence Drive, San Diego, CA 92121, USA
Tel: +1 858 458 1800 **Fax:** +1 858 657 5403 Email
sales@radynecomstream.com
Website: www.radynecomstream.com

Vocality
Vocality International Ltd. Unit 1, Ramsden Grange, Hambledon Road, Godalming, Surrey GU7 1XQ, UK
Tel: +44 1483 861 999 **Fax:** +44 1483 861 888 **Email:** sales@vocality.com
Website: www.vocality.com

RF – Up-/Downconverters

Advent Communications Ltd
Preston Hill House, Nashleigh Hill, Chesham, Bucks HP5 3HE, UK
Tel: + 44 1494 774400 **Fax:** + 44 1494 791127
Email: sales@advent-comm.co.uk
Website: www.advent-comm.co.uk

CML
Continental Microwave Ltd., 1 Crawley Green Road, Luton, Beds LU1 3LB, UK
Tel: +44 1582 424 233 **Fax:** +44 1582 455 273
Website: www.continental-microwave.co.uk

Mitec
Mitec Telecom Inc., 9000 Trans-Canada Highway, Pointe Claire, Quebec H9R 5Z8, Canada
Tel: +1 514 694 9000 **Fax:** +1 514 694 3814
Website: www.mitectelcom.com

MITEQ
MITEQ Inc., 100 Davids Drive, Hauppauge, NY 11788, USA
Tel: +1 631 436 7400 **Fax:** +1 631 436 7430
Website: www.miteq.com

Peak
Peak Communications Ltd., Kirklees House, 22 West Park St, Brighouse, West Yorkshire HD6 1DU, UK
Tel:+44 1484 714200 **Fax:**+44 1484 723666 **Email:** sales@peakcom.co.uk
Website: www.peakcom.co.uk

Tripoint Global Communications
Electronic Products Division, 101 Round Hill Drive, Rockaway, NJ 07866, USA
Tel: +1 973 627 5981 **Fax:** +1 973 627 0932
Website: www.tripointglobal.com

RF – Components

Com Dev
Com Dev Wireless, 2435 North Central Express Way, Palisades Central 11 Suite 255 Richardson, Dallas TX 75080 USA
Tel: +1 972 231 4979 **Fax:** +1 972 231 7576 **Email:** sat-sales@phase.com
Website: www.comdev.ca

Mitec
Mitec Telecom Inc., 9000 Trans-Canada Highway, Pointe Claire, Quebec H9R 5Z8, Canada
Tel: +1 514 694 6666 **Fax:** +1 514 694 3814 **Website:** www.mitectelcom.com

RF – High Power Amplifiers

CPI Satcom
CPI Satcom Division, 811 Hansen Way, PO Box 51625, Palo Alto, CA 94303, USA
Tel: +1 415 846 2800 **Fax:** +1 415 424 1744
Website: www.cpii.com

Marconi Applied Technologies Ltd.,
106 Waterhouse Lane, Chelmsford, Essex, CM1 2QU, UK
Tel: +44 1245 493493 Fax +44 1245 492492
Website: www.marconitech.com

ETM
ETM Electromatic Inc., 35451 Dumbarton Court, Newark, CA 94560, USA
Tel: +1 510 797 1100 **Fax:**: +1 510 797 4358 **Email:** support@etm-inc.com
Website: www.etm-inc.com

MCL
MCL Inc., 501 South Woodcreek, Bolingbrook, IL 60440–4999, USA
Tel: +1 630 759 9500 **Fax:** +1 630 759 5018 www.mcl.com

Vertex
VertexRSI , 2600 North Longview Street, Kilgore, TX 75662–6842
Tel: +1 903 984 0555 **Fax:** +1 903 984 1826
Website: www.rsicom.com

Xicom
Xicom Technology , 3550 Bassett Street, Santa Clara, CA 95054, USA
Tel: +1 408 213 3000 **Fax:** +1 408 213 3001
Website: www.xicomtech.com

RF: Measurement instrumentation

The following are the principal suppliers of RF power and spectrum measurement equipment.

Agilent Technologies
395 Page Mill Road, PO Box 10395, Palo Alto, CA 94303, USA
Tel: +1 650 752 5000
Website: www.agilent.com

Tektronix
Tektronix Inc., 14200 SW Karl Brown Drive, PO Box 500 Beaverton, OR 97077, USA
Tel: (US only) 800–835 9433
Website: www.2.tek.com

Inmarsat Satphone Manufacturers

Globecomm
Globecomm Systems Inc., 45 Oser Avenue, Hauppauge, NY 11788–3816, USA
Tel: +1 516 231 9800 **Fax:** +1 516 231 1557
Website: www.globecommsystems.com

Glocom
Glocom Inc, 20010 Century Blvd, Germantown, MD 20874, USA
Tel: +1 301 916 2100 **Fax:** +1 301 916 9438
Website: www.glocom-us.com

IN-SNEC
IN-SNEC Groupe Intertechnique, 2 rue de Caen, Bretteville L'Orgueilleuse 14740, France
Tel: +33 2 31 29 49 49 **Fax:** +33 2 31 80 65 49 **Email:** insntsi@wanadoo.fr
Website: www.insnec.com

MTI
Mobile Telesystems Inc., 205 Perry Parkway, Suite 14, Gaithersburg, Maryland 20877, USA
Tel: +1 301 590 8500 **Fax:** +1 301 590 8558 **Website**: www.mti-usa.com

NEC
NEC Corporation, 4035 Ikebe-cho, Tsuzuki-ku, Yokohama 224–8555, Japan
Tel: +81 45 939 2538 **Fax:** +81 45 939 2230

Nera
Nera Satcom, Kokstadveien 25, PO Box 6010, 5020 Bergen, Norway
Tel: +47 55 225 100 **Fax:** +47 55 225 651
Website: www.nera.no

Ottercom
Ottercom Ltd, Shannon Way, Tewkesbury, Gloucestershire, GL20 8ND , UK
Tel: +44 (0)1684 290020 **Fax:** +44 (0)1684 295535
Website: www.otter.co.uk

OSS
OSS Satellite Systems Inc., 1 Brandywine Drive, Deer Park, NY 11729, USA
Tel: +1 631 586 0030 **Fax:** +1 631 586 7032
Website: www.oss-sat.com

STN Atlas
STN ATLAS Elektronik GmbH, Sebaldsbruecker Heerstrasse 235, Bremen D-28305, Germany
Tel: +49 421 457 0 **Fax:** +49 421 457 2900
Website: www.stn-atlas.de

Thrane & Thrane
Thrane & Thrane A/S, Lundtoftegardsvej 93D DK 2800 Lyngby, Denmark
Tel: +45 39 55 88 00 **Fax:** +45 39 55 88 88
Website: www.tt.dk

Inmarsat Equipment Specialists – Broadcast

7E
7E Communications Ltd, Swan House, 203 Swan Road, Hanworth, Feltham, Middx TW13 6LL, UK
Tel: +44 208 744 8500 **Fax:** +44 208 744 8501

Comrex
Comrex Corporation, 65 Nonset Path, Acton, MA 01720, USA
Tel: +1 978 263 1800 **Fax:** +1 978 635 0401 **Email:** info@comrex.com

Glensound
Glensound Electronics Ltd., 5+6 Brooks Place, Maidstone, Kent ME14 1HE, UK
Tel: +44 1622 753662 **Fax:** +44 1622 762330 **Email:** office@glensound.co.uk
Website: www.glensound.co.uk

Livewire
Livewire Digital Ltd, Units 14+15, First Quarter, Blenheim Road, Epsom, Surrey KT19 9QN, UK
Tel: +44 1372 731400 **Fax:** +44 1372 731420 Email enquiry@livewire.co.uk
Website: www.livewire.co.uk

SCOTTY
SCOTTY Tele-Transport Corp., Telslastrasse4 8074 Grambach, Austria
Tel: +43 316 409426–0 **Fax:** +43 316 409426–53 **Email:** scotty@scotty.co.at
Website: www.scotty.co.at

Miscellaneous Products

Cabling systems

Belden
Belden Wire & Cable Co., PO Box 1980, 2200 US Highway 27 South, Richmond, IN 47374–1980, USA
Tel: +1 765 983 5200 **Fax:** +1 765 983 5257
Website: www.belden.com
(Multi-core cable systems suited for ENG/SNG)

Camplex
Concept W Corp., 3302 West 6th Avenue, Emporia, KS 66801, USA
Tel: +1 316 342 7743 **Fax:** +1 316 342 7405 **Email:** sales@camplex.com
Website: www.camplex.com
(Video / audio multiplexers for cabling)

Telecast Fiber
Telecast Fiber Inc., 102 Grove Street, Worcester, MA 01605, USA
Tel: +1 508 754 4858 **Fax:** +1 508 752 1520 **Email:** sales@telecast-fiber.com
(ENG/SNG Fibre Optic Systems)
Website: www.telecast-fiber.com

Flight Cases

EDAK
EDAK AG, Rheinauerweg 17, Dachsen 8447, Switzerland
Tel: + 41 52 647 2111 **Fax:** + 41 52 647 2230 **Email:** edak@edak.ch
Website: www.edak-cases.com

Hardigg
Hardigg Industries Inc., 147 North Main Street, South Deerfield, MA 01373–0201, USA
Tel: +1 413 665 2163 **Fax:** +1 413 665 4801
Website: www.hardigg.com

Zero
APW Zero Cases Europe, Unit 5 Alpha Park, Bevan Way, Dartmouth Road, Smethick, West Midlands, B66 1BZ, UK
Tel: +44 121 558 2011 **Fax:** +44 121 565 2115
Email:ZERO.cases.europe.sales@dial.pipex.com
Website: www.zerocases.com

Satellite Link Software

Satmaster Pro
Arrowe Technical Services, 58 Forest Road, Heswall, Wirral CH60 5SW, UK
Tel: +44 151 342 4846 **Fax:** +44 151 342 5142 **Email:** www.arrowe.com

SATfinder
Design Publishers, 800 Siesta Way, Sonoma, CA 95476, USA
Tel: +1 707 939–9306 **Fax:** +1 707 939–9235 **Email:** design@satnews.com
Website: www.satnews.com
Satellite database and link budget software

AcuSat
Blue Ravine Software, PO Box 6477, Folsom, CA 95763–6477, USA
Website: www.acusat.com
Uplink pointing calculation software

SNG Training Courses

BBC
BBC Training & Development, Wood Norton, Evesham, Worcs, UK
Tel: +44 1386 420216 **Email:** woodnorton@bbc.co.uk
Website: www.bbctraining.co.uk

This is the BBC's own training college, and courses are run on SNG operations throughout the year – small class sizes maximise 'hands-on' experience of the facility's own SNG uplink equipment. Courses include Digital Satellite Communications *and* Operational Satellite Broadcasting

NAB
National Association of Broadcasters, 1771 N Street, NW Washington, DC 20036, USA
Tel: +1 202 429 5300, **Fax:** +1 202 775 3520
Website: www.nab.org
Satellite Uplink Training Course (annual 4 day course in October)

Trade Exhibitons

The following trade shows feature SNG equipment.

NAB
National Association of Broadcasters Annual Convention & Exhibition, Las Vegas, USA
Website: www.nab.org
Held annually in April in Las Vegas, USA

IBC
International Broadcasting Convention & exhibition, RAI Centre, Amsterdam, The Netherlands
International Broadcasting Convention, Aldwych House, 81 Aldwych, London WC2B 4EL, UK
Tel:+44 (0)207 611 7500 **Fax:**+44 (0)207 611 7530 **Email:** show@ibc.org
Website: www.ibc.org
Held annually in September in Amsterdam, The Netherlands

BroadcastAsia
Asia-Pacific Sound, Film & Video Exhibition & Conference, Singapore
Website: www.broadcast-asia.com
Held biennially in June in the Far East

Satellite <year> 200X
Satellite convention & exhibition, Washington DC, USA
Website: www.satellite200X.com
Held annually in Spring in Washington DC, USA.

Information sources

Publishers – Print

INMARSAT – see previous entry
Publishes Via Inmarsat – *specialist magazine on Inmarsat-related subjects*
Email: via.inmarsat@inmarsat.org

Intertec Publishing
PO Box 12901, Overland Park, KS 66282–2901, USA
Tel: +1 913 341 1300 **Fax:** +1 913 967–1898
Website: www.intertec.com
Publishers of Satellite Broadband, Broadcast Engineering, *and* World Broadcast Engineering

Satellite Broadband, 5680 Greenwood Plaza, Suite 100, Englewood, CO 80111, USA
Tel: +1 303 741 2901 **Fax:** +1 720 489 3253
Website:
www.thebroadbandspace.com

Phillips Satellite Group
Phillips International, Inc., 7811 Montrose Road, Potomac, MD 20854, USA
Tel: +1 301 340 2100
Website: www.satellitetoday.com
Publishers of Satellite News, Via Satellite, Satellite Industry Directory

TLA Publishing
TLA Publishing Ltd., 28 – 30 St. John's Square, London EC1M 4DN, UK
Tel: +44 207 426 8900 **Fax:** +44 207 426 8909 **Email:**
Move-IT@TLAgroup.com
Publishes MOVE-IT – a specialist magazine on mobile satcomms

Publishers – Electronic

American Journalism Review
Website:http://.ajr.newslink.org

MLESAT Publishing
MLE Inc., PO Box 159, Winter Beach, FL 32971,USA
Fax: +1 734 433 0935 **Email:** mlesat@hotmail.com
Website: www.mlesat.com
Publishers of satellite systems information – books, CD-ROMs, software, videotapes & study packages

The Satellite Encyclopedia
www.tbs-satellite.com
Publishers of satellite systems information on the web and by email newsletter, both on subscription

SpaceDaily Express
Website: www.spacedaily.com
Daily digest of space and satellite news

SatNews Online
Website: www.satnews.com

Miscellaneous interesting sources of information

US Government Regulations
Website: www.access.gpo.gov
includes FCC documents

ISOG
WBU Inter-Union Satellite Operations Group
Website: www.tvo.org/naba/isog.html

Lloyds Satellite Constellations
www.ee.surrey.ac.uk/Personal/L.Wood/constellations
satellite information

Mobile Satellite Users Association (MSUA)
Mobile Satellite Users Association, 1350 Beverly Road Suite 115–341, McLean, VA 22101, USA
Tel: / Fax: +1 302 664 1213
Website: www.msua.org

John Walker's 'Index Librorum Liberorum
Website: www.fourmilab.ch
interesting astronomical information

Glossary of terms

16-QAM Sixteen-state quadrature amplitude modulation.

29 – 25 $\log_{10}\theta$ Mathematical characteristic describing performance of an SNG uplink antenna to meet the 2° satellite spacing requirement.

2B+D Basic rate ISDN, with 2 × 64 kbps bearers and a 16 kbps data channel.

3G Third generation mobile system: digital cellular telephony platform for beyond 2000.

4:2:0 Describes a picture where the chrominance is sampled at half the rate relative to luminance, in both the horizontal and vertical planes.

4:2:2 Describes a picture where the chrominance is sampled at half the rate relative to luminance in the horizontal plane, but at the same rate as luminance in the vertical plane.

4:4:4 Describes a picture where the chrominance and luminance are sampled at the same rate in both the horizontal and vertical planes: effectively chrominance is given the same bandwidth as the luminance.

601 See ITU-R BT.601.

64-QAM Sixty-four-state quadrature amplitude modulation.

8-PSK Eight-phase shift key modulation.

A Ampere (or amp): measurement of electrical current.

ABC American Broadcasting Company: US network formed in 1945 from the Blue Network Company, which was originally NBC's Blue Network.

AC Alternating current: type of electrical current.

AC-3 See Dolby AC-3.

Access The general term for the ability of a telecommunications user to make use of a network.

ACTS Advanced Communication Technologies and Services (EU project).

ADC Analogue-to-digital conversion: process of converting analogue signals to a digital representation. DAC represents the reverse translation.

Ad hoc capacity Capacity available for bookings on a temporary basis; see Occasional capacity.

ADPCM Adaptive differential pulse code modulation.

Algorithm Refers to the computational code used to compress or decompress a signal.

AM Amplitude modulation.

Amplifier Device used to increase the power of an electronic signal.

Amplitude modulation Process where a baseband message signal modulates (alters) the amplitude and frequency of a high-frequency carrier signal, which is at a nominally fixed frequency, so that the carrier signal varies in amplitude.

Analogue Signal which can take on a continuous range of values between a minimum and a maximum value; method of transmitting information by continuously variable quantities, as opposed to digital transmission, which is characterized by discrete 'bits' of information in numerical steps.

ANIK Canadian domestic satellite system.

ANSI American National Standards Institute (US).

Antenna Device for transmitting and receiving radio waves; in SNG, the directional parabolic antenna used for satellite transmissions.

AOR Atlantic Ocean Region: describes coverage area of a satellite.

Aperture Cross-sectional area of a parabolic antenna.

Apogee The point in a satellite's orbit when it is furthest away from the Earth.

A-pol INTELSAT circular polarization definition.

Artifacts Imperfections in a digital signal caused by the compression process.

ASI Asynchronous serial interface: DVB compliant signal.

AT&T American Telephone and Telegraph (US).

ATLANTIC Advanced Television at Low bit-rate And Networked Transmission over Integrated Communication systems (EU ACTS project).

ATM Asynchronous transfer mode: division of digital signals into small packets and transmitted in small, fixed-size data 'cells'.

Atmospheric losses Losses caused by the travel of the signal through the atmosphere – encountered on both the uplink and the downlink.

Attenuation Loss in power of electromagnetic signals between transmission and reception points.

Attitude control Orientation of the satellite in relationship to the Earth and the Sun.

Audio sub-carrier An extra carrier typically between 5 and 8 MHz carrying audio information on top of a video carrier.

Automatic frequency control AFC: circuit which automatically controls the frequency of a signal.

Automatic gain control AGC: circuit which automatically controls the gain of an amplifier so that the output signal level is virtually constant for varying input signal levels.

Availability The percentage of time over a year that a satellite link will be received successfully. Not particularly applicable to SNG uplinks.

Azimuth Angle of rotation (horizontal) that a parabolic antenna must be rotated through to point to a specific satellite in a geosynchronous orbit; the compass bearing of the satellite in the horizontal plane from a point on the Earth's surface.

Backhaul Contribution circuit: traditionally a terrestrial communications path linking an earth station to a local switching network; now also refers to satellite contribution paths.

Back-off Process of reducing the input and output power levels of a travelling wave tube to obtain more linear operation and hence minimize the risk of distortion. Expressed in dB.

Bandpass filter Active or passive circuit which allows signals within the desired frequency band to pass but prevents signals outside this pass band from getting through.

Bandwidth Measure of spectrum (frequency) use or capacity; the amount of information that can be passed at a given time over a certain span of frequency.

BAPT Bundesamt für Post und Telekommunikation: former German licensing authority, replaced by RTP.

Baseband Basic direct output signal obtained directly from a camera or videotape recorder or other video source.

Basic rate ISDN ISDN service of 2 × 64 kbps bearers and a 16 kbps data control channel. Also referred to as 2B+D.

Baud Rate of data transmission based on the number of symbols transmitted per second; see Symbol.

BBC British Broadcasting Corporation: UK network formed in 1927 from the British Broadcasting Company established in 1922.

Beacon Low-power carrier transmitted by a satellite which supplies the satellite operations control centre on the ground with a means of monitoring telemetry data and tracking the satellite.

Beam Signal from uplink or satellite.

Beamwidth Angle or conical shape of the beam the antenna projects.

Bearer Generic term for a carrier signal.

BER Bit error rate: a measurement of the quality of a data link.

B-frame Bi-directional compressed frame, encoded as a combination of forward and backward motion prediction information. B-frame can only be encoded after processing the requisite previous and future frames. It can be considered as the average of two P-frames – one that has already been encoded and sent (i.e. in the past) and one that has been encoded but not yet sent (i.e. in the future) – used as predictors for the current input frame to be processed.

Binary See Digital.

Bird Colloquial term for satellite.

Bit Single digital unit of information.

Bit error rate See BER: the fraction of a sequence of message bits that are in error. A bit error rate of 10^{-6} means that there is an average of one error per million bits.

Bit-rate Speed of a digital transmission, measured in bits per second.

Bit-stream Continuous stream of data bits transmitted over a communication channel with no separators between the character group.

Bizphone Land-line, cellular or satellite phone used for satellite co-ordination at the SNG uplink.

Block downconverter Device used to convert the C- or Ku-band signal down to L-band.
Block A group of pixels in a compression process.
Blockiness Picture degradation where the macroblocks are seen.
Bonding Method of combining a number (n) of discrete data channels of data bandwidth d to provide the user with a single combined data channel of $n \times d$.
Boresight Main focus of an antenna; the direction of maximum gain of an antenna.
Box Area in which the satellite has to be maintained while in station-kept geostationary orbit; see Station-kept box.
B-pol INTELSAT circular polarization definition.
bps Bits per second.
BPSK Binary phase shift key: modulation technique.
BRI Basic rate ISDN; Basic Rate Interface (for ISDN).
Broadband Referring to a bandwidth greater than the baseband signal; a single path that has multiple independent carriers multiplexed onto it; a large bandwidth digital 'pipe'.
Broadbeam See Widebeam.
BSS Broadcast-Satellite Service: ITU service designation.
Buffer Digital memory store on the input and output of a compression device.
BZT Bundessamt für Zülassengun in Telekommunikation: former German national telecommunication standards body, now privatized as CETECOM GmbH.
C/N See Carrier-to-noise ratio.
CAA Civil Aviation Authority (UK).
Cable television System which receives transmissions from programme sources and distributes them to users (usually homes) via fibre-optic or cable, usually for a fee.
Capacity Satellite bandwidth or channel.
Carnet Customs document used for international transportation of goods.
Carrier; carrier frequency Continuous (usually high-frequency) electromagnetic wave which can be modulated by a baseband signal to carry information; main frequency on which a voice, data or video signal is sent.
Carrier-to-noise ratio C/N: the ratio of the received carrier power and the noise power in a given bandwidth, expressed in dB.
Cascading Concatenating connection of devices.
C-band Frequencies in the 4–6 GHz range, used both for terrestrial microwave links and satellite links.
CBS Columbia Broadcasting System: US network formed in 1928 by William Paley from the struggling United Independent Broadcasters company.
CCIF Comité Consultatif International des Fils à Grande Distance : International Telephone Consultative Committee; part of the ITU.
CCIR Comité Consultatif International des Communications Radio à Grande Distance: International Radio Consultative Committee; part of the ITU.
CCIR-601 Recommendation developed by the CCIR for the digitization of colour video signals; now ITU-R BT.601

CCIT Comité Consultatif International des Communications Telegraphes à Grande Distance: International Telegraph Consultative Committee; part of the ITU.

CCITT Comité Consultatif International des Communications Téléphoniques et Telegraphes à Grande Distance: International Consultative Telegraph and Telephone Committee; part of the ITU established in 1956 when the CCIF and the CCIT merged.

CDL Commercial driver's license (US).

CDMA Code division multiple access: refers to a multiple-access scheme where the transmission process uses spread-spectrum modulations and orthogonal codes to avoid the signals interfering with one another.

CENELEC Comite Europeen de Normalisation Electrotechnique (European Committee for Electrotechnical Standardisation).

Centre frequency Frequency of the centre of a channel on a satellite transponder.

CEPT Conference Europeane des Administrations des Postes et des Telecommunications.

CES Coast earth station: alternative term for LES in INMARSAT system.

CETECOM CETECOM GmbH: successor to BZT; privatized German telecommunications standards entity.

CFR Code of Federal Regulations (US).

Chain Transmission path; usually in an SNG uplink referring to the equipment that makes up the transmission channel.

Channel Specific frequency range in which a signal is transmitted in a satellite transponder.

Check-bits Data bits added for error detection.

Chip-set Set of integrated circuit devices which form the heart of a compression encoder.

Chrominance Colour signal.

CIF Common intermediate format: developed so that computerized video images could be shared from one computer to another. An image that is digitized to CIF format has a resolution of 352 × 288 or 352 × 240, which is essentially one-half of CCIR-601.

Circular polarization Geometric plane in which the electromagnetic signals are transmitted. Circular polarization can be either clockwise (left-hand circular polarization, or LHCP) or counter-clockwise (right-hand circular polarization, or RHCP).

Clamp Video processing circuit that removes the energy dispersal signal component from the video waveform.

Clarke Belt; Clarke Orbit Geostationary orbit; named in honour of Arthur C. Clarke.

Clean carrier Unmodulated carrier signal.

Cleanfeed Studio audio output transmitted to the remote site so that the reporter is able to conduct a two-way dialogue with the studio presenter. Also called IFB, mix-minus, return sound or return audio.

Cliff-edge effect Describes the fact that digital signals suffer from achieving either perfect results or nothing: as a digital signal passes a particular threshold, the signal fails catastrophically with no warning.

Clock Term for electronic timing signal or circuit.

CNN Cable News Network: US network.

Coax Coaxial cable, used for carrying signals at frequencies from video baseband up to IF.

Codec COder/DECoder: a device which converts an analogue signal into or from a digital signal.

Codeword Descriptor of a block of data.

Coding delay See Latency.

Coding order Order in which a group of pictures is coded in a digital compression process.

Coefficient Mathematical term for a number with a particular algebraic significance.

Co-located Where a number of satellites share the same geostationary orbital assignment.

Colour difference Signals representing the difference in brightness between the luminance and the relative colour signal.

Colour sub-carrier Sub-carrier that is added to the main video signal to convey the colour information.

Common carrier Regulated telecommunications company which will carry signals for a fee; common carriers include telephone companies as well as owners of communications satellites.

Companding COMpressing/exPANDING: a noise-reduction technique that applies compression at the transmitter and complementary expansion at the receiver.

Component video Video presented as three separate signals: luminance (Y) and two colour difference signals (C_b and C_r), with line and field timing synchronization pulses.

Composite video TV signal with multiplexed luminance and chrominance signals with line and field timing synchronization pulses.

Compression Process of removing redundant data to minimize the use of bandwidth, ideally without visibly or audibly degrading the programme.

COMSAT US Signatory to INTELSAT.

CONUS CONtiguous US: description of arrangement of satellite coverage beams to cover the whole of the US except Hawaii and Alaska.

Convolutional code Type of error correction code used in digital coding.

CPB Constrained parameters bit-stream: subset of MPEG-1 valid bit-streams.

CPC Central product classification (UN).

Cross-modulation Form of signal distortion in which modulation from one or more RF carrier(s) is imposed on another carrier.

Cross-polar discrimination XPD: measurement of immunity from interference from signals on the opposite polarization; describes the ratio of (wanted) signals on the desired polarization compared to the (unwanted) signals on the opposite polarization.

Cross-polar interference XPI: interference from signals in the same transponder (IPs) or adjacent transponders, and signals at similar frequencies but on the opposite polarization – related to XPD

Cross-strapping Process on-board a satellite where an uplinked signal in one frequency band is converted to a different frequency band for the downlink, e.g. C-band uplink to Ku-band downlink.

Cross-talk Unwanted leakage of signal from one channel to another.

CSC Centre for Satellite Control, EUTELSAT, Paris.

CTR Common Technical Regulation (EU).

CTR-030 EU standard defining operation of SNG systems in Europe; formerly TBR-030 and ETS 300 327.

CTU Central telemetry unit: central satellite on-board system, feeding back confirmation signals confirming that ground control instructions have been acted on.

DAB Digital audio broadcasting.

DAC Direct access customer: a customer who has been authorized by a Signatory or DATE to directly access the INTELSAT space segment.

DAMA Demand-assigned multiple access: highly efficient means of instantaneously assigning channels in a transponder according to immediate traffic demands.

Data communication Transfer of digital data by electronic or electrical means.

DATE Duly authorized telecommunications entity: a country which is not party to the INTELSAT Agreement may designate an entity to transact business with INTELSAT.

dB Decibel: logarithmic unit of measurement of electronic signals.

dBA Measure of sound pressure level.

dBi Power relative to an isotropic source expressed in decibels; the amplification factor ('gain') of an antenna with reference to the gain of a theoretical 'isotropic' antenna (the 'i' in dBi).

DBS Direct broadcasting by satellite (also DTH): service that uses satellites to broadcast multiple channels of television programming direct-to-home to small consumer antennas.

dBW Measurement of the ratio of the power to 1 W (0 dBW) expressed in decibels.

DC coefficient First value in a DCT 8 × 8 pixel sample block which describes the average brightness of the whole block.

DC Direct current.

DCE Data communications equipment; equivalent to the TA (Terminal Adapter) in ISDN.

DCT Discrete cosine transform: refers to the coding methodology used to reduce the number of bits for actual data compression.

Decibel dB: expression of ratios of two power levels in logarithmic form, to express either a gain or loss in power between the input and output devices.

Decoder Device that reconstructs an encrypted signal so that it can be clearly received.

Decompression Process of decoding information and reconstructing the data that was removed in the compression process. Both the compression and decompression processes are often referred to together overall as 'compression'.

Dedicated Satellite capacity committed for a particular purpose, e.g. DTH, business data services, occasional video services.

De-emphasis Reinstatement of a uniform baseband frequency response following demodulation; the inverse of pre-emphasis.

Delay Time taken for a signal from the uplink to travel through the satellite to the downlink; transmission delay for a single 'hop' satellite connection is approximately one-quarter of a second.

Demodulation; demodulator The inverse process of modulation to recreate the original baseband message signal; the device which extracts or demodulates the 'wanted' signals from the received carrier.

Demultiplex Process of recovering the individual messages from a single communications channel that carries several different messages simultaneously.

De-orbit Process of removing a satellite from its geostationary orbit; see Graveyard orbit.

Deviation Frequency range over which the carrier is altered to carry the information signal.

Differential coding Simplest form of compression, sending only the data that describes the difference between two frames.

Digital Information expressed as data; signal state expressed as on (1) and off (0) states; a binary signal.

Digitizing Process of converting a signal to digital form.

Direct-access agreement Agreement between an entity and INTELSAT or EUTELSAT allowing purchase of satellite capacity directly without involving a Signatory.

Discrete cosine transform Specific implementation of a mathematical process called a Fourier transform, involving trigonometry; the intra-frame coding technique that takes advantage of spatial redundancy, representing a sequence of numbers as a waveform.

Dish Colloquial term for a satellite antenna; also generic term for an SNG uplink operation.

Dolby AC-3 Proprietary multi-channel perceptual digital audio coding technique providing five full audio range channels in three front channels (left, centre and right) plus two surround channels, used for HDTV transmissions in the US.

Domsat Domestic satellite system.

DOT Department of Transportation (US).

Downlink Ground station for receiving satellite communications; communications satellite link to Earth.

Drive Information signal applied to an HPA or upconverter.

DTE Data terminal equipment; for example, an audio codec.

DTH Direct-to-home satellite broadcasts, also called direct broadcasting by satellite, where the satellite broadcasts a very high power so that signals can be received by small consumer antennas.

DTMF Dual-tone multi-frequency: in-band telephone signalling commonly known as 'touch-tone'.

DTT Digital terrestrial television.

DTV Digital television.

Dual path Ability to simultaneously transmit two signals from one uplink system to a single satellite.

Dual polarity Ability to transmit or receive on horizontal and vertical polarization planes simultaneously.

Dummy load Device that will absorb all the power from an HPA for test or standby purposes.

Duplex Simultaneous two-way signal; a mode in which there exists two-way communication between the users.

DVB Digital Video Broadcasting: the European-backed project to harmonize standards of digital video.

DVB-S Digital Video Broadcasting – Satellite.

Earth station Generic term for any transmitting or receiving station for satellite transmissions.

E_b/N_0 Energy per bit to noise density ratio.

ED See Energy dispersal.

EDF Energy dispersal frequency: typically ±2 MHz for analogue.

Edge of coverage Limit of a satellite's defined service area. In many cases, the EOC is defined as being 3 dB down from the signal level at beam centre, though reception may still be possible beyond the –3 dB point.

EESS EUTELSAT earth station standard.

Efficiency Property of an antenna: typically 55–70% for an SNG antenna.

EIRP Effective isotropic radiated power: defines the total system power in dBW; describes the strength of the signal leaving the satellite antenna or the uplink antenna, and is used in determining the C/N and S/N.

ELCB Earth leakage circuit breaker.

Elevation Upward tilt of an antenna measured in degrees, required to point accurately the antenna at the satellite; angle of tilt (inclination) of an uplink antenna; the angle of the direct path to the satellite above the horizon.

EMC Electromagnetic compatibility: generic term for a Directive of the EU covering electromagnetic interference (EMI).

EMI Electromagnetic interference.

Encoder Device used to digitally compress the video and audio signals for transmission.

Encryption Process of encoding or scrambling television signals so that unintended audiences are unable to view the signal.

Energy dispersal ED: signal added to analogue modulation to disperse energy nodes that occur coincident to television line and field synchronizing signals; in digital systems, pseudo-random signal added to reduce energy peaks.

ENG Electronic news gathering.

Entropy coding Process where values are looked up in a fixed table of 'variable length codes', where the most probable (frequent) occurrence of coefficient values is given a relatively short 'code word' and the least probable (infrequent) occurrence is given a relatively long one.

Entropy The useful information content of a signal.

EOL End of life – of a satellite.

Ephemeris data Data regularly published by satellite operators describing cyclical influences of gravity from the Sun and the Moon, as well as solar radiation. These all affect the positioning of the satellite.

ERC European Radiocommunications Committee: governed by the CEPT. Administers spectrum for Europe.

ERO European Radicommunications Office: agency of the ERC. A planning agency for pan-European frequency management.

Error correction Process where bits are added in the modulation of a digital signal to correct for errors in transmission.

ESA European Space Agency.

ESOG EUTELSAT System Operations Guide.

ETS 300 327 European technical standard for SNG, now superseded; see CTR-030.

ETS European telecommunication standard.

ETSI European Technical Standards Institute: established in 1988 on the joint initiative of the EU and the CEPT to produce technical standards for telecommunications across the EU.

Eureka International consortium of broadcasters, network operators, consumer electronic industries and research institutes that developed digital audio standards; included the Eureka 147 project which devised MUSICAM.

Eurobeam European widebeam.

EUTELSAT European Telecommunications Satellite Organisation.

Exciter Generic term for the combination of an analogue modulator and upconverter, which provides the drive signal to the input of an HPA.

Fade margin The difference between the calculated or actual performance of a link and the threshold of operation; see Threshold margin.

Far-field Zone of non-ionizing radiation.

FCC Federal Communications Commission (US): national telecommunications regulatory body.

FDMA Frequency division multiple access: refers to the use of multiple carriers within the same transponder where each uplink has been assigned a frequency slot and bandwidth, usually employed in conjunction with frequency modulation.

FEC Forward error correction: data correction signal added at the uplink to enhance concealment of errors that occur on the passage of the signal via the satellite to the downlink.

Feed Generic term for the feedhorn, or the feedhorn and the antenna; a generic term for a transmission.

Feeder link Backbone Earth-satellite links in contribution and distribution networks.

Feedhorn In SNG, the assembly on the end of the arm extending out from the antenna from which signals are transmitted into the focus of the antenna.

Fibre-optic Transmission process using mono-frequency sources of light modulated with a signal or group of signals transmitted through a glass fibre.

Field Partial scan of a frame; there are two fields in a single TV frame.

Figure of merit G/T: a measure of the performance of a downlink station expressed in units of dB/K, depending on the receive antenna and low noise amplifier combination G (in dBi), and the amount of receiving system noise T expressed as a temperature in kelvin (K). The higher the G/T number, the better the system.

Figure-of-eight orbit See Inclined orbit.

Firewire IEEE 1394: data transfer standard for consumer and semi-professional audio-visual devices.

Flange power The output power of the HPA delivered to the output port for delivery to the uplink antenna.

Flight A satellite series, as in EUTELSAT 2 Flight 2.

Flight-case Electronics transit case built for ruggedness, typically made from aluminium alloy or polypropylene, with internal shockproofing, to withstand the rigours of shipping by air.

Flux-gate compass Electronic compass.

Flyaway Transportable earth station used for SNG which can be broken down and transported in cases to location.

FM Frequency modulation.

Focal point Area toward which the antenna reflector directs a signal for transmission, or where the signal is concentrated for reception.

Footprint Arrangement of satellite coverage beams to cover a particular area; a map of the signal strength of a satellite transponder showing the EIRP contours of equal signal strengths as they cover the Earth's surface.

Forward error correction FEC: inner convolutional coding that adds unique codes to the digital signal at the source so errors can be detected and corrected at the destination.

fps Frames per second.

Frame A single TV picture is referred to as a 'frame' and there are two 'fields' in every frame.

Frequency The number of times that an electromagnetic wave goes through a complete cycle in one second. One cycle per second is referred to as 1 Hertz (Hz) and is the basic measurement of frequency.

Frequency band Specific span of electromagnetic spectrum.

Frequency co-ordination Process to analyse and eliminate frequency interference between different satellite systems or between terrestrial microwave systems and satellites.

Frequency modulation Process whereby a carrier signal is shifted up and down in frequency ('deviates') from its 'at rest' (centre) frequency in direct relationship to the amplitude of the baseband signal.

FRR First right of refusal: method of reserving satellite capacity before that capacity is available.

FSA; FSL Free space attenuation; free space loss: the ionospheric and atmospheric attenuation of the signal.

FSS Fixed-Satellite Service: ITU designation.

G.711 ITU-T specification for audio data coding using 64 kbps of bandwidth to provide 3 kHz telephone quality audio.

G.722 ITU-T specification for audio data coding using 64 kbps of bandwidth to provide 7.5 kHz audio over a 64 kbps circuit.

G.728 ITU-T specification for audio data coding using 16 kbps of bandwidth to provide 3 kHz audio of bandwidth.

G/T See figure of merit.

Gain Measure of amplification expressed in dB.

Gain step The gain setting of a transponder, which can be varied in discrete steps from the TT&C ground station according to operational requirements.

Galactic noise Thermal noise (interference) from space.

GAN Global Area Network: see M4.

GATS General Agreement on Trade and Services (UN).

GATT General Agreement on Tariffs and Trade (UN).

Gb Gigabyte: equivalent to one thousand million (1×10^9) bytes of information.

GEO Geostationary Earth Orbit.

Geostationary Earth orbit GEO: describes a special geosynchronous orbit which is symmetrical above the Equator such that the satellite appears to remain stationary relative to a location on the surface of the Earth.

Geosynchronous orbit Describes a circular orbit around the Earth, with an average distance from the centre of the Earth of about 42 000 km (26 000 miles) in which the satellite orbit has a period equal to the rotation period of the Earth.

Get-away The path towards the desired satellite from the uplink or downlink to obtain a clear look-angle.

GFI; GFCI Ground fault interrupt; ground fault circuit interrupter.

GHz Gigahertz: unit of frequency equal to one thousand million cycles per second, or 1×10^9 Hz (1 000 000 000 Hz).

Glitch Imperfections in a digital signal.

Global beam C-band satellite antenna pattern which effectively covers one-third of the globe, typically aimed at the centre of the Atlantic, Pacific and Indian Oceans.

GMDSS Global Maritime Distress and Safety System: operated by INMARSAT as a life-saving service.

GMPCS Global Mobile Personal Communication by Satellite.

Good-night Verbal confirmation between uplink and satellite control centre of end of a transmission.

GOP Group of pictures.
GPRS General packet radio services: wireless packet-data service; part of the third generation of digital cellular telephony (3G).
GPS Global Positioning System: satellite-based navigation system.
GR Guaranteed reservation: where a reservation will be accepted up to one year in advance on currently operational satellite capacity, and up to three years in advance for planned satellite capacity (INTELSAT).
Graveyard orbit Orbital positions at 105°W and 75°E where satellites are moved or allowed to drift to at the end of their operational lives; thus this space junk can then be monitored to reduce the risk of debris damaging any active spacecraft.
Ground station Generic term for satellite communications earth station.
Group of pictures (GOP) Digital picture compression sequence that contains a combination of I-, B- and P-frames.
GSM Global System for Mobile Communications: digital cellular telephony standard.
GTO Geostationary transfer orbit: see Transfer orbit.
Guard-band; guard-channel Separation of signals in a transponder to minimize the risk of interference with each other.
H.221 Framing protocol used as part of H.320 (ITU-T standard).
H.230 Control and indication protocol used as part of H.320 (ITU-T standard).
H.242 Communications procedure protocol used as part of H.320 (ITU-T standard).
H.261 Audio compression algorithm standard (ITU-T standard).
H.263+ Further refinement of H.320 to improve performance (ITU-T standard).
H.320 Specification for video-conferencing over ISDN (ITU-T standard).
Half-transponder Method of transmitting two TV signals within a single satellite transponder through the reduction of each TV signal's deviation and power level.
HDTV High definition television: television system with approximately twice the horizontal and twice the vertical resolution of current 525-line and 625-line systems, with component colour coding, picture aspect ratio of 16:9 and a frame rate of at least 24 Hz.
Headend Control centre located at the antenna site of a cable television system, for processing of received signals for transmission to the cable system subscribers.
Hemi-beam C-band beam that typically covers a continent, or straddles parts of two continents to provide inter-continental connectivity.
Hertz Hz: term for the basic measure of an electromagnetic wave completing a full oscillation from its positive to its negative pole and back again in what is known as a cycle; thus 1 Hertz is equal to one cycle per second.
HGV Heavy goods vehicle (UK).
HIVITS HIgh quality VIdeotelephone Television System: European project.
Hop A single Earth-satellite-Earth path.
Horizontal blanking interval Contains the horizontal timing and synchronization information for the television display.

Horn See Feedhorn.

Hot-spot Burn points on a waveguide or feed-assembly.

HPA High-power amplifier.

HSCSD High-speed circuit switched data: digital cellular data standard.

HSD High-speed data.

HSE Health and Safety Executive: UK safety agency.

HT High-tension: refers to voltages over a thousand volts, often found in HPAs.

Hz Unit of frequency, where 1 Hz describes one cycle per second; 1 kHz = 1000 Hz (1×10^3 Hz); 1 MHz = 1000 000 Hz (1×10^6 Hz); 1 GHz = 1000 000 000 Hz (1×10^9 Hz).

IAEA International Atomic Energy Agency.

IATA International Air Transport Association: international civil aviation industry body.

IBO Input back-off: see Back-off.

IBS INTELSAT Business Services: describes specific data service of INTELSAT.

IDR framing See Overhead framing.

IDR Intermediate data rate: refers to INTELSAT's general data services.

IEC International Electrotechnical Commission (European).

IEEE Institute of Electrical and Electronics Engineers (US).

IEEE 1394 See Firewire.

IESS INTELSAT earth station standard.

IESS-306 INTELSAT earth station standard for the modulation of analogue TV frequency-modulated carriers.

IESS-308 INTELSAT earth station standard for the modulation of QPSK-modulated digital carriers.

IF Intermediate frequency signal, between stages in an uplink or downlink transmission chain – typically at frequencies of 70 or 140 MHz, or in the L-band (at around 1000 MHz or 1 GHz).

IFB Interruptible foldback or interruptible feedback: allows a producer, director etc. to communicate with the reporter during a 'live' remote transmission.

I-frame Intra-compressed frame: only has compression applied within (intra) the frame, and has no dependence on any frames before or after.

IFRB International Frequency Registration Board: part of the ITU.

IGO Intergovernmental organization.

ILS Instrument landing system.

IMN INMARSAT mobile number.

IMO International Maritime Organisation: founder of INMARSAT.

IMT-2000 International mobile telecommunications for the second millenium: ITU equivalent to European UMTS.

Inclination Angle between the orbital plane of a satellite and the equatorial plane of the Earth.

Inclined orbit Satellite status near the end of its operational life, when North–South station-keeping manoeuvres are largely abandoned; this type of capacity is often used for SNG.

Information rate Fundamental video data rate.

INIRC International Non-Ionizing Radiation Committee.

Injection orbit See Transfer orbit.

In-line fed prime focus Antenna where the feedhorn is directly in front of the vertex of the antenna.

INMARSAT International Maritime Satellite Organisation: operates a network of satellites for international transmissions for all types of international mobile services including maritime, aeronautical and land mobile.

Inner code Error correction code; see Forward error correction.

Input back-off IBO: where the input signal to the HPA is reduced so that the TWT is operating on the linear part of its power transfer curve; expressed in dB.

INTELSAT International Telecommunications Satellite Organisation.

Interference Energy which interferes with the reception of desired signals.

Interlace In a television frame of two TV fields, where each TV field is offset by a line, and each pair of TV fields is sent in an alternate sequence of odd- and even-numbered lines in the field.

Interleaving Where consecutive bits are interleaved amongst other bits in the bit-stream over a certain period of time to improve ruggedness of data.

Intermodulation products IP: spurious signals generated typically in the HPA which cause interference to other adjacent signals.

Interoperability The ability of one manufacturer's equipment (e.g. an MPEG-2 encoder), compliant to an agreed standard, to operate in harmony with the same type of equipment from another manufacturer.

Interview mode Low-latency mode (typically less than 250 ms) on an MPEG-2 encoder.

Intra-frame coding Form of compression of looking for spectral and spatial redundancy inside each frame without reference to any other frame, and sending such frames in a continuous sequence.

Inverse DCT Reverse process to DCT coding.

Inverter Electrical powering device that produces 110/240 V from a 12/24 V source, e.g. a vehicle battery.

IOC INTELSAT Operations Center: master control operation for INTELSAT.

Ionizing radiation Electromagnetic radiation of X-rays, gamma radiation, neutrons and alpha particles, that can have sufficient energy to permanently alter cell structures.

IOR Indian Ocean Region: describes coverage area of a satellite.

IP Intermodulation product; Internet protocol.

IPFD Incidental power flux density: a measure of the illumination of the satellite by the uplink.

IRD Integrated receiver decoder: an integrated MPEG-2 digital receiver which combines a downconverter, a demodulator and a compression decoder.

IRPA International Radiation Protection Association.

ISDN Integrated services digital network: ITU-T term for integrated transmission of voice, video and data on the digital public telecommunications network.

ISO International Standards Organization.

ISOG Broadcasting industry lobby group of the WBU, which seeks to influence particularly in the areas of SNG and contribution circuits.

Isotropic antenna A hypothetical omni-directional point-source antenna that radiates equal electromagnetic power in all directions – the gain of an isotropic antenna is 0 dBi – that serves as an engineering reference for the measurement of antenna gain.

ITU International Telecommunications Union: undertakes global administration of spectrum. Founded in 1865 as the International Telegraph Union, it changed its name to the International Telecommunication Union in 1934.

ITU-D ITU development sector: provides technical assistance to developing countries.

ITU-R BT.601 Formerly CCIR-601; recommendation developed for the digitization of colour video signals.

ITU-R ITU radiocommunication sector: establishes global agreements on standards for international use of radio frequencies, broadcasting, radiocommunication and satellite transmission standards.

ITU-T ITU telecommunication sector: establishes global agreements on standards for telecommunications.

JPEG ISO Joint Picture Expert Group standard for the compression of still pictures that shrinks the amount of data necessary to represent digital images from 2:1 to 30:1, depending on image type.

K See Kelvin.

Ka-band Frequency range from 18 to 31 GHz; in satellite communications, the frequency band 17.75–19.50 GHz is used for transmission.

kbps Kilo-bits per second (1000 bits per second).

Kelvin K: the temperature measurement scale used in the scientific community, where 0 K represents absolute zero and corresponds to –273 degrees Celsius (–459 degrees Fahrenheit); thermal noise in an electronic system is measured in Kelvin.

kHz Kilohertz: Hz $\times$ 10^3 (1000 Hz).

Kilohertz See kHz.

kph Kilometres per hour.

Ku-band Frequency range from 10.5 to 17 GHz; in satellite communications, the frequency band 13.75–14.50 GHz for transmission and 10.7–12.75 GHz for reception.

kVA Kilovolt amperes: measurement of absolute maximum electrical power into an electrical load, which is in reality usually reduced due to the electrical characteristics of the load.

LAN Local area network (computers).

Latency Delay through a compression process.

Launcher Generic term for the feedhorn; vehicle – typically a rocket – used to carry a satellite into orbit.

Layer MPEG group of audio coding algorithms.

L-band In SNG terms, the frequency band 1.0–2.0 GHz.

Leasing The rental of satellite capacity for a fixed term, typically one week to five years.
Left-hand circular polarization See Circular polarization.
LEO Low Earth orbit: below 5000 km.
LES Land earth station (fixed); ground station; INMARSAT ground station.
Level In MPEG, the level defines the image resolution and luminance sampling rate, the number of video and audio layers supported for scaleable profiles, and the maximum bit-rate per profile.
LHCP See Circular polarization.
Licence Formal authority permitting radio transmissions, usually incurring a fee.
Linear polarization Geometric plane in which the electromagnetic signals are transmitted. Linear polarization is divided into vertical (X) and horizontal (Y) polarization.
Linearizer Part of an HPA design that effectively compensates for the non-linearity near the top of the power transfer characteristic.
Line-up Period of time before the beginning of a transmission when operational technical parameters are checked and confirmed with satellite control centre.
Link budget Calculation to ascertain the performance of a complete satellite link system.
Lip-synch error Where the audio is out of synchronism with the video – commonly seen on compressed transmissions due to system errors.
Live shot Short live transmission, typically for a brief news update.
Live stand-up See Stand-up.
LNA Low noise amplifier.
LNB See Low noise block-downconverter.
Look-angle Azimuth and elevation angles that an uplink or downlink has to be orientated along to point towards the desired satellite.
Loopback; Lookback Ability at the uplink to receive its own signal downlinked from the satellite for verification.
Lossless Compression process that creates a perfect copy of the original signal when it is decompressed at the end of the signal chain.
Lossy Compression process that relies on the fact that the human eye and ear can tolerate some loss of information yet still perceive that a picture or sound is of good or adequate quality.
Low-noise amplifier LNA: pre-amplifier between the antenna and the earth station receiver, located as near the antenna as possible for maximum effectiveness, usually attached directly to the antenna receive port. The LNA is especially designed to contribute the least amount of thermal noise to the received signal.
Low-noise block downconverter LNB: combination of a low-noise amplifier and downconverter built into one device attached to the feed; it amplifies the weak signal from the satellite and frequency shifts it down from Ku-band (or C-band) to L-band, as required at the input to the satellite receiver.
LPG Liquefied petroleum gas.
Luminance Brightness of a video picture.

M4 Multi-Media Mini-M: INMARSAT successor to Inmarsat-B System, offering smaller sized terminals with enhanced data capabilities. Also called Global Area Network.

MAC (A, B, C, D2) Multiplexed analogue component: colour video transmission system; subtypes refer to the various methods used to transmit audio and data signals.

Macroblock Number of 8 × 8 pixels luminance blocks, used as part of motion prediction in the digital compression process.

Magnetic North direction towards which a compass needle will ordinarily point when influenced by the magnetic field of the Earth.

Margin Amount of signal in dB by which the satellite system exceeds the minimum threshold required for operation.

Mask Antenna radiation performance pattern with respect to a particular specification, e.g. to meet 2° spacing of satellites.

Matrix Switching router that can connect any input source to any output destination.

MB Megabyte: 1 million bytes per second (1×10^6 bytes per second).

Mbaud Megabaud: 1 million symbols per second (1×10^6 symbols per second).

Mbps Mega-bits per second: 1 million bits per second (1×10^6 bits per second).

MCPC Multiple channels per carrier.

MCR Master control room.

MEO Medium Earth orbit: 10 000–20 000 km above the Earth.

MES Mobile earth station.

Metadata Programme content description information that is carried as an auxiliary data channel within the MPEG-2 signal.

MHz Megahertz: unit of frequency equal to 1 million cycles per second or 1×10^6 Hz (1000 000 Hz).

Microwave Refers to frequencies typically in the range of 2–30 GHz.

Mix-minus The transmission to the remote location of the output of the studio *minus* the contribution from the remote location; see Cleanfeed.

Modem MODulator/DEModulator: device that transmits and receives data.

Modulation rate Modulated symbol rate.

Modulation; modulator Process of manipulating the frequency or amplitude of a carrier in relation to a signal; device which combines a 'message' signal with a high-frequency carrier signal for transmission.

Motion JPEG Method of using JPEG coding techniques to compress and transmit television pictures.

Motion prediction; estimation Process of calculating the position of an object from one frame to the next and sending only the co-ordinates of the block of data that describes that object from one frame to the next by simply placing it in the new position.

Motion vector The difference co-ordinates of the block of data that describes an object that has moved in position between one frame to the next.

MoU Memorandum of understanding: interorganizational or intergovernmental agreement establishing fundamental principles and understandings.

Mount Mechanical assembly upon which the SNG antenna is mounted, complete with all necessary mechanical adjustment controls.

MP Main profile – MPEG.

MPE Maximum permissible exposure.

MPEG Motion Pictures Expert Group.

MPEG-1 Video compression standard for non-interlaced, computer-type data streams. Typical MPEG-1 video compression goes up to 100:1 for images composed of 352 pixels (picture elements) × 288 lines at a refresh rate of up to 30 frames per second.

MPEG-2 Standard covering the compression of data (coding and encoding) for digital television, derived from MPEG-1 and defined for interlaced broadcast TV. It provides improved picture quality, higher resolution and additional features such as scaleability.

MPEG-2 MP@HL MPEG-2 main profile at high level: the higher bit-rate system adopted to provide high-definition television in widescreen format.

ms Millisecond: one thousandth of a second (1×10^{-3})

MSD Medium speed data: usually for INMARSAT purposes; equivalent to a data rate of 9.6 kbps.

MSps Mega-symbols per second: measurement of data rate – 1000 000 symbols per second (1×10^{6} symbols per second).

MSS Mobile-Satellite Service: ITU designation.

Multilateral Transmission from one origin to a number of destinations simultaneously.

Multiple channels per carrier MCPC: where multiple programme signals are combined onto one carrier signal and uplinked.

Multiplexer Combines a number of separate signals into one signal for simultaneous transmission over a single circuit.

MUSICAM Masking pattern adapted Universal Sub-band Integrated Coding And Multiplexing: digital audio compression standard.

NABA; NANBA North American Broadcasters Association: formerly known as North American National Broadcasters Association – a trade association based in Canada.

NASA National Aeronautical Space Administration (US).

NBC National Broadcasting Company: US network formed in 1926 by RCA.

NBC Nuclear, biological, chemical: describes a method of protection used in warfare.

Near-field Zone of non-ionizing radiation.

Newsgathering Journalistic and technical process of gathering news for broadcast.

NGO Non-governmental organization.

Noise Unwanted and unmodulated energy that is always present to some extent within any signal; interference or unwanted signals.

Noise figure Figure of merit of a device, such as an LNA or receiver, expressed in dB, which compares the device with a perfect one.

Noise floor Level of constant residual noise in a system.

Non pre-emptible Satellite capacity that is guaranteed and cannot be pre-empted for restoration or other operational reasons.

Non-dedicated Satellite capacity that is not dedicated for any particular purpose, e.g. telephony, Internet, DTH.

Non-ionizing radiation Electromagnetic radiation encompassing the spectrum of ultraviolet radiation, light, infrared radiation and radio frequency radiation, with low energy, insufficient to permanently alter cell structures.

NRPB National Radiological Protection Board (UK).

NSS New Skies Satellites N.V.: privatized part of INTELSAT.

NT1 Network termination device – ISDN.

NT2 point at which a second network termination device such as an ISDN switchboard is connected to the NT1 at the T-interface.

NTSC National Television Systems Committee: generic term for US analogue colour television system used in North America and Japan, based upon 60 fields per second and 525 lines.

Oblate Describes the shape of the Earth, which is not perfectly circular but slightly flattened in the Polar regions.

OBO Output back-off.

Occasional capacity; Occasional use Capacity available for ad hoc bookings.

Occupied bandwidth Absolute bandwidth occupied by the signal within a transponder or allocated channel.

OET Office of Engineering and Technology, FCC (US).

Offset prime focus Type of parabolic antenna where the focus of the antenna is displaced: instead of being at the centre of the parabola (the 'vertex') the focus is shifted up.

OHF See Overhead framing.

OMT Orthogonal mode transducer.

On-axis In line with the main beam of an antenna.

OOC Orion Operations Center (Loral), Maryland, US.

Orbital period Time taken for a satellite to go through one complete orbit.

Orthogonal mode transducer OMT: multi-port microwave device that allows the transmission of signals on one polarization while receiving signals on the opposite polarization.

Oscilloscope Electronic instrument for viewing and measuring electrical and electronic signals.

OSHA Occupational Safety and Health Administration: US national safety agency.

OTS Orbital Test Satellite: early UK satellite programme of the 1970s.

Outage Failure of satellite link.

Outer code Adds a number of parity (check) bits to blocks of data.

Output back-off OBO: the amount the output signal of an HPA is reduced as a result of input back-off being applied. Expressed in dB.

Overhead framing Additional data signal added to the data stream from the encoder to carry auxiliary information data – used in wider telecommunications applications. Often generically referred to as IDR framing.

Over-run Last-minute satellite booking extension.

Packet-switching Data transmission method that divides messages into standard-sized packets for greater efficiency of routing and transport through a network.

PAL Phase alternate line: the European CCIR analogue TV standard based upon 50 fields per second and 625 lines.

Panning-up Process of accurately aligning satellite uplink antenna to the satellite.

Paper satellites Orbital allocation applications made well in advance of the actual placing of a satellite in position; many organizations hold allocations for orbital slots that may never be used, as a 'spoiling' tactic.

Parabolic antenna Antenna with a reflector described mathematically as a parabola; essentially an electromagnetic wave lens which focuses the radio frequency energy into a narrow beam.

Parking orbit The initial orbit a satellite is placed into after launch, before being placed into the transfer orbit leading to the final geostationary orbit.

Payload Load carrying capacity; on a rocket launch vehicle, the satellite is the payload.

PCM Pulse code modulation.

PDU Power distribution unit.

Perceptual coding A coding technique which takes advantage of the imperfections of the human visual and auditory senses.

Perigee The point of a satellite's orbit where it is closest to the Earth's surface.

Permission Agreement from a national government for SNG transmissions.

PES Packetized elementary stream.

Petal Segment of a flyaway antenna.

PFD Power flux density: a measure of the satellites illumination of the Earth's surface.

P-frame Predicted compressed frame: calculated by applying motion prediction to a previous frame and deriving a difference signal (inter-coding).

Phase noise Phase interference in the wanted digital signal which can cause the digital signal to fail to be demodulated correctly.

Phase stability Characteristic that is important in digital transmission components – lack of phase stability can cause the digital signal to fail to be demodulated correctly.

Phase-combiner Multi-port microwave device which ensures that the individual outputs from two HPAs are combined together so that the outputs are added in-phase.

Phase-locked loop PLL: type of electronic circuit used to demodulate satellite signals.

Piece-to-camera Where a reporter gives a news report face to camera.

Pipe Generic term for a data circuit.

Pixel Smallest element that is discernible (quantified and sampled) in a picture.

Pixellation Aberration of a digitally coded television picture where the individual digitally coded samples can be seen.

Pointing loss Loss due to misalignment of boresight of antenna towards satellite, and movement of satellite within its station-kept 'box'.

Polarization Geometric plane in which the electromagnetic signals are transmitted.

Polarization rotator Device that can be manually or automatically adjusted to select one of two orthogonal polarizations on an antenna.

Polarization skew An uplink or downlink waveguide is rotated to compensate for the angular difference between the antenna position on the Earth's surface and the satellite position above the Equator.

Pool Grouping of newsgathering entities, usually for a limited period, to pool resources to cover a news event.

POP Point of presence.

POR Pacific Ocean Region: describes coverage area of a satellite.

Power balancing The operational adjustment of two signal powers to minimize the production of intermodulation products; the adjustment of carriers to minimize small signal suppression.

Power density Density of radiated power from an antenna, measured in W/m^2 (mW/cm^2).

Power meter A measurement instrument which can be connected directly to the sample port of the HPA to give a constant readout of the power being produced from the HPA.

pp Peak-to-peak.

Precipitation loss Loss due to rain attenuating the signal, particularly in the Ku-band, and further dispersing the signal as it passes through the drops of water.

Precipitation zone To aid in calculating the effect of precipitation loss, the world is divided into precipitation zones or rain climatic zones, each of which has a numerical value defined by the ITU, used in the calculation of a link budget.

Pre-emphasis Type of frequency-dependent level boosting/cutting or filtering.

Pre-emptible Satellite service that may be interrupted or displaced by another, higher-priority service, typically (but not exclusively) during a satellite contingency situation.

PRI Primary rate ISDN.

Primary rate ISDN PRI: primary rate ISDN provides 30 × 64 kbps (B) data channels and 1 × 64 kbps (D) signalling channel.

Profile Profiles define the colour space resolution in MPEG compression and also the scaleability of the bitstream; the combination of a profile and a level produces an architecture which defines the ability of a decoder to handle a particular bit-stream.

Progressive scan Alternative to interlaced scanning, where each line is scanned in sequence in the frame.

PSK Phase shift key modulation: digital phase modulation.

PSTN Public Switched Telephone Network.

PSU Power supply unit.

Psycho-acoustic; psycho-visual Description of the models of the brain's perception of vision and sound.

PTC Piece-to-camera.

PTO Public telecommunications operator; power take-off: where a generator may be driven through a vehicle's road engine and transmission system while the vehicle is at rest.

PTT Post, telephone and telegraph administration: national telecommunications agencies directly or indirectly controlled by governments, in charge of telecommunications services in most countries of the world.

Pulse code modulation PCM: time division modulation technique in which analogue signals are sampled and quantized at periodic intervals into digital signals – typically represented by a coded arrangement of 8 bits.

QAM Quadrature amplitude modulation: modulation technique.

QCIF Quarter common interchange format: 176 × 144 pixels.

QPSK Quadrature phase shift key: modulation technique.

Quantization The process that assigns a specific number of bits (resolution) to each frequency coefficient after DCT coding.

RA Radiocommunications Agency (UK): national radiocommunications regulatory body.

Rain climatic zone See precipitation zone.

Rain outage Loss of signal due to absorption and increased sky-noise temperature caused by heavy rainfall.

RCCB Residual current circuit breaker.

RCD Residual current device: a device that measures the difference in current flowing between the supply wires; if a difference exists then a fault is assumed and the supply is rapidly disconnected.

Receiver Rx: device which enables a particular satellite signal to be separated from all others being received by an earth station, and converts the received signal into video and audio.

Redundancy Protection against failure of either a part or the whole of the system that would thus make the entire system inoperable. In an SNG uplink that is redundant, there are two transmitters and two modulators.

Reed-Solomon code Type of outer convolutional error correction code; for each given number of symbols forming a block, an additional parity check block of data is added to make up a complete block of data.

Reg. TP Reguliersrungbehorde für Telekommunikation und Post: German national licensing body.

Resolution Sharpness of the picture.

Restoration The action of a satellite system operator to sustain defined service level in the event of satellite failure or anomaly.

Return audio; reverse audio See Cleanfeed.

RF; RF bandwidth Radio frequency: refers to radio transmissions or relating to that part of the electromagnetic spectrum used for radio transmissions.

RHCP Right-hand circular polarization.

RJ-11; RJ-45 Standard types of telephone line connector.

Router Network device that determines the optimal path along which network traffic should be forwarded.

Routing matrix See Switching matrix.
RR Radio Regulations: instruments of the ITU.
RS Reed-Solomon code.
RS.232 Common serial data interface.
RU Nineteen-inch rack-height unit; See U.
Rx Receive; receiver.
S point See S-interface.
S/N Signal-to-noise ratio – typically expressed in dB.
S/T bus; S/T interface See S0 bus/interface.
S0 bus; S0 interface Converts the ISDN data from the format used between the exchange and the socket to the type used in the subscriber's premises.
Sampling frequency (rate) Rate at which samples are taken.
Sampling Instantaneous measurement of the picture at a particular point in time; the value of the signal at that instant is typically converted to a binary number.
SAR Specific energy absorption rate: the biological effects of non-ionizing radiation expressed in terms of the rate of energy absorption per unit mass of tissue.
Satellite link Microwave link between a transmitting earth station and receiving earth station through a satellite.
Satellite newsgathering SNG: the means by which news is gathered for television and radio broadcast using satellite communications.
Satellite system operator Organization which operates and controls one or more satellite spacecraft.
Satellite, artificial Electronic communications relay station orbiting the Earth.
Satellite, geostationary An electronic communications relay station orbiting 35 785 km (22 237 miles) above the Equator, moving in a fixed orbit at the same speed and direction as the Earth, i.e. approximately 11 000 kph (7000 mph).
Satphone Generic term for satellite telephone, usually used on the INMARSAT system.
Saturated flux density SFD: the power required to achieve saturation of a single channel on a satellite.
Saturation Where a high power amplifier is driven into the non-linear part of its power transfer characteristic, such that an increase in input power results in little or no increase in output power.
Scintillation Rapid fluctuation in amplitude and phase caused by the signal travelling a longer path through the atmosphere and ionosphere.
Scope See Oscilloscope
SCPC Single channel per carrier.
Screening Process of checking material for political content by censor before transmission.
SDI Serial digital interface.
SDTV Standard definition television.
SDV Serial digital video.
SECAM SEquentiel Coleur Á Memoire: analogue colour television broadcast standard of France, Russia and a number of other French-speaking nations

(typically former colonies) which, though based on 50 fields per second and 625 lines, is incompatible with PAL.

Sensitivity Ability of an electronic device (typically a receiver) to detect and process very small signals satisfactorily.

Serial digital video Digital video transmitted serially, typically at a data rate of 270 Mbps.

Shannon theory Information theory showing that signals will always have some degree of predictability; data compression commonly uses the principle of statistical analysis of a signal to predict changes.

Shore power Power from an external source.

Side-chain An auxiliary transmission path on an SNG uplink.

Sidelobes A parabolic antenna does not produce a completely perfect radiation pattern, which would be a single focused beam, but has a main 'lobe' (centred on the boresight) and a number of 'sidelobes' radiating out from the antenna.

Sidereal cycle The Earth rotates once in every 23 hours, 56 minutes and 4.1 seconds, rather than exactly 24 hours – equivalent to one sidereal day.

SIF Standard interchange format: format for exchanging video images of 240 lines × 352 pixels for NTSC, and 288 lines × 352 pixels for PAL and SECAM.

Signal-to-noise ratio S/N: ratio of the signal power and noise power in analogue systems; a video S/N of 54–56 dB is considered to be of full broadcast quality.

Signatory National administration or organization which forms part of the establishment of an international treaty organization such as INTELSAT or EUTELSAT.

Simplex Transmission in one direction only between sending and receiving station.

Single channel per carrier SCPC: a transmission of single programme signal on one carrier.

S-interface The ISDN subscriber interface where the connection to the ISDN network is through another network termination device such as an ISDN switchboard.

SIS Sound-in-sync.

Skew An adjustment that compensates for slight variance in angle within the same sense of polarity.

Slant range The length of the path between a satellite and the ground earth station.

Slice Sequence of macroblocks across the picture in a digital compression process.

Slot, orbital Longitudinal position in the geosynchronous arc at which a communications satellite is located – allocated by the ITU.

Small signal suppression Suppression effect seen on a relatively low-power carrier signal when a low-power carrier and a high-power carrier share the same transponder on a satellite.

SMPTE Society of Motion Picture and Television Engineers (US).

SMPTE-125M US standard for 4:2:2 component video signals – equivalent to CCIR-601.

SMPTE-259M US standard for 270 Mbps serial digital video.

SMS Satellite Multi Services: EUTELSAT's business data services.

SNG Satellite newsgathering.

Snow Form of noise seen on analogue pictures caused by a weak signal from the satellite, characterized by alternate dark and light dots appearing randomly on the picture tube.

SNV Satellite newsgathering vehicle.

Solar eclipse Event when the Earth moves between the satellite and the Sun, preventing the satellite from receiving energy from the Sun to provide solar power.

Solar noise Description of the electromagnetic 'noise' generated from the Sun.

Solar outage Event during the equinoxes when the satellite is on that part of its orbit where it is between the Sun and the Earth, causing a downlink receiving antenna pointed at a satellite to become 'blinded' as the Sun passes behind the satellite.

Solid state power amplifier SSPA: a high-power amplifier using solid-state electronics as the amplifying element, as opposed to a vacuum tube device.

Sound-in-sync SIS: audio signal carried as a digitized signal in the sync (line timing) pulses of the analogue TV video signal – European standard.

Space segment Generic term for the part of the transmission system that is in the sky.

Sparklies See Threshold noise.

Spatial redundancy The similarity between pixel values in an area of the picture which can be exploited to save the amount of information transmitted: more often than not, two neighbouring pixels will have very nearly the same luminance and chrominance values.

Spectral redundancy The similarity between frequency values in an area of the picture which can be exploited to save the amount of information transmitted.

Spectrum The full range of electromagnetic radio frequencies.

Spectrum analyser A radio receiving instrument with the ability to repeatedly and automatically tune across a band of electromagnetic spectrum, showing the amplitude of signals present in that band on a display screen.

Spike A voltage or energy surge; irregularities in the power supply waveform.

Spill-over Satellite signal that falls on locations outside the beam pattern's defined edge of coverage.

Splitter Passive device (one with no active electronic components) which distributes a signal into two or more paths.

Spotbeam High-power satellite signal with a focused antenna pattern that covers only a small region.

Spreading loss As a signal is transmitted by an antenna, the signal spreads out to cover a very wide area and gets weaker as it travels further away from the antenna.

Spurious radiation Any radiation outside a defined frequency band; potential source of interference.

SQIF Sub-quarter common intermediate format: 128 × 96 pixels.

SSOG Satellite Systems Operations Guide – INTELSAT.

SSPA Solid-state power amplifier.

Stand-up Position where a reporter gives a news report; generic term for a short news report from the field.

Station-keeping Small orbital adjustments regularly made to maintain the satellite's orbital position within its allocated 'box' on the geostationary arc.

Station-kept box The space in which the satellite is maintained while in station-kept geostationary orbit.

Store and forward Process by which high-quality video is digitized and compressed at the high bit-rate, stored, and later transmitted in non-real time through a low data rate circuit, e.g. INMARSAT.

Sub-carrier An additional signal to carry additional information piggybacked onto a main signal, and which is applied at a frequency above the highest frequency of the main carrier.

Sub-sampling Processing of chrominance components to remove spectral redundancy.

Sub-satellite point Point on the Equator directly beneath the satellite; if a line was drawn from the centre of the Earth to the satellite, the point at which the line passes through the Earth's surface.

Sun-out See Solar outage.

Superbeam See widebeam.

Switched-56 56 kbps digital telephony data service in the US, similar to ISDN, and now rapidly being superseded by ISDN.

Switching matrix Router that can connect different source inputs to different destination outputs; also called a routing matrix.

Symbol; symbol rate Used in connection with the transmitted data, where a defined number of bits represents a 'symbol' – a unit of information. One symbol per second is termed a 'baud'.

Synchronization Process of achieving the same timing relationship at the transmitter and the receiver in order that information can be correctly conveyed.

System noise Unwanted signals and artifacts generated within a receive system, typically expressed as a power figure to balance against the signal power.

T point See T-interface.

TA ISDN terminal adapter: provides interface conversion between the S0 bus and the serial communications interface on the subscriber data terminal equipment (DTE).

Tachograph Device fitted to a vehicle (typically a goods vehicle) which records vehicle activity over a 24-hour period onto a circular chart. Used for statutory reasons.

Talkback Bi-directional circuit to allow studio-remote conversations, instructions and information to be exchanged, between the remote location and the control room of the studio.

Talking-head Colloquial term for a single head shot of a reporter.
Tape feed Video or audio tape transmission.
TBR Technical basis for regulation (ETSI).
TBR-030 Precursor to CTR-030.
TDMA Time division multiple access: refers to a form of multiple access where a single carrier is shared by many users so that signals from earth stations reaching the satellite consecutively are processed in time segments without overlapping.
Teleport Fixed earth station; typically has a large number of satellite antennas.
Telstar Early US satellite.
Temporal redundancy Similarity in pixel values between one frame of a picture and the next frame, which can be exploited to save the amount of information transmitted; redundancy in the data from one frame to the next can be removed.
TES Transportable earth station: mobile microwave radio transmitter used in satellite communications, including SNG systems.
Threshold extension In an analogue FM signal system, analogue receivers have filters that track the modulated bandwidth to maximize the use of the allocated bandwidth, rapidly adjusting ('tracking') in response to the incoming FM deviation.
Threshold margin In an analogue uplink: the level of uplink power that can be lost before the onset of threshold noise affects the picture. In a digital uplink, the margin that can be lost before complete failure of the signal.
Threshold noise In an analogue FM link, noise occurs in the picture as black and white spikes – sparklies – where noise spikes instantaneously exceed the carrier level.
Threshold Point at which a signal will fail to be correctly demodulated: a digital receive system will fail abruptly with either a 'frozen' or black picture when the input signal fails to achieve threshold; an analogue system will suffer increasing degradation to the point of failure the further below threshold the signal level drops.
Thruster Small booster rocket motor typically found on a satellite, used to maintain correct station-keeping.
T-interface ISDN subscriber interface at the NT1.
Tonne One thousand kilograms.
Tracking, Telemetry and Control (or Command) Management of in-orbit satellite primary systems from the ground; sometimes referred to as tracking, telemetry and command.
Transfer orbit Transitional orbit by which a satellite is moved from the parking orbit into geostationary orbit.
Transmission order Order in which the group of pictures is transmitted.
Transmission rate Aggregate data rate – including RS, convolutional encoding and the information rate.
Transmitter Electronic device consisting of oscillator, modulator and other circuits which produce an electromagnetic signal for radiation into the atmosphere by an antenna.

Transponder Combination receiver, frequency converter and transmitter package, physically on-board the satellite.

Transport stream Multiplex of programme channels in a DVB system.

Travelling wave tube TWT: powerful microwave amplifying vacuum tube used in a high-power amplifier, commonly employed in SNG systems as well as on satellites.

True North Theoretical North Pole: geographical designation of North Pole at 90°N latitude, 0° longitude.

Truncation noise On an analogue uplink, if the receiver IF filter bandwidth is too narrow, high-order 'sidebands' of the vision signal, corresponding to areas of highest deviation, are severely attenuated.

TS Transport stream.

TT&C See Tracking, telemetry and control (or command).

Turnaround Act of downlinking a satellite signal, altering it and instantaneously uplinking it again; often used to provide a multi-hop path around the globe, or to change the signal from Ku-band to C-band.

Turnkey System or installation provided complete and ready for operation by a manufacturer or supplier.

TVRO Television receive only: description of a small receive-only facility, using a small antenna; in SNG often used for off-air check or cueing purposes.

TVSC TV Service Centre: INTELSAT satellite booking centre.

Tweeking Engineering slang for adjustment to optimize performance.

Two-way Studio-remote interview in a news broadcasts.

TWT Travelling wave tube.

Tx Transmit; transmission.

Type-approval Official process of obtaining technical approval for an earth station to be used with a particular satellite system.

U Rack-height unit: a unit of measurement of height for equipment racks, where 1 U = 1.75 in (44.5 mm); also referred to as RU.

U-interface The point at which the ISDN telephone line terminates at a subscriber's premises.

UMTS Universal Mobile Telecommunication System: ETSI system.

Unilateral Transmission from one origin to one destination only.

Upconverter Earth station equipment to convert from IF to RF frequency: transforms the modulated IF signal from the modulator up to the desired transmit frequency by a process of frequency shifting or conversion.

Uplink Earth station used to transmit signals to a satellite; the transmit earth station-to-satellite connection.

UTC Universal time co-ordinated: measurement of time, locked to an atomic clock reference, particularly used in space science and engineering; virtually the same as Greenwich Mean Time.

V Volt: measurement of electrical potential difference.

V.35 Protocol used for communications between a network access device and a packet-data network (ITU-T standard).

VA Volt amperes: a measurement of power.

VAC Volts, alternating current.

Variable length coding Process where further efficiencies are used to reduce the amount of data transmitted while maintaining the level of information.

VAST-p Video and audio storage and transmission – portable: store and forward video unit for newsgathering manufactured by TOKO.

VBI Vertical blanking interval.

VDC Volts, direct current.

Vertex Geometric centre of the parabola on an antenna.

Vertical blanking interval Contains the vertical timing and synchronization information for the television display.

Video-conferencing The process of visually and aurally linking two or more groups across a telecommunications network for (typically) business meeting purposes.

Viterbi Decoding algorithm for FEC.

VLC Variable length coding.

VPC Variable phase combiner: particular type of phase combiner which has a variable adjustment to match the phase of the two inputs.

VSAT Very small aperture terminal: refers to small fixed earth stations, usually with antennas up to 2.4 m, used in a 'star' or 'mesh' network for business communications.

VSWR Voltage standing wave ratio: measurement of mismatch in a cable, waveguide or antenna system.

W Watt: measurement of power.

WARC see WRC.

Waveguide Metallic microwave conductor, typically rectangular in shape, used to connect microwave signals with antennas; generic term for the feedhorn assembly; the connecting piece (or pieces) between the output of the HPA and the feed-arm of the uplink antenna.

Waveguide loss Loss due to waveguides introducing some attenuation to the signal.

WBU World Broadcasting Union.

Wideband Large frequency bandwidth: see broadband.

Widebeam Ku-band satellite antenna pattern offering broader coverage, up to 8000 km (5000 miles) in diameter, but only available at lower power levels because the power is spread over a wider area.

WP 4-SNG ITU Study Group which undertakes studies specifically relating to satellite newsgathering.

WRC World Administrative Radiocommunication Conference: regular summit conference for the ITU-R. After 1992, renamed World Radiocommunications Conference.

WTDC World Telecommunication Development Conference: the regular summit conference for the ITU-D.

WTO World Trade Organisation (UN).

WTSC World Telecommunication Standardisation Conference: regular summit conference for the ITU-T.

X polarization Vertical polarization – the linear geometric plane in which the electromagnetic signals are transmitted.

X.21 Serial communications interface standard (ITU-T standard).

X.25 Data packet-switching standard (ITU-T standard).

XPD Cross-polar discrimination.

XPI Cross-polar interference.

Y polarization Horizontal polarization – the linear geometric plane in which the electromagnetic signals are transmitted.

Z Relative time reference to the start of a transmission.

Zone beam C-band equivalent of Ku-band widebeam, with high gain covering a smaller area.

Bibliography

General interest: Newsgathering

Arnett, P. (1994) *Live From The Battlefield*. Bloomsbury Press.
Gall, S. (1994) *News From The Front*. Heinemann.
Simpson, J. (1999) *Strange Places, Questionable People*. Pan.
Bell, M. (1996) *In Harm's Way*. Penguin Books.

Satellite communications engineering

Ranked in order of relevance to the author.

Pratt, T. and Bostian, C. (1986) *Satellite Communications*. John Wiley.
Morgan, W. and Gordon, G. (1989) *Communications Satellite Handbook*. John Wiley.
Evans, B. (ed.) (1999) *Satellite Communication Systems*. Inspec/IEE.
Maral, G., Bousquet, M. and Bousquet, M. (1998) *Satellite Communications Systems: Systems, Techniques and Technology*. John Wiley.
Gordon, G.D. and Morgan, W.L. (1993) *Principles of Communication Satellites*. John Wiley.
Elbert, B.R. (1987) *Introduction to Satellite Communications*. Artech House.
Roddy, D. (1995) *Satellite Communications*. McGraw-Hill.

Digital compression

Watkinson, J. (1999) *MPEG-2*. Focal Press.

Papers

Ely, S.R. (1996) MPEG Video Coding: A Basic Tutorial Introduction. BBC RD 1996/3; www.bbc.co.uk/rd.

Tudor, P.N. (BBC) (1995) MPEG-2 video compression. *IEE Electronics and Communications Engineering Journal*, December, 257–264; www.bbc.co.uk/rd.

Strachan, D. (1996) SMPTE Tutorial: Video Compression, SMPTE; www.smpte.org.

Reference works

Gallagher, B. (ed.) (1989) *Never Beyond Reach*. Inmarsat.

Inmarsat (1997) *Inmarsat-B High Speed Data Reference Manual.*

INTELSAT (1998) *Satellite Newsgathering Handbook.*

INTELSAT (1998) *The Video Book.*

Index